Insulating Films on Semiconductors, 1979

Invited and contributed papers from the Conference on Insulating Films on Semiconductors (INFOS 79) held at the University of Durham, 2–4 July 1979

Edited by G G Roberts and M J Morant

Conference Series Number 50

The Institute of Physics
Bristol and London

CODEN IPHSAC 50 1–293 (1980)

British Library Cataloguing in Publication Data

Conference on Insulating Films on Semiconductors,
University of Durham, 1979
Insulating films on semiconductors, 1979.
–(Institute of Physics. Conference series; no. 50 ISSN 0305-2346).
1. Semiconductors–Congresses
I. Title II. Roberts, G G III. Morant, Martin John IV. Series
621.3815′2 TK7871.85

ISBN 0-85498-141-1
ISSN 0305-2346

The First European Insulating Films on Semiconductors Conference was organised by the Institute of Physics (Thin Films and Surfaces Group) in collaboration with the Institution of Electrical Engineers, the International Union of Pure and Applied Physics and the Solid State Physics sub-committee of the Institute of Physics.

Organising Committee
G G Roberts (*Chairman*), C Hilsum, J C Bass, M J Morant

Programme Committee
M Cardwell (*Caswell*), A Goetzberger (*Freiburg*), C Hilsum (*Malvern*), M Pepper (*Cambridge*), G G Roberts (*Durham*), J F Verwey (*Eindhoven*)

Sponsors
The Organising Committee is glad to acknowledge generous financial support from the European Research Office of the US Army, the United States Office of Naval Research, London, Imperial Chemical Industries Ltd, Plessey Research (Caswell) Ltd, Standard Telecommunication Laboratories Ltd, Philips Research Laboratories and Ferranti Electronics Ltd.

Honorary Editors
G G Roberts and M J Morant

Published by The Institute of Physics, Techno House, Redcliffe Way, Bristol BS1 6NX, and 47 Belgrave Square, London SW18 8QX, England.

Typeset by DJS Spools Ltd, Horsham, West Sussex, and printed in Great Britain by J W Arrowsmith Ltd, Bristol.

Insulating Films on Semiconductors, 1979

Preface

The development of the MOS transistor in the 1960's opened up a new field of study in solid state physics, that of insulating films on semiconductors or 'MIS physics'. Analysis of the electrical properties of the metal–oxide–silicon structure showed that it could yield fundamental information about charge states in the insulator and at the insulator–semiconductor interface. Understanding and control of these charges was also of technological concern and essential for the production of reliable MOS devices. The subject has therefore continued to be of absorbing interest to a large number of physicists and engineers, some concerned with basic physical processes and some with device applications which now include memory devices, microwave transistors, MIS solar cells and electroluminescent and switching devices. Considerable effort is now being devoted to MIS structures based on materials other than Si; particular attention is being focused on the group III–V compounds because of their favourable transport characteristics.

The first of what it is hoped will become a biennial series of conferences on the topic of Insulating Films on Semiconductors (INFOS 79) was held at the University of Durham on 2–4 July 1979; it brought together many enthusiasts for MIS physics and its applications. This book, which contains all the invited and contributed papers, enables recent research in the subject to be digested at leisure by a larger number of readers. Thanks are due to the referees and the editorial staff of the Institute of Physics for its rapid production.

The control of interfacial disorder in MOS devices has also opened up another area of solid state physics research, namely that of space-charge layers, a subject which concentrates on the electronic properties of the semiconductor rather than the insulator. The conference organisers decided at an early stage not to include papers dealing exclusively with fundamental studies of this topic. However, as it is an area of general interest to those involved in MIS science, a special evening lecture was arranged. In this, Professor R A Stradling discussed various effects consequent upon the decrease in dimensionality of the electron gas due to its confinement at the semiconductor–insulator interface. These effects include the appearance of minigaps in the electronic band structure and the disappearance of scattering terms in the conductivity associated with the mobility edge. The interplay of collective effects such as density waves and Wigner crystallisation with disorder was also emphasised. Professor Stradling concluded his lecture with a discussion of a proposed tunable source in the far infrared for spectroscopic applications that relied on the coupling of electromagnetic radiation to plasma modes in the inversion layer. This provided a timely illustration of the fundamental research on the properties of space-charge layers in MOS structures that is now generating further devices on which future technologies might be based.

More than half of the conference sessions were devoted to the properties of silicon oxide on silicon, which is still the 'prototype' for all MIS studies because of its apparent simplicity and its technological importance. The conference showed that in spite of many hundreds of papers over the last 15 years, there is still a great deal that is not fully understood about charge states in MOS structures.

Chapter 1 contains papers that are concerned primarily with the properties of SiO_2 itself, starting with two on the mechanism of thermal oxidation. The good control of oxide thickness in the manufacture of devices is based on data that are largely empirical and it is surprising how little is understood about the actual growth mechanism. Processes in the thermal oxidisation of silicon in oxygen were discussed in an invited paper by Sir Nevill Mott. As the established theories of oxidation apply strictly only to crystalline films, vitreous SiO_2 is attractive for developing theoretical ideas of the growth of non-crystalline oxides in general. One of the main problems is to explain the experimental fact that molecular oxygen can pass through SiO_2 without atomic exchange in the parabolic region of growth. After discussing several possible models, Professor Mott emphasised that the detailed understanding of oxidation requires still more experimental data, particularly on thin oxides grown in really dry oxygen. With even small concentrations of water the thermal growth rate of SiO_2 on silicon can be greatly increased, and D R Wolters described some of the complicated additional processes that probably occur under these conditions.

Although SiO_2 is an almost perfect insulator, it may contain trapped electrons and semimobile ions when forming part of a device structure and this often causes slow deterioration of the device properties. As the channel length of MOST's is reduced, unintentional electron injection into the oxide, with the subsequent build-up of space charge, becomes an increasing problem. In memory devices, where electrons are deliberately injected through the oxide, unwanted space charge must also be avoided. Hence there has been extensive work in recent years on electron traps in SiO_2, as described in the invited paper by D R Young. With controlled injection, the MOS capacitor is a particularly elegant structure for investigating the charging and discharging of traps, enabling their location, energy, concentration and cross section to be determined from comparatively simple experiments. The formation of additional traps by the controlled implantation of specific ions further enhances the technique, as shown in the paper by R F DeKeersmaecker and D J DiMaria for As ion-implanted oxides. Detailed measurements of trap energies and densities in comparatively thick wet oxides were described by D D Rathman, F J Feigl and S R Butler. Oxide films on polysilicon show enhanced high-field conduction due to electron injection from asperities, the characteristics of which were described by M Conti, G Corda and R Gastaldi. Ageing effects may again be explained by the build-up of space charge, although a very high concentration of traps is indicated in these oxides.

The second source of charge in SiO_2 is ionic. It is well known that the movement of positive ions causes threshold voltage instability in MOST's, a problem which has been largely overcome industrially by improved techniques for the growth of 'ion-free' films. However, there continues to be great interest in the behaviour of ions such as Na^+ and K^+ in SiO_2, as shown in the papers by J F Verwey and by J P Stagg and M R Boudry, because the long-term wear-out of MOS structures is probably controlled by the clustering of residual ions. The electrical properties of larger defects in SiO_2, which still reduce the initial yield of devices, were described in the paper by K J Dean, D A Baglee, R A Stuart and W Eccleston.

The defect structure of the $Si{-}SiO_2$ interface is discussed in Chapter 2. As this remains one of the main problem areas in MOS technology, it naturally attracted a number of papers describing techniques aimed at understanding the origin of the defects or methods designed to reduce their number. The keynote address was given by M Schulz who described the various measurement techniques for analysing the densities of interfacial states, their capture cross sections and their energy locations. The merits and limitations of the CC–DLTS method are highlighted in both his review and in the contribution by E Klausmann. A feature of this method of analysis is the variation of the capture cross section of the states with energy position in the band gap. Although the evaluation of these cross sections is complex, available evidence points to their magnitude being dependent on processing technology but always decreasing to very low values near the conduction-band edge. To explain this effect, Professor Schulz suggested a model involving tunnelling to a discrete level in the SiO_2 where the tunnelling depth is variable due to round-off near the interface. The cross section theme was carried further by D W Greve and W E Dahlke. These authors have used a photoemission technique to show that the optical cross section increases monotonically with photon energy and that the density of interface states in an n-type device increases towards the conduction band.

Generally speaking, the residual low density of states at the Si–SiO_2 interface is achieved only by empirical methods of oxidation and annealing. For example, the density of fast surface states in Si devices may be reduced by annealing in hydrogen at comparatively low temperatures. A variant of this technique which yields extremely low surface-state densities for CCD's was described by L Risch, E Pammer and K Friedrich. It involves depositing layers containing hydrogen followed by an annealing step. It is also well known that sodium can have an important influence on surface-state properties. However, fewer data are available on interface states due to group III and group V impurity atoms. J Snel presented evidence that such acceptors and donors can induce interface states and oxide charge at the Si–SiO_2 interface. He concluded that donor atoms like phosphorus and arsenic induce far fewer interface states than acceptor atoms. The well-known slow-state instability was the subject of the paper by C Werner, H Bernt and A Eder. By measuring fluctuations of the surface potential before and after an electrical bias stress was applied to the gate electrode at elevated temperatures, these authors measured the increase in the inhomogeneity of the slow surface states.

A novel method which might give information about surface sub-bands was described in the paper by W Hönlein, K von Klitzing and G Landwehr. It is based on measurement of the tunnelling characteristics of p^+–n^+ junctions in which one side is formed by a surface inversion layer. At present the technique requires improved device structures, but the expected characteristics should be obtained in future work.

Chapter 3 contains papers on structures in which the oxide is sufficiently thin for tunnelling to occur. In his review of MOS tunnelling H C Card explained the importance of using experimental techniques that can distinguish between the contributions of the majority and minority carriers to the tunnelling current. The injection ratio was also the central theme of a theoretical paper by G Kamarinos, G Pananakakis and P Viktorovitch that described a self-consistent model of an MIS tunnel structure. Dr Card also summarised the practical importance of devices that rely on thin films of tunnelling dimensions deposited onto semiconductors. Papers describing two such applications are included in in this chapter: G Sarrabayrouse, J Buxó, A Muñoz-Yagüe, A E Owen and J-F Sebaa presented analytical solutions for a MISS tunnel switching device while F A Abou-Elfotouh and M A Almassari described the preparation conditions (with associated surface characterisation measurements) for minimising the surface-state density x thickness product in order to achieve optimum conversion efficiency in MOS solar cells. A feature emphasised in all the tunnelling contributions is the fine control that is required of the insulator thickness. One possible way of achieving the desired accuracy was alluded to in a paper by M C Petty and G G Roberts, which described an MIS structure fabricated by depositing monomolecular organic layers onto the surface of CdTe.

Another important application of tunnelling occurs with double-dielectric films as used in MNOS memory devices. The development of successful models for the charge–discharge characteristics of such films was reviewed in the invited paper by M Pepper. However, much less is known about the problems of charge decay and fatigue which may occur in practical memory devices and there is also great need for more study of the chemical composition and physical properties of the individual films. One step towards this was given in the paper by N Lieske and R Hezel where Auger profiling of an MNOS structure shows that the 'oxide' is, in fact, a silicon oxynitride. Simultaneous electron energy-loss spectroscopy shows the presence of electron states in this film which are affected by annealing, and obviously these might affect the charge transfer. It was most unfortunate that Dr Lieske was unwell during the conference so that he could not present this paper.

MIS structures based on GaAs or InP are not being commercially exploited at present. However, both of these materials are attractive for high-frequency, low-noise applications. From the evidence heard at the two conference sessions dealing with group III–V semiconductors, presented in Chapter 4, InP looks more encouraging than GaAs in this regard and it may be superior to Si for very high-frequency discrete and integrated circuit MOSFET technology. For example, D Fritzsche reported characteristics for a p-type InP, chemical-vapour-deposited SiO_2 MISFET which demonstrated that inversion has been achieved. Moreover, mobilities up to 1000 $cm^2 V^{-1} s^{-1}$ can be obtained in the n-channel inversion mode. The main drawback of the device is thought to be electrical drift due to tunnelling of thermally excited electrons into the native oxide. C W Wilmsen, J F Wager and J Stannard, in their paper describing ESCA profiles for CVD oxides on InP, also stressed how dependent device performance is on the stoichiometry of the inner layer between the SiO_2 and the semiconductor. A prototype n-channel MOSFET using semi-insulating InP and silicon oxide grown by a plasma decomposition technique was described by A J Grant, D C Cameron, L D Irving, C E Greenhalgh and P R Norton.

In two contrasting review articles W E Spicer and H H Wieder both emphasised that the phenomena of oxygen adsorption and growth are very different in Si and the group III–V compounds; this difference is reflected in the densities of interface states and the degree of Fermi level pinning in the two types of material. Professor Spicer has relied on synchrotron radiation to study the surface electronic structure and the chemistry of III–V semiconductors by examining preferentially the last few atomic layers through the use of a large range of photon energies. His own work and that of his colleagues shows that Fermi level pinning takes place before a true continuous overlayer is formed. He also provided convincing evidence that the levels that pin the Fermi level are not induced directly by the adatom but are induced indirectly, probably by disruption of the III–V lattice near the surface. His measurements were made on bulk (110) material. However, confirmation that a similar state of affairs exists on (100) epitaxial layers was provided by Dr Wieder in his summary of dielectric–semiconductor interfaces of binary and quaternary alloy III–V compounds. He showed, for example, that the properties of

GaAs MIS structures are not dependent on either the crystallographic orientation of the semiconductor or the type of insulating layer and its method of preparation. In his view, the high surface-state density inherent in using this material holds little promise for its eventual use in MIS microwave digital integrated circuits. However, several papers presented at the conference indicated that good MIS characteristics may be achieved with this material, provided some provision is made to eliminate the natural oxide layer before depositing the thick insulator film. An attractive feature of the deposition technique used by B Bayraktaroglu, W M Theis and F L Schuermeyer is the ability to grow good-quality insulators at relatively low temperatures. The wet anodisation method described by S Hannah and B Livingstone also suggests that pinning of the surface potential can be avoided. The paper by J Nishizawa and I Shiota explained how gallium oxynitride films could be optimised to produce a stable interface with GaAs for surface passivation.

Discussions at the III–V semiconductor sessions emphasised present-day problems, but also reflected guarded optimism for the future. That is, they stressed the difficulties inherent in making MIS or MOS structures on this class of material, but included promising results for p-type InP and a hint that GaAs is not as dead as some would think. It will be interesting to see whether these advances will eventually enable the III–V compounds to compete with submicron Si technology.

Overall, the conference showed the need for more research on almost every topic discussed. It is planned to hold a second INFOS conference at Erlangen in 1981, at which some of the questions may be answered, and, doubtless, others raised. The intention of the organisers will be the same as that attempted by the organisers of INFOS 79 – to bring together physicists, electronic engineers, chemists and other scientists to share their thoughts on the fundamental aspects of the field of insulating films on semiconductors.

G G Roberts
M J Morant

Contents

The physics of space-charge layers

R A Stradling
Physics Department, University of St Andrews, North Haugh, St Andrews, Fife, UK

Abstract. In contrast to other papers presented at this conference, this review concentrates on the electronic properties of the semiconductor rather than of the insulator in MOS structures. One of the features of the space-charge layer that is of interest to physicists is the decrease in the dimensionality of the electron gas due to its confinement at the interface. The change to a quasi-two-dimensional system gives rise to a number of new effects such as the appearance of minigaps in the electronic band structure for silicon surfaces cut on low symmetry planes and to a disappearance of scattering terms in the conductivity associated with the Mott–Anderson mobility edge and with the peak conductivity measured in the Shubnikov–de Haas effect. The competing effects of disorder and electronic correlation are reviewed. An example of where correlation becomes dominant in a two-dimensional system is provided by the recent observation of Wigner crystallisation of the electron gas at the surface of liquid helium.

1. Introduction

The delay of nearly 40 years between the initial patents and the production of the first commercially important device based on the field effect induced by an insulated-gate structure arose because of an inability to control the density of interfacial states prior to the development of present-day techniques for the thermal oxidation of silicon. When control of interfacial charge density was achieved, a whole new area of semiconductor physics was opened up. Physicists are still greatly interested in the experimental and theoretical aspects of disorder, and the quasi-two-dimensional electron gas formed by the space-charge layer in MOS structures forms an excellent vehicle for such studies. In addition to the interest in the change of dimensionality in the MOS studies, their capacitor structure has the unique experimental advantage that the density of carriers and hence the Fermi energy can be varied over a very wide range, simply by changing the voltage applied to the device. This feature has been exploited in studies of interfacial disorder as well as investigations of the carrier transport within the space-charge layer. The change in dimensionality has resulted in the observation of a number of new effects, including 'minigaps' due to the lifting of valley degeneracy of the conduction band by low-index surfaces, and density-wave interactions.

Thermally activated conduction was reported some time ago close to threshold. Similar thermal activation has now been reported away from threshold in a magnetic field. Disorder effects certainly play a role in this observation, but the relative contributions of other effects such as density waves, Wigner crystallisation and impurity banding remains unclear. An impetus to resolving the roles of these contributory mechanisms has been given by the recent observation of Wigner crystallisation of the electron gas at the surface of liquid helium.

0305-2346/80/0050-0001$02.00

2. The quantised nature of the space-charge layer

The space-charge density is controlled simply by the voltage applied to the gate. The potential confining the induced charge is formed by all the charge near to the interface and therefore includes the depletion charge and the charge on the gate, insulator and interface as well as the space charge itself. In addition, the dielectric images produced by the sandwich structure of the device itself should be considered. Hence a solution of the Schrödinger equation must be found self-consistently from Poisson's equation. For a free-electron system, the solutions separate into two independent parts: one for motion perpendicular to the interface, where the energies are split into a series of sub-bands separated by energies of the order of 10 meV which depend on the level of inversion; the solution for motion parallel to the surface retains the parabolic dispersion relation for free electrons. The splitting of the sub-bands is such that only the lowest state is occupied for motion perpendicular to the surface except at high temperatures or levels of inversion. It is this localisation of the space charge to the region of the interface and the confinement of the carriers to the lowest quantum state which gives rise to the two-dimensional nature of the free-electron gas. The localisation that increases with increasing space-charge density may be further controlled by application of an electrical bias between the space charge and the substrate across the depletion layer. By changing the interaction with the surface, substrate bias is an extremely valuable experimental parameter in the study of disorder effects.

3. Threshold studies in zero magnetic field

MOS structures provide a particularly interesting system for studying Anderson localisation and the associated metal–insulator phase transition. Experiments with these devices differ in two important respects from measurements with impure bulk or amorphous semiconductors. Firstly, the MOS structure allows the Fermi level to be varied continuously in the same device over a large range of energies above and below the critical energy E_c. Secondly, the loss of dimensionality means that the equation for the minimum metallic conductivity no longer contains a factor of the dimensions of length, which is related to the mean-free path, and becomes:

$$\sigma_{mm} = 0{\cdot}1(e^2/\hbar) = 2\times10^{-5}\ \mathrm{S}.$$

Thus the minimum metallic conductivity for a two-dimensional system is expected to be completely independent of the range and details of the potential fluctuations in the system. Pepper *et al* (1974a, b) first exploited the special features of MOS structures to study the metal–insulator transition. The initial experimental results were consistent with the independent-particle model of Anderson localisation and both imperfect MNOS interfaces and good-quality MOS devices gave minimum metallic conductivities of the order of 10^{-5} at the onset of metallic conduction. At lower densities and temperatures, variable-range hopping was observed which, because of the two-dimensional feature of the system, gave an exponent proportional to $T^{-1/3}$ instead of the $T^{-1/4}$ dependence expected for bulk semiconductor material.

More extensive investigations by several groups of workers (reviewed by Pepper 1978 and Adkins 1978) showed that the independent-particle model does not apply for all devices and that, when thermal excitation above a mobility edge is expected, values of σ_{mm} considerably greater than $0{\cdot}1\, e^2/\hbar$ are frequently observed. The Hall effect, which is expected to become activated with the same energy as the conductance when the Fermi energy is below the mobility edge, frequently shows anomalous behaviour that is not correlated with the value of σ_{mm} derived for the device concerned. An electron-liquid model has been proposed to explain this anomalous behaviour in which gross inhomogeneities are present in the interfacial region and where correlation plays a dominant role near to any metal–insulator transition (Adkins 1978).

By introducing Na^+ ions near to the interface, it has also proved possible to generate an impurity band separated from the band edge. Three modes of conduction have been suggested for this band, activation above a mobility edge, nearest-neighbour hopping and variable-range hopping (Hartstein and Fowler 1978).

4. Shubnikov–de Haas measurements on space-charge layers

When a strong magnetic field is applied normal to the interface, the energy spectrum of the space-charge layer becomes discrete because of the quantisation of the orbital motion parallel to the interface. Fowler *et al* (1966) demonstrated that it was possible to observe a well-developed Shubnikov–de Haas effect in the electrical conductivity as the Fermi energy was varied by changing the voltage on the gate. These quantum oscillations showed fine structure demonstrating valley and spin splitting. In the absence of scatterers, the density of states would be a series of δ-functions at each Landau-state energy. In the presence of scatterers each level is broadened by an amount proportional to the scattering. In the high-field limit, however, the transverse conductivity (σ_{xx}) is proportional to the diffusion of the orbit centres and hence to the number of scatterers and the density of states which, at the centre of each level, is inversely proportional to the number of scatterers. Hence, as has been verified experimentally, the conductivity maxima are independent of the scattering and are simply given by:

$$\sigma_{xx}^{\max} = (e^2/\pi^2\hbar)(N + \tfrac{1}{2})$$

where N is the Landau index (Ando and Uemura 1974). In contrast to the conductivity maxima which do not depend on the mobility or the effective mass, the conductivity minima are determined by the overlap between adjacent Landau levels and hence are very sensitive to the degree and type of scattering.

It is the appearance and subsequent growth of these conductivity minima with decreasing temperature which gives rise to the Shubnikov–de Haas effect in space-charge systems. A detailed analysis of the temperature and mobility dependence of the minima has been performed in order to determine the processes involved (Nicholas *et al* 1977, 1978, Englert and von Klitzing 1978, Pepper 1978). The standard method of demonstrating that a process involves thermal activation is by making an Arrhenius plot where the quantity of interest is plotted on a logarithmic scale against reciprocal temperature. When such plots are made for the values of minimum conductivity observed when the Fermi energy (E_F) lies between two Landau states for particular values of magnetic field and Landau index, two distinct slopes are found (Nicholas *et al* 1977). The lower temperature of the two sections either has a lower activation energy than the higher-temperature region or the conductivity becomes independent of temperature. The higher-temperature region is due to the excitation of carriers into the first Landau level above E_F so that these carriers and the 'holes' left below E_F dominate the conductivity under these conditions. As the temperature is reduced, conduction can only take place close to the Fermi energy.

The region where the conductivity becomes independent of temperature, which occurs at relatively weak magnetic fields, can be interpreted as arising from the overlap of low density-of-state regions from adjacent Landau levels. If a Gaussian function is assumed for the density of states, an analytical relation can be deduced for the minimum conductivity observed, which contains the low-field mobility in an exponential function. Excellent agreement is found between the experimental values of conductivity and those predicted by this relation. As the magnetic field is raised, the saturation region where the minimum conductivity is independent of temperature changes over to a temperature range where the conductivity is activated. The straight line sections on the Arrhenius plots then extrapolate back to a common intercept on the conductivity axis. This behaviour is reminiscent of the activation expected for excitation above a mobility edge. On the assumption that such an edge occurs in the tails of the Landau levels and that a metal–insulator transition can be induced by increasing the magnetic field, the values found for the minimum metallic conductivity are remarkably close to the predicted value of $0{\cdot}1e^2/\hbar$ for the first Landau levels, although the values for higher Landau levels are an order of magnitude smaller.

As was found near to threshold in the absence of a magnetic field, it became clear on investigating a wider range of devices that this 'ideal' behaviour with a single intercept was only characteristic of certain devices. Nevertheless, it is thought that the three distinct mechanisms of thermal excitation into the centre of the Landau levels which occurs at higher temperatures, followed at lower temperatures by either conduction between extended states in the Landau state tails or thermal excitation from localised states in the tail region, provide a qualitatively correct picture of the conduction near to the Shubnikov–de Haas minima. A numerical study of localisation in a strong magnetic field confirms the general features of the proposed mechanisms (Aoki 1977), although the variation from device to device may well involve correlation, as was proposed for the zero-field threshold studies discussed in §3.

The conductivity maxima which, at least for intermediate values of magnetic field, are reasonably well fitted by the relation $e^2/\pi^2\hbar(N+\frac{1}{2})$, were found by Kawaji *et al* (1978) to be thermally activated for the two valley components of the lowest spin state for the $N=0$ Landau level (i.e. in the region slightly above threshold where the space-charge density is low). As was found for the conductivity minima, two distinct regions of thermal activation were observed at fields up to 15 T. A 'phase' diagram was constructed showing the temperature dependence of the critical concentration of carriers where the conductivity became thermally activated and the concentration where the break occurred between the two types of activation. It was postulated that there were two phases of the conduction electrons apart from the independent-particle gaseous phase and that these phases were states of a 'Wigner glass' where the electrons are localised under mutual repulsive interactions which are stronger than, but comparable to, the local potentials.

Nicholas *et al* (1977, 1978) demonstrated that, at very high magnetic fields (up to 35 T), conductivity maxima well away from threshold become thermally activated and large regions of the Landau level spectra are suppressed completely for Landau indices up to $N = 3$. However, the relative importance of collective phenomena such as Wigner crystallisation and localisation induced by random potentials is not yet clear and both further experiments with a range of well-characterised samples and a more quantitative theoretical description of the conductivity under these conditions are required.

5. Wigner crystallisation observed for a two-dimensional sheet of electrons bound to the surface of liquid helium

The previous two sections demonstrate how the unique features of the space-charge layer, the density of which is controlled by a capacitor structure, can be exploited to study localisation effects. At the same time, the experiments discussed highlight the difficulties in achieving results which change from device to device in a systematic or comprehensible manner. The source of the present difficulties in obtaining results which are reproducible from sample to sample certainly arises from the problems associated with the preparation of a near-ideal semiconductor interface with well-characterised properties. The same problems held back the commercial development of the field-effect principle for some 30 years. There is, however, another two-dimensional system of electrons with many features in common with the carriers in the conducting channels of MOS structures, but which has a much higher degree of perfection and longer mean-free paths from the electrons. The system concerned consists of the two-dimensional gas of electrons outside the surface of liquid helium and bound to the vacuum–helium interface by the image potentials that arise from the dielectric discontinuity between the two media (Crandall and Williams 1972). The lack of immobile defect states at the surface of the liquid has led to the first unambiguous detection of a Wigner crystal formed on the ordering of the free charge (Grimes and Adams 1979).

Apart from the much greater perfection of the helium surface compared with an interface between two solids, the other significant difference between the two low-dimensional systems is that the concentrations of surface charge that can be deposited onto the surface of liquid helium are much smaller, with the consequence that the electron gas remains classical down to temperatures of the order of 0·1 K. The electrons are introduced from a filament at the centre of a capacitor plate near the surface of the helium. Another capacitor plate is placed below the surface of the helium. The separation of the plates is some four orders of magnitude greater than the oxide thickness in MOS structures and the space-charge density of electrons is correspondingly reduced. The filament and top plate are negatively biased, thereby causing the electrons to accumulate at the helium surface.

At low densities the attractive potential for a single charge located at a distance Z above the helium surface is simply:

$$V = -(e^2/4\pi\epsilon_0)[(\epsilon - 1)/4(\epsilon + 1)](1/Z)$$

where ϵ is the dielectric constant of liquid helium. Thus the motion in the z direction becomes quantised in a series of one-dimensional hydrogen-like levels given by:

$$E_n = -R^*/n^2 = -(13{\cdot}6)[(\epsilon - 1)/4(\epsilon + 1)]^2(1/n^2)\ \text{eV}.$$

Because of the very small difference in dielectric constant between the liquid and the gas ($\epsilon - 1 = 0{\cdot}057$), the effective Rydberg R^* is reduced to ~0·6 meV, which corresponds to the microwave region of the spectrum. Hydrogen-like transitions between the bound states can be observed at frequencies close to 150 GHz. In order to avoid changing the frequency of the klystron in this experiment, a field-plate is located close to the surface of the liquid helium and the levels are Stark-tuned through the photon energy involved. In this respect and in general principle, these experiments are very similar to spectroscopic experiments involving transitions between sub-bands formed in inversion layers discussed in §6. By application of a magnetic field perpendicular to the surface of the helium, motion in the *xy* plane becomes quantised. Cyclotron resonance from the bound carriers can be observed in experiments analogous to the cyclotron resonance measurements in MOS devices. By applying electric fields parallel to the surface of the helium, hot-electron effects can also be generated, in the same manner as in MOS experiments.

In the experiments which detect the ordering of the electron gas into a Wigner crystal, the lower capacitor plate contains an RF element that generates a spatially varying electric field parallel to the surface of the helium which induces motion of the trapped electrons in this plane. Both capacitor plates are surrounded by a cylindrical container for the helium. At temperatures of the order of 0·5 K sharp resonances are seen in the impedance of the driven element as the RF region is swept (see figure 1). The amplitude of these resonances grows rapidly with decreasing temperature below a critical temperature for their appearance. The critical temperature is strongly dependent on the density of the trapped charge (N_s) and obeys a relation of the form $N_s^{1/2} \propto T$ as may be seen from figure 2. This relation is suggestive of the ratio of the Coulomb potential energy to the kinetic energy per particle for a classical electron gas, which is given by:

$$\Gamma = \pi^{1/2} N_s^{1/2} e^2/kT.$$

Thus data shown in figure 2 demonstrate that the ordering takes place at a constant value of Γ.

The interpretation of the observed resonances has been provided by Fisher *et al* (1979) who obtained an almost perfect fit to the observed frequencies on calculating the mode spectrum for the vibrations of the electron crystal coupled to the capillary waves (ripplons) of the helium surface. The results of their calculations of the frequencies

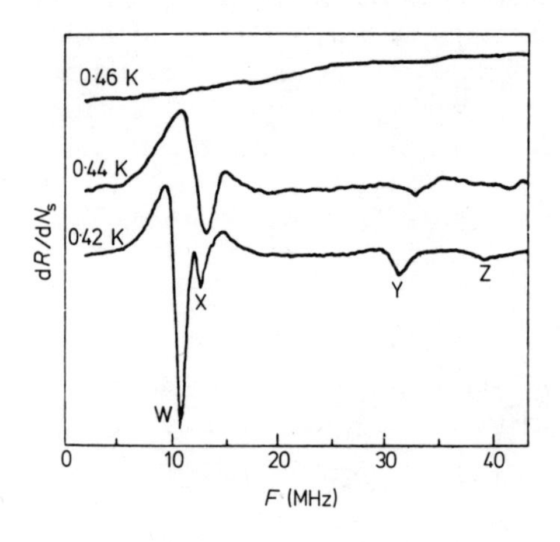

Figure 1. The derivative resistance of an RF element coupled to the space-charge layer of electrons bound to the surface of liquid helium as a function of the driving frequency. Four resonance dips appear below 0·457 K where the sheet of electrons has crystallised into a triangular lattice (Grimes and Adams 1979).

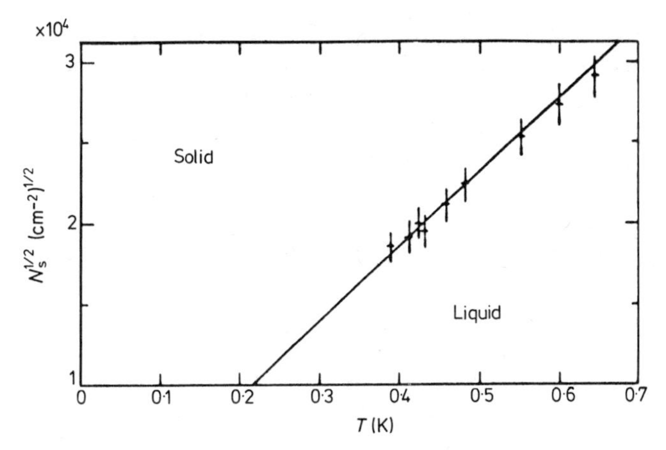

Figure 2. Variation of the critical temperature for the formation of a Wigner lattice of electrons on the surface of liquid helium with the concentration of surface charge (Grimes and Adams 1979).

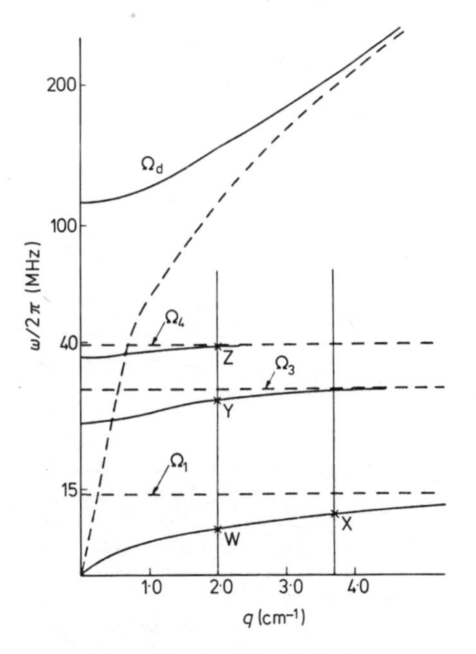

Figure 3. The dispersion relation calculated by Fisher *et al* (1979) for the vibrational modes of a two-dimensional lattice of electrons coupled to the ripplon modes of the surface of liquid helium. The dips displayed in figure 1 for the derivative resistance of the RF element used to excite the electrons correspond to the wavevectors marked with vertical lines. The excitations are dimensional resonances and therefore the wavevectors are determined by the dimensions of the container of the liquid helium. $n_s = 4{\cdot}55 \times 10^8$ cm^{-2}; $E_L = 415$ V cm^{-1}.

of the coupled-mode system are shown as full curves in figure 3. The unperturbed modes which are shown in figure 3 by the broken curves consist of the two-dimensional plasmon for the electrons that has the form $\omega_p^2 \alpha N_s q$ and the ripplons, the frequencies of which are given by $\omega_r^2 = (\alpha/\rho)k^3$ where α and ρ are the surface tension and density of the helium. Ripplons will be resonantly excited when an integral number of ripplon wavelengths equals the spacing between rows of electrons, i.e. when $k \sim G_n$ where G_n is a reciprocal lattice vector for the electron crystal. The lowest three of the unperturbed ripplon frequencies are shown in figure 3. In the experiment, standing waves of the coupled system are excited with wavevectors determined by the dimensions of the cylindrical container and the boundary conditions that the current flow be zero at the walls. The wavevectors concerned are shown as vertical lines in figure 3. The excellent agreement between theory and experiment demonstrates the formation of a Wigner crystal on the surface of the helium. This unambiguous result will stimulate further investigations into the role of similar correlation effects in MOS structures at low temperatures.

6. Submillimetre spectroscopy of space-charge layers

The understanding of the interfacial disorder which brought about present-day MOS technology produced an improvement in the mobility of the carriers in the space-charge layer: surface mobilities in excess of $1\ m^2V^{-1}s^{-1}$ can be achieved at low temperatures. Even though these mobilities are still well below those of the highest-quality bulk silicon, they are perfectly adequate for the formation of discrete size-quantised sub-bands and for the observation of cyclotron resonance from the two-dimensional electron gas. At the same time the experimental techniques for the generation and detection of far-infrared radiation have advanced and considerable improvement in the understanding of space-charge layers has resulted from spectroscopy at submillimetre wavelengths.

For hole carriers in the most commonly investigated silicon MOS system, the cyclotron resonance is complicated by the degeneracy of the valence bands. The detailed interpretation of the observed spectrum is not completely clear, although the main features can be explained qualitatively in terms of admixtures of the three interacting bands concerned. When the space-charge layer consists of electrons, the interpretation should be much easier as the conduction band is much simpler than the valence band in silicon. With the ⟨100⟩ surface, the electron mass at high levels of inversion is close to the value found in

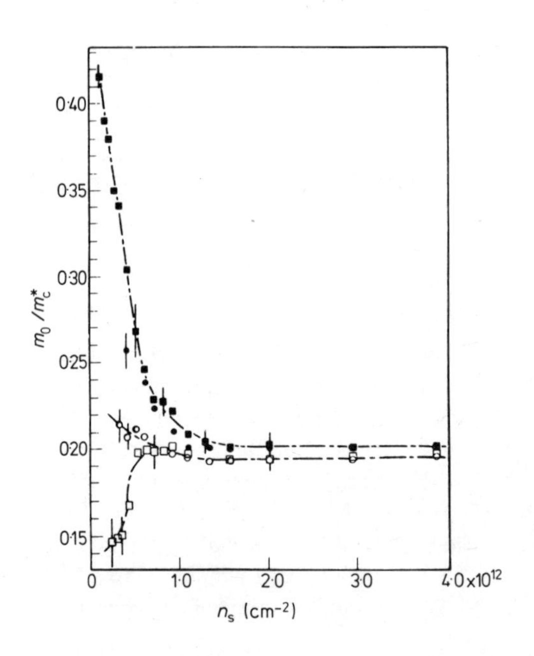

Figure 4. The cyclotron mass observed with n-type inversion layers on silicon ⟨100⟩ surfaces ($T = 10$ K, $f = 891$ GHz) as a function of surface charge at zero stress and at a uniaxial stress such that the sub-bands of cyclotron mass of $0{\cdot}42m_0$ lie below the normal ground-state sub-bands of cyclotron mass $0{\cdot}20m_0$ at low levels of inversion. At higher levels of inversion these two sets of sub-bands change order but, as this figure demonstrates, only a single cyclotron line of intermediate mass is observed as the different valleys cross. This observation can be explained by postulating a density-wave interaction that couples the two valleys together (Stallhofer *et al* 1976). Squares, sample K36; circles, sample W14; open symbols, $S = 0$; full symbols, $S \sim 1{\cdot}5 \times 10^9$ dyn cm^{-2}.

the bulk. However, at high temperatures or under suitable uniaxial compression where a second set of valleys of different mass should become occupied, a shift in mass is observed instead of the appearance of a second resonance line, as can be seen from figure 4 (Stallhofer *et al* 1976). This is believed to result from a density-wave interaction between the two sets of valleys which lowers the total energy of the multivalley electronic system (Kelly and Falicov 1977). Here the unambiguous nature of the cyclotron resonance experiment has been of assistance in revealing a new physical interaction. However, the remarkable narrowing by almost a factor of 10 of the cyclotron resonance line that accompanies the shift in mass awaits a quantitative explanation, although electron correlation and possibly Wigner effects are likely to be involved. The general feature observed in the threshold region, namely an abrupt increase in the cyclotron mass from a value well below the bulk mass to a value slightly above the bulk, followed by a slow decrease with increasing levels of inversion which can be seen for the zero-stress results of figure 4, has also been found with other semiconductors (von Ortenberg *et al* 1979).

For some time it has been known that the electric dipole transitions between the size-quantised sub-bands formed in inversion layers can be detected as the emission of submillimetre radiation, as well as by the more normal absorption techniques, provided that the carriers in the dipole space-charge layer are heated out of equilibrium by electric fields applied between the source and drain contacts (Gornik and Tsui 1976, 1978). The energies of the transitions involved can be varied by means of a voltage applied to the gate of the MOS structure, which gives rise to the possibility of a tunable source for spectroscopic applications. However, the power levels detected from this mechanism to date have been quite weak. Very recently a new type of emission has been reported from MOS devices which appears to be more promising for spectroscopy (Tsui and Gornik 1978). Not only are the power levels two orders of magnitude greater than those found with interband emission, but the tuning range involved is considerably wider. This type of emission can be observed when the interfacial plane concerned is not the common high-symmetry ⟨100⟩ plane, but the original semiconductor surface used in fabricating the MOS device is cut a few degrees off the ⟨100⟩ plane. When the inversion layer is formed on such a high-index plane, the twofold degeneracy of the silicon conduction band found with the high-symmetry ⟨100⟩ plane is lifted by valley–valley interactions at the crossing points of the two conduction bands in the two-dimensional Brillouin zone. 'Minigaps' therefore occur at these points, the magnitude depending on the strength of the valley–valley interaction. This in turn depends on the occupancy of the bands and hence on the gate voltage. Furthermore, the matrix elements are high for transitions across the minigaps. Consequently readily detectable amounts of power are observable and the predominant emission frequency is voltage-tunable, as may be seen from figure 5.

An interesting feature of the two-dimensional nature of the space-charge layers is that the first term in the plasma dispersion relation is proportional to the wavevector (**k**) to the power of a half as discussed in the previous section. As the phase velocity of the plasma excitations is very much less than the velocity of light, electromagnetic radiation does not normally couple to the plasma modes and only the energy loss from the free carriers is observed with MOS devices. However, coupling can be induced by the fabrication of a periodic slow-wave structure with a grid spacing of value a on the surface of a semi-conductor. Then interaction between the photons and the plasmons occurs at values of **k** given by $2\pi n/a$ which gives rise to the possibility of a number of device applications based on the coupled modes concerned. Figure 6 shows the plasmon absorption peak for $\mathbf{k} = 2\pi/a$ superimposed on the free-carrier loss (Allen *et al* 1977).

7. Conclusions

The control of interfacial disorder which has resulted in the ever increasing importance of MOS technology has also opened up one of the most active and interesting areas of

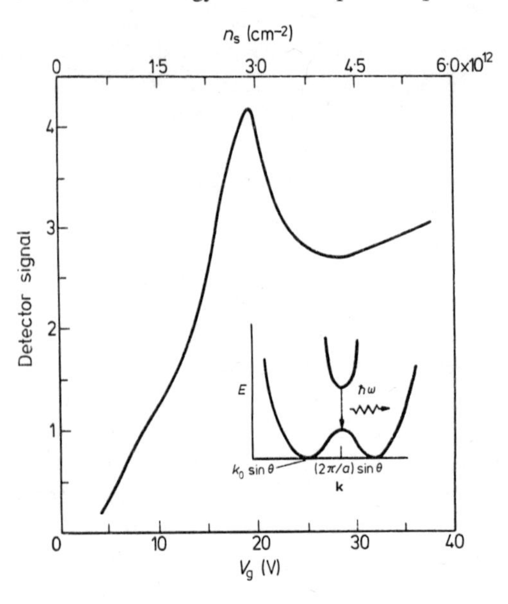

Figure 5. The signal detected by a narrow-band detector centred at 4·4 meV and emitted from a MOS device fabricated on a surface cut a few degrees off a ⟨100⟩ plane (9° Si MOSFET). The recording demonstrates the reasonably monochromatic nature of the radiation emitted from transitions across the minigaps produced by valley–valley interactions in these 'vicinal' planes (Tsui and Gornik 1978). $T = 4{\cdot}2$ K.

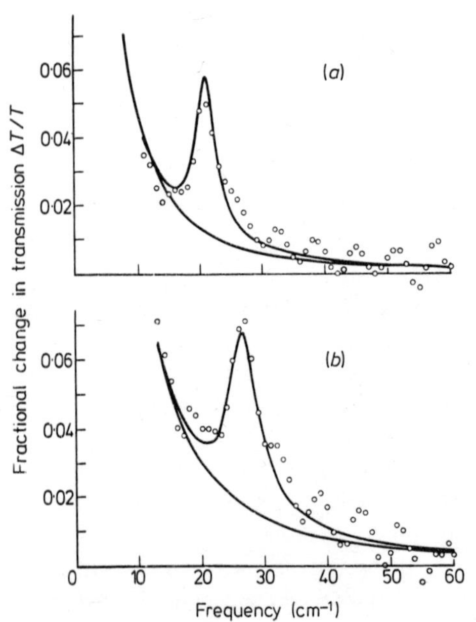

Figure 6. Plasmon absorption peaks superimposed upon free carrier loss from inversion layers in silicon MOS devices. The coupling to the plasmons is induced by a grid structure fabricated on the surface of the device (Allen *et al* 1977). (*a*) $n_s = 1{\cdot}55 \times 10^{12}$ cm^{-2}, $m^* = 0{\cdot}2$, $1/\tau = 0{\cdot}74 \times 10^{12}$ s^{-1}; (*b*) $n_s = 2{\cdot}51 \times 10^{12}$ cm^{-2}, $m^* = 0{\cdot}2$, $1/\tau = 1{\cdot}1 \times 10^{12}$ s^{-1}.

solid state physics. Subsequent to the development of techniques for reducing the degree of disorder, the MOS device has been employed as a unique vehicle for the study of interfacial defects. Collective effects such as density waves and Wigner crystallisation have been demonstrated and the interplay of these effects with disorder is presently attracting considerable attention. As exemplified by the development of a tunable source in the far-infrared, fundamental research on the properties of space-charge layers is in turn generating further devices on which future technologies may be based.

References

Adkins C J 1978 *Phil. Mag.* **B38** 535
Allen S J, Tsui D C and Logan R A 1977 *Phys. Rev. Lett.* **38** 980
Ando T and Uemura Y 1974 *J. Phys. Soc. Jap.* **36** 959
Aoki A 1977 *J. Phys. C: Solid St. Phys.* **10** 2573
Crandall R S and Williams R 1972 *Phys. Rev.* **A5** 2183
Englert T and von Klitzing K 1978 *Surface Sci.* **73** 70
Fisher D S, Halperin B I and Platzman P M 1979 *Phys. Rev. Lett.* **42** 798
Fowler A B, Fang F F, Howard W E and Stiles P J 1966 *Proc. 8th Int. Conf. on Physics of Semiconductors, Kyoto* p371
Gornik E and Tsui D C 1976 *Phys. Rev. Lett.* **39** 1425
—— 1978 *Solid St. Electron.* **21** 139
Grimes C C and Adams G 1979 *Phys. Rev. Lett.* **42** 795
Hartstein A and Fowler A B 1978 *Surface Sci.* **73** 19
Kawaji A, Wakabayashi J, Namiki M and Kusuda K 1978 *Surface Sci.* **73** 121
Kelly M J and Falicov L M 1977 *Phys. Rev.* **B15** 1974
Nicholas R J, Stradling R A, Askenazy S, Perrier P and Portal J C 1978 *Surface Sci.* **73** 106
Nicholas R J, Stradling R A and Tidey R J 1977 *Solid St. Commun.* **23** 341
von Ortenberg M, Tuchendler J, Silbermann R and Thuillier J C 1979 *Phys. Rev.* **B19** 2276
Pepper M 1978 *Surface Sci.* **73** 40
Pepper M, Pollitt S and Adkins C J 1974a *J. Physique* **74** 1273
Pepper M, Pollitt S, Adkins C J and Oakley R E 1974b *Phys. Lett.* **47A** 71
Stallhofer P, Kotthaus J P and Koch J F 1976 *Solid St. Commun.* **20** 519
Tsui D C and Gornik E 1978 *Appl. Phys. Lett.* **32** 365

Mechanisms for the thermal growth of vitreous oxide layers on silicon

N F Mott

University of Cambridge, Cavendish Laboratory, Madingley Road, Cambridge

Abstract. The rate of growth of vitreous oxide layers on silicon heated in oxygen is examined. For thick films the law proposed by Deal and Grove describes the observations, whereas for thin films, that given by Cabrera and Mott applies. In the parabolic region, the rate of oxidation is proportional to the partial pressure of O_2 and marker experiments with ^{18}O show that the oxygen molecule travels through the network without exchanging oxygen atoms with it. The implications of these results for the mechanism of transport are examined, both for thermal and anodic oxidation.

When oxygen or some other chalcogen reacts with a metal or with silicon in such a way as to form a compact film, either the oxygen or the metal must pass through the film. When the film is crystalline, the mechanism by which this occurs is fairly well understood. Normally the rate of transport of one constituent is much greater than that of the other. If metal is the mobile species, then metal atoms dissolve in the oxide at the interface, in the form of electrons and interstitial metal ions or anion vacancies. If the equilibrium concentration of the defect at the interface is c, a concentration gradient c/x is set up across a film of thickness x, leading to a growth rate given by:

$$\mathrm{d}x/\mathrm{d}t = 2cD\Omega/x \qquad (1)$$

where D is the diffusion coefficient of the defect which is assumed to be smaller than that of the electrons and Ω is the volume of oxide per cation. It is assumed that the densities of electrons and cations are such as to ensure charge neutrality and large enough to ensure a strong space-charge field if this were not so. Thus electrons and defects move together.

A review of the derivation of equation (1) is given by Mott and Gurney (1948); it integrates to give the parabolic law $x^2 = At$. The rate of oxidation then does not depend on oxygen pressure, unless this is very low. That this is so for the oxidation of zinc was shown by Wagner and Grünewald (1938). For the oxidation of copper in the parabolic range, however, copper vacancies are formed at the interface between oxygen and Cu_2O. Thus one oxygen molecule produces four vacant Cu sites and four positive holes, so a rate of oxidation proportional to $p^{1/8}$ is expected, where p is the oxygen pressure. The authors quoted found $p^{1/7}$, the small discrepancy being ascribed to partial ordering of the charged particles.

The present author (1947a, b), Cabrera and Mott (1948/49) and Fehlner and Mott (1970) have discussed low-temperature oxidation under conditions such that the growth law is logarithmic in the time. Here the densities of electrons and ions moving through

0305-2346/80/0050-0012$01.00

the oxide are supposed to be so small that no appreciable space charge is set up and their rates of transport can be considered independently of each other. Oxygen ions are supposedly absorbed on the surface of the oxide, giving rise to surface states. The next assumption is that electrons can be transported through the oxide, either by tunnelling or in some other way, such as motion in an impurity band, and fast enough to ensure that electrons in these surface states raise the Fermi energy there to that of the metal or silicon. A potential difference V that is independent of the thickness x is therefore set up across the film resulting in a field $F = V/x$, as shown in figure 1. If metal is trans-

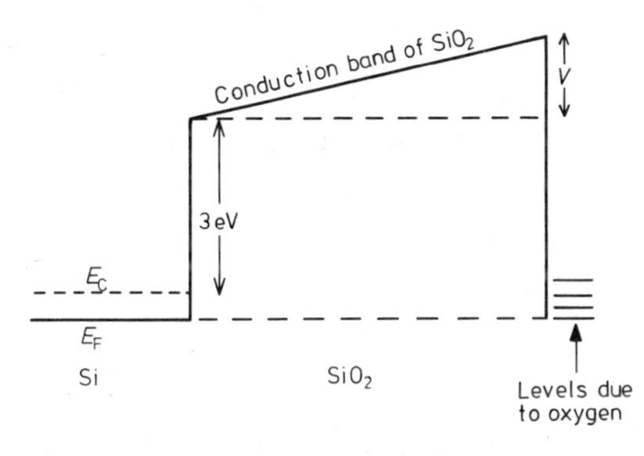

Figure 1. Potential energy diagram for a film of SiO_2 formed on Si. E_C is the bottom of the conduction band of silicon; E_F is the Fermi energy.

ported from metal to the oxide/oxygen interface, the rate-determining step is the removal of a metal ion from a 'kink site' on the metal surface into an interstitial site in the oxide. For this a barrier of height W must be surmounted, and if a is the distance of its summit from an equilibrium position in the kink site in the metal, the chance per unit time of this occurring is:

$$\omega \exp[-(W - qaF)/kT] \tag{2}$$

where ω is of the order of the phonon frequency and q is the charge on the ion. Integration of the resultant rate equation gives a logarithmic growth law. If oxygen diffuses through the oxide or if a metal vacancy is generated at the oxide/oxygen interface, similar considerations can be applied to the formation of an interstitial ion or vacancy there; the electrons in the surface form a two-dimensional metal and the field $F = V/x$ is uniform across the surface (Fehlner and Mott 1970). Equation (2) can also be applied to anodic oxidation, as was first proposed by Verwey (1935).

In the limit of large thickness $x (qaF \ll kT)$, equation (2) leads to a parabolic growth rate, if the exponential is replaced by:

$$\exp(-W/kT) \sinh(qaF/kT) \simeq \exp(-W/kT) aqF/kT. \tag{3}$$

The rate of generation of defects is now inversely proportional to x. However, the pressure dependence in this regime is different from that in the 'classical' parabolic regime, because electrons do not contribute to the entropy. Thus in the oxidation of copper we should expect an oxidation rate proportional to $p^{1/4}$, where p is the pressure of O_2. As far as we know, this regime has not been observed for copper.

Although logarithmic growth laws have had some success in describing the growth of protective and anodic films, the observed films, for instance on aluminium, tantalum or silicon, are not normally crystalline. As far as we know, no model has yet been proposed to describe the passage of 'network-forming' ions or atoms through such films and we do not know whether the concept of a 'defect' is still appropriate. The purpose of this paper is to discuss models for the growth of films of vitreous SiO_2 on silicon heated in oxygen. Vitreous SiO_2 is much better understood than other non-crystalline oxides, such as Al_2O_3, which is why we choose it for a first attempt at a theory.

The most striking fact about the oxidation of silicon is that the (parabolic) rate of oxidation is proportional to the pressure of O_2 and shows a remarkable similarity to the rate of oxygen diffusion through vitreous silica (Motzfeld 1964). Revesz and Zaininger (1968) pointed out that this appears to imply that a neutral or charged O_2 molecule is the diffusing specimen. That oxygen in some form is the diffusing specimen seems almost certain from tracer work with ^{18}O. Thus Rosencher *et al* (1979) found that, when a film first grown to a thickness X_0 in O_2 of natural isotopic constitution is then grown to a thickness $X_0 + dX_0$ in highly enriched O_2, 93% of the heavy isotope is found near the Si–SiO_2 interface, 7% near the SiO_2–O_2 surface and very little in the bulk.

The classical work on the oxidation of silicon is that of Deal and Grove (1965), who found for both dry and wet oxygen a growth law:

$$x^2 + Ax = B(t + \tau) \tag{4}$$

although τ is zero for wet oxygen. B (the parabolic coefficient) is proportional to the partial pressure of O_2 or H_2O and A is independent of this pressure. For the linear range, it is thought that the film contains an almost constant concentration of the diffusing species and the rate is determined by the reaction at the Si–SiO_2 interface. Perhaps surprisingly, A decreases with increasing temperature. Deal and Grove introduced the concept of the Debye length:

$$X_D = [(kT/q)/(\kappa\epsilon_0/2qC)]^{1/2}$$

where C is the concentration of excess oxygen, and obtained a value of 15 Å for dry oxygen at 1000 K, but much less for wet oxygen, because C is 10^3 times larger for H_2O. Their quantity τ is the result of some rapid growth process for $X < X_D$ and for wet oxygen τ is observed to be negligible. The Cabrera–Mott process, whether parabolic or logarithmic, is of course valid only for $X < X_D$. Moreover, if neutral O_2 is the diffusing species, the concept of a Debye length cannot be meaningful. In this connection it is satisfactory that Kamigaki and Itoh (1977) have found that the Cabrera–Mott logarithmic relationship is valid for dry air oxidation for thicknesses below about 200 Å. Also they found an activation energy of 43·9 kcal mol^{-1}, which they compared with the value of 42·2 kcal mol^{-1}, the energy needed to break a Si–Si bond. They appear to assume that the mechanism works only for the movement of ions from the substrate. According to Fehlner and Mott (1970), this is not so, but for this process movement of silicon ions might be preferred because of the high charge on the ion, in spite of the smaller activation energies (28·5 kcal mol^{-1} for dry oxygen, 16·3 kcal mol^{-1} for wet oxygen), reported for the constant B in equation (4). There is no *a priori* reason why the same species should move in the diffusion-controlled and in the Cabrera–Mott regimes. On the other hand I would expect in the latter the same mechanism as for anodic oxidation and, according to Mackintosh and Plattner (1977), the use of noble gases as markers shows that the oxygen is mobile here, although earlier work by Schmidt and Owen (1964) came to contrary conclusions.

We turn now to the nature of the diffusing species. According to the experimental work of Shackleford and Masaryk (1978), the interstitial volume available in SiO_2 is not large and it seems surprising that O_2 and H_2O can be absorbed with a heat of absorption $-W$ and diffuse with an activation energy U, so that $W+U$ is only of the order of 1 eV. We will therefore examine other models. One possibility is that the network deforms so as to form a sort of cavity to admit the molecule. The surface energies of network glasses are believed to be small if the surface does not involve broken bonds, as will presumably be the case for surfaces formed above the glass transition temperature. For liquid SiO_2 the surface energy is about 0·1 eV per bond, so we might expect a small value of W. On the other hand such a mechanism would give a very large value of U, namely that for the viscosity η. For network glasses the Vogel (1921)–Fulcher (1925) equation

$$1/\eta = A \exp\left[-a/(T-T_0)\right]$$

does not apply; the equation was derived by Cohen and Turnbull (1959) from a model in which free volume in a random assembly of hard spheres allows diffusion. According to D Turnbull (private communication; see also Turnbull and Bagley 1975) the viscosity of SiO_2 obeys an Arrhenius equation:

$$1/\eta = A \exp(-B/kT)$$

with B near to the Si–O bond energy ($\sim$4 eV). This mechanism, then, is ruled out.

We next consider defects in SiO_2, which could diffuse and carry oxygen through the network. For SiO_2 the most familiar defect is the 'non-bridging oxygen'; the oxygen should be bonded to one silicon and, if neutral, to one oxygen. One oxygen molecule will form four of these so if they are the relevant defects, the rate of oxidation should go as $p^{1/4}$. If a hole in the valence band can form a polaron by allowing two oxygens to attract each other, as assumed by Mott (1977), *a fortiore* two D^O should be able to form a pair. But even if these were stable at 1300 K, the rate of oxidation would still be proportional to $p^{1/2}$ instead of p as observed, so these defects cannot be the ones that transport oxygen.

To form a silicon vacancy requires the absorption of two oxygen atoms, so if this is the diffusing defect, the dependence on oxygen pressure would be as observed. The first problem, however, is whether the energy to form such a defect can be as low as the observed activation energy for oxidation (1·24 eV for dry O_2, 0·71 eV for wet O_2: Revesz 1967), in view of the large energy ($\sim$4 eV) of the Si–O bond. The second is whether this model is consistent with transport by oxygen. An electric field accelerates oxidation if it is in the direction to drive negative charge towards the silicon (Jorgensen 1962) and it appears therefore that the defects must be negative. Jorgensen also showed by marker experiments that oxygen is the diffusing species. He came to the conclusion that oxygen ions are responsible, and not molecules, but his work is earlier than that of Motzfeld. As regards the transport of electrons, in contrast to Ta and Al, a large electronic current is observed to flow through the oxide during the anodisation of silicon (Schmidt and Mickel 1957, Schmidt 1965). It is thought that 'holes' are generated at the Si–SiO_2 interface, but it is unlikely that these are in the SiO_2 valence band. Mott (1977) proposed that some kind of impurity band, attributable to defect states, allows the passage of current either at strong fields or high temperatures. These states might be the 'slow recombination states' of silicon technology and must have energies near the Fermi energy of the silicon, say 3–4 eV below the SiO_2 conduction band. Assuming that the energies of these states are spread over an energy much greater than kT, with the lowest occupied and the upper ones empty, taking an electron from this impurity band will not significantly change the entropy.

The rate-determining process might then be the formation of a silicon vacancy carrying (probably) a negative charge $2e$. To form such a vacancy an O_2 molecule would combine with one silicon to form an additional molecule of SiO_2 at the surface of the network, two electrons being taken from the impurity band. To do this four Si–O bonds must be broken, but four more are formed at the surface. In the vacancy two oxygen dangling bonds pointing towards its centre could form an O–O bond which would compensate for much of the energy of dissociating O_2. The other two oxygen p orbitals would be fully occupied and carry the negative charge. These oxygen orbitals are similar to the lone-pair orbitals in the valence band, so if the band gap is 9 eV (Mott 1978) the impurity band is 4 eV below the conduction band and the lone-pair orbitals are 2 eV above the valence band, we gain:

$$(2 \times 3 - U_{\mathrm{H}})\ \mathrm{eV}$$

where U_{H} is the interaction energy (Hubbard U) for the pair of electrons in the centre, perhaps 10 eV. Then finally the doubly charged centre will gain energy by polarising its surroundings, by an amount:

$$[(2e)^2/2r]\,[(1/\kappa_0) - (1/\kappa)]$$

where r is the radius of the centre and κ_0 and κ are the high-frequency and static dielectric constants respectively. We estimate this to be:

$$(8/r)\ \mathrm{eV}$$

if r is in Å. The energy to charge the centre is then made up of the following parts:

(*a*) For two electrons to drop from the Fermi energy of the silicon into the lone-pair oxygen orbitals	−6 eV
(*b*) The Hubbard U	+10 eV
(*c*) Polarisation energy if $r = 2A$	−4 eV.

With these values the energy adds up to zero. But obviously all these estimates are crude in the extreme and all we can maintain is that a value between 1 and 2 eV is not unreasonable for the energy to form the charged defect.

However, this model seems to be ruled out by the results of Rosencher *et al* (1979) to which we have already referred. These show that $^{18}O_2$ enters the network, diffuses through it and reacts with the silicon without any exchange of oxygen ions with those already in the network. This fact forces us back to the assumption that the network does contain enough space to admit O_2 and H_2O, and if the diffusing specimen is negative, to allow them to form negative ions. Perhaps particularly surprising is the fact that $W + U$ is less than the value of U for Na^+ ions in vitreous silica according to the results of Owen and Douglas (1959), who obtained a value of ~1·2 eV.

Finally we discuss the effect of n- and p-type dopants on the rate of oxidation. If the temperature is low enough for the dopants to affect the Fermi energy of the silicon, an effect would be expected in the Cabrera–Mott logarithmic and parabolic ranges, but not in the diffusive range (B in equation 4). On the other hand in the linear range of the Deal–Grove equation, an effect of pile-up of the dopant at the interface is likely, and we believe has been observed.

To summarise, it seems that in the parabolic region O_2 diffuses through the network without exchanging atoms with it. In anodic oxidation and for thin films (Cabrera–Mott regime), negative oxygen ions move, but whether in molecular or atomic form or as some defect is not yet clear.

References

Cabrera N and Mott N F 1948/49 *Rep. Prog. Phys.* **12** 163
Cohen, M H and Turnbull D 1959 *J. Chem. Phys.* **31** 1164
Deal B E and Grove A S 1965 *J. Appl. Phys.* **36** 3770
Fehlner F and Mott N F 1970 *Oxidat. Metals* **2** 59
Fulcher G S 1925 *J. Am. Ceram. Soc.* **6** 339
Greaves G N 1978 *The Physics of SiO_2 and its Interfaces* ed S T Pantelides (New York: Pergamon) p268
Jorgensen P J 1962 *J. Chem. Phys.* **37** 874
Kamigaki Y and Itoh Y 1977 *J. Appl. Phys.* **48** 2891
Mackintosh W D and Plattner H H 1977 *J. Electrochem. Soc.* **124** 317
Mott N F 1947a *J. Chim. Physique* **44** 172
—— 1947b *Trans. Faraday Soc.* **43** 429
—— 1977 *Adv. Phys.* **26** 363
—— 1978 *The Physics of SiO_2 and its Interfaces* ed S T Pantelides (New York: Pergamon) p390
Mott N F and Gurney R W 1948 *Electronic Processes in Ionic Crystals* 2nd edn (Oxford: Oxford UP)
Motzfeld K 1964 *Acta Chem. Scand.* **18** 1596
Owen E A and Douglas R W 1959 *J. Glass Technol.* **43** 159
Pepper M 1977 *7th Conf. on Amorphous and Liquid Semiconductors, University of Edinburgh* eds A E Owen and W E Spear p 477
Revesz A G 1967 *Phys. Stat. Solidi* **19** 193
Revesz A G and Zaininger K H 1968 *RCA Rev.* **29** 1
Rosencher E A, Straboni A, Rigo S and Amsel C 1979 *Appl. Phys. Lett.* **34** 254
Schmidt P F 1965 *IEEE Trans. Electron Devices* **ED-12** 102
Schmidt P F and Mickel W 1957 *J. Electrochem. Soc.* **104** 230
Schmidt P F and Owen A E 1964 *J. Electrochem. Soc.* **111** 682
Shackleford J F and Masaryk J S 1978 *J. Non-Cryst. Solids* **30** 127
Street R A and Mott N F 1979 *Phys. Rev. Lett.* **35** 1293
Turnbull D and Bagley B C 1975 *Treatise on Solid State Chemistry* vol 5 ed N B Hannay (New York: Plenum) p573
Verwey E J 1935 *Physica* **2** 1059
Vogel H 1921 *Phys. Z.* **22** 645
Wagner C and Grunewald K 1938 *Z. Phys. Chem.* **B40** 455

The role of water in the oxidation of silicon

D L Wolters
Philips Research Laboratories, Eindhoven, The Netherlands

Abstract. The existing theory on thermal oxidation kinetics of silicon is extended by accounting for the solubility and transport behaviour of water in silica. From the solubility data of water in silica reported in the literature it can be shown clearly that the dissolution of water is a two stage process at temperatures up to 1200 °C: during the first stage silanol groups are formed (slow process); and in the second stage these silanol groups are subsequently hydrolysed forming hydronium ions and non-bridging oxygen atoms in the silica network (fast process).

Consistent with this incorporation model it is shown that the transport of water occurs by ambipolar diffusion of hydronium and hydroxyl ions.

The incorporation model and the transport of effectively two water molecules (H_3O^+ + OH^-) leads consequently to a parabolic rate constant proportional to water pressure.

The formation of silanol groups (first stage) is a retarded process and must consequently lead to a time dependent diffusion coefficient. Incorporation of this time dependence in the transport equation results in a linear–parabolic expression for oxide growth.

The catalytic role of water when present in trace amounts in oxygen requires a model which accounts for the interaction of the water and oxygen diffusion. Such a model is presented and it properly describes the growth data in mixed ambients. It is shown that the so called 'initial growth process' is no more than an implication of the model.

1. Introduction

The well-known linear–parabolic growth theory for oxide layers on silicon from Deal and Grove (1965) is expressed by the equation

$$k_{par}^{-1}x_0^2 + k_{lin}^{-1}x_0 = t + t^*$$

or

$$x_0^2 + Ax_0 = B(t + t^*)$$

where k_{par} $(= B)$ is the parabolic rate constant, and k_{lin} $(= B/A)$ is the linear rate constant, x_0 is the oxide thickness, and t is time.

For oxidation in pure steam the theory assumes Henry's law to be valid (equilibrium concentration of water or solubility $\sim P_{H_2O}$). However, this is contradicted by the solubility data for silica (Revesz and Evans 1969). The solubility varies with $P_{H_2O}^{1/2}$. The parabolic rate constant should vary with the solubility (i.e. $\sim (P_{H_2O})^{1/2}$).

For oxidation in O_2/H_2O mixtures it is assumed that oxygen and water diffuse independently and as a consequence the linear rate constant should vary proportionally with the partial pressure of water (Deal *et al* 1978).

The parameter t^* is introduced in the expression to correct for what is called the 'initial growth process'.

0305-2346/80/0050-0018$02.00

We shall present a model which makes clear why $k_{par} \sim P_{H_2O}$. It will also point out that $k_{lin} \sim (P_{H_2O})^{1/2}$ in mixed ambients and that traces of water can enhance the growth rate. There is no need to introduce a separate initial growth process. It will be shown that there is an excellent fit of the derived expression with published data.

2. The incorporation of water in silica

The solubility of water in silica is not unambiguously determined by temperature and pressure as can be seen in figure 1(*a*). Different samples with various thermal pre-treatments have different solubilities. The thermal pre-treatment or equilibration at a certain high temperature is usually characterised by a so-called 'fictive temperature'. A plot of the solubility at a certain (low) temperature (e.g. 750 °C) against the recip-

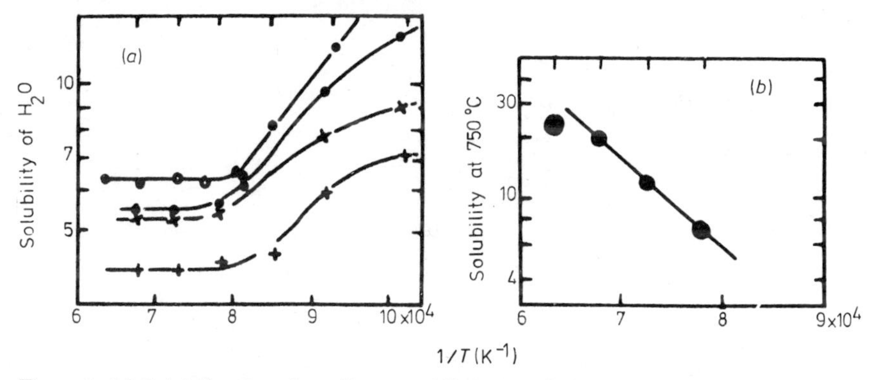

Figure 1. (*a*) Solubility data from Drury and Roberts (1963). Variation of the solubility of water in silica with the reciprocal absolute temperature. The parameter is the thermal history. (*b*) Variation of the solubility with the reciprocal fictive (!) temperature (Roberts and Roberts 1964).

rocal fictive temperature shows that there is an activated high temperature process which determines the low temperature dissolution (see figure 1*b*).

Water dissolved at high temperatures ($\gg$1500 °C) was shown to be incorporated as free and hydrogen-bonded silanol groups (Walrafen 1975). The proposed equilibrium reaction is:

$$\vdots\text{Si–O–Si}\vdots + H_2O \rightleftharpoons 2\vdots\text{SiOH}. \tag{1}$$

Water dissolved at relatively low temperatures ($\ll$1200 °C) shows a much broader absorption spectrum in the infrared region, indicating other incorporation mechanisms. Therefore we propose a second mechanism in which the silanol groups of (1) are hydrolysed:

$$\vdots\text{SiOH} + H_2O \rightleftharpoons \vdots\text{SiO}^- + H_3O^+. \tag{2}$$

In silanol formation the solubility (here defined as the equilibrium concentration of water related species) varies proportionally with the square root of water pressure:

$$[\vdots\text{SiOH}] \sim (P_{H_2O})^{1/2}. \tag{3}$$

In the second (hydrolisation) process the concentration is given by:

$$[H_3O^+] = [\vdots\text{SiOH}]\,[\text{SiO}^-]^{-1} P_{H_2O}. \tag{4}$$

The electroneutrality condition requires $[H_3O^+] = [SiO^-]$ and when it is assumed that the silanol concentration is frozen in, the solubility is given by:

$$\text{solubility} = [H_3O^+] = [SiOH]^{1/2}\,(P_{H_2O})^{1/2}. \tag{5}$$

3. Transport of water in silica

The transport of water can be investigated by inspection of the diffusion profiles of hydrogen incorporated during silicon oxidation. Breed and Doremus (1976) concluded from these profiles that water diffused in molecular form through the silica layer while it reacted with the silica to form silanol groups. Here, we will derive the slope that is to be expected in a lg(concentration)–lg(distance) plot, and will then prove that bimolecular transport of water is more probable than monomolecular transport.

The hydrogen, formed from the reduction of water by silicon, diffuses back to the ambient but meanwhile undergoes reaction. The proposed reaction is:

$$\vdots\,\text{Si–O–Si}\,\vdots + H_2 \rightleftharpoons \,\vdots\,\text{SiH} + \,\vdots\,\text{SiOH}. \tag{6}$$

For a growing layer, the steady-state condition may be assumed and thus the flux of oxidising species is constant and a linear concentration gradient will be set up. The growth rate $\mathrm{d}x_0/\mathrm{d}t$ is then proportional to the concentration gradient:

$$\mathrm{d}x_0/\mathrm{d}t = +M_0^{-1}|J| = M_0^{-1}D(\partial c/\partial x)_{x=0} = M_0^{-1}D(\Delta c/x_0) \tag{7}$$

where D is the diffusion coefficient, J is the flux density, Δc is the concentration difference over the layer and x_0 is the layer thickness. M_0 is the number of molecules needed to grow unit volume of oxide. $x_0 = 0$ at the Si/SiO_2 interface.

When for instance H_2O is the diffusing species (water in molecular form as concluded by Breed and Doremus 1976) the water concentration gradient is constant but the silanol concentration must vary with the square root of distance:

$$[H_2O] \sim [\,\vdots\,\text{SiOH}]^2 \sim x. \tag{8}$$

or

$$\ln[\,\vdots\,\text{SiOH}] = \tfrac{1}{2}\ln(x) + \text{constant}.$$

The flux of hydrogen formed at the silicon must be equal and opposite to the water flux, or:

$$J_{H_2} = -J_{H_2O}. \tag{9}$$

The concentration gradient of hydrogen is constant but opposite to that of the water:

$$[H_2] \sim x_0 - x. \tag{10}$$

From equation (6) we can state that:

$$\ln[H_2] = \ln[\text{SiOH}] + \ln[\,\vdots\,\text{SiH}] + \text{constant}. \tag{11}$$

Substitution of equations (8) and (10) in logarithmic form into equation (11) gives:

$$\ln[\,\vdots\,\text{SiH}] = -\tfrac{1}{2}\ln(x) + \ln[(x_0 - x)/x_0] + \text{constant} \tag{12}$$

$$\ln[(x_0 - x)/x_0] \approx 0 \qquad \text{when } x \approx 0.$$

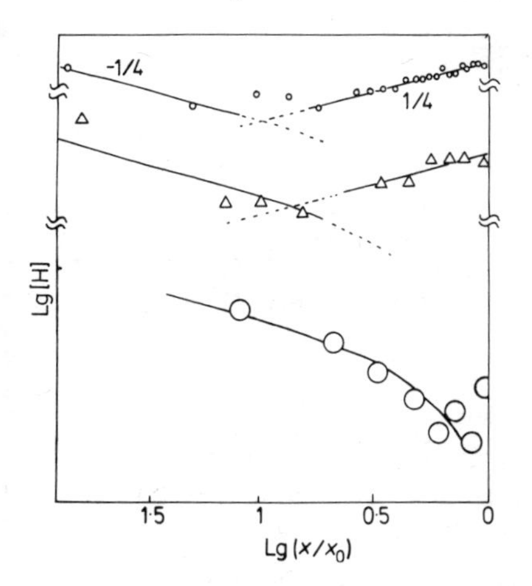

Figure 2. Compilation of H profiles in silica from Burkhardt (1967) to demonstrate the slopes $-\frac{1}{4}$ and $\frac{1}{4}$ on a lg(concentration)–lg(x/x_0) plot. The curves drawn correspond to equations (13) and (14). ○ $x_0 = 6750$ Å; △ $x_0 = 4400$ Å; ◯ $x_0 = 820$ Å.

The hydrogen profile consists of silane groups at the silicon side and silanol groups at the ambient side.

The same method can be used to show that when $H_3O^+ + OH^-$ is transported instead of H_2O, the coefficients in the logarithmic expressions (8) and (12) are divided by two (compare equation (8) with (13) and equation (12) with (14)):

$$\ln[\vdots\text{SiOH}] = \tfrac{1}{4}\ln(x) + \text{constant} \tag{13}$$

$$\ln[\vdots\text{SiH}] = -\tfrac{1}{4}\ln(x) + \tfrac{1}{2}\ln[(x_0 - x)/x_0] + \text{constant} \tag{14}$$

for instance $[H_3O^+ + OH^-] = [(H_2O)_2] \sim [H_2O]^2 \sim [\vdots\text{SiOH}]^4 \sim x$.

The experimental data reported by Burkhardt (1967) are replotted in figure 2 to show that the slopes are more likely to be ¼ and $-$¼ than ½ and $-$½. From these profiles we can conclude that the water species are probably transported by ambipolar diffusion of H_3O^+ and OH^-. Other forms of dimeric transport $(H_2O)_2$ are also possible.

4. The parabolic oxidation rate

The parabolic rate constant can be found from equation (7):

$$\mathrm{d}x_0/\mathrm{d}t = k''D[(C_s - C_i)/x_0]. \tag{15}$$

C_s and C_i are the concentrations at the outer and inner interfaces respectively and k, k' and k'' are constants (k and k' to be used below).

Assuming $C_i \ll C_s$, equation (15) becomes, after integration:

$$x_0^2 = k'DC_st. \tag{16}$$

Here C_s is the concentration of the diffusant $(H_3O^+ + OH^-)$ at the surface. It can be related to the partial pressure in the ambient by:

$$C_s = [H_3O^+ + OH^-]_s = [(H_2O)_2]_s \sim [H_2O]_s^2 = (\text{Solubility})^2$$

and by substituting equation (5) we get

$$C_s \sim (\text{Solubility})_s^2 \sim P_{H_2O}[\vdots\text{SiOH}]_s. \tag{17}$$

Substitution of equation (17) into (16) gives:

$$x_0^2 = kDP_{H_2O}[\vdots SiOH]_s t \equiv k_{par} t. \tag{18}$$

From equation (18) it follows that k_{par} varies linearly with P_{H_2O}. This is, however, not a consequence of Henry's law, as was supposed by Deal and Grove (1965), but it results from the dependence of the solubility on the square root of the water pressure and the coupled transport of two water molecules ($H_3O^+ + OH^-$).

5. Time-dependent transport

When the solubility of water on silica changes slowly with time, this must affect the effective diffusion coefficient. The effective diffusion coefficient is equal to the product of a mobility and a concentration. When the latter reaches equilibrium with a large relaxation time, the diffusion coefficient D_{eff} will be time-dependent (Crank 1975).

We derive a general growth equation in which the time dependence of the diffusion coefficient is accounted for. The treatment is similar to the derivation given by Deal and Grove (1965) when a new time parameter T is introduced. With this parameter a steady-state approximation can be made and the derivation results in an expression in which the real time parameter is replaced by T. The limiting cases for the expression will be illustrated.

We propose an expression such as:

$$D_{eff} = D_e[1 - \exp(-t/\tau)] \tag{19}$$

where D_e is the equilibrium diffusion coefficient at $t = \infty$. We define a parameter T by:

$$dT \equiv (D_{eff}/D_e)\,dt = [1 - \exp(-t/\tau)]\,dt \tag{20}$$

or:

$$T \equiv D_e^{-1}\int D_{eff}\,dt = t + \tau\exp(-t/\tau) - \tau. \tag{21}$$

With the aid of equations (19) and (20), Fick's law

$$\partial c/\partial t = (\partial/\partial x)[D_{eff}(\partial c/\partial x)] \tag{22}$$

can be rewritten as

$$\partial c/\partial T = (\partial/\partial x)[D_e(\partial c/\partial x)] \tag{23}$$

(cf Crank 1975 chap 7). Now the right-hand side of equation (23) is time-independent and in the steady state approximation $\partial c/\partial T = 0$.

The growth of the oxide layer is described by equation (7) and substitution of D by D_e with equation (20) gives:

$$dx_0/dT = M_0^{-1}D_e(\partial c/\partial x)_{x=0}. \tag{24}$$

Furthermore, we assume for simplicity that the boundary-reaction rate constants are time-independent. At any moment the reaction rate at the boundary must equal the transport by diffusion:

$$\begin{aligned} D_e(\partial c/\partial x) &= h(C^* - C_s) \qquad & x &= x_0 \\ D_e(\partial c/\partial x) &= kC_i \qquad & x &= 0. \end{aligned} \tag{25}$$

C^* is the concentration in the ambient and is taken to be constant; h and k are the reaction rate constants at $x = x_0$ and $x = 0$ respectively.

The condition for using the steady state approximation is:

$$\mathrm{d}x_0^2/\mathrm{d}T \ll 6D_e.$$

The solution of equations (24), (25) and (26) is quite similar to the case with constant D given by Deal and Grove (1965):

$$\frac{x_0}{A/2} = \left(1 + \frac{T}{A^2/4B}\right)^{1/2} - 1 \tag{28}$$

with $A = 2D_e[(1/k) + (1/h)]$ and $B = 2D_e C^*/M_0$. The only difference is that we replace t by T. If we use expression (21) for T and substitute it into equation (28) we get:

$$\frac{x_0}{A/2} = \left[1 + \left(\frac{t + \tau \exp(-t/\tau) - \tau}{A^2/4B}\right)\right]^{1/2} - 1. \tag{29}$$

The limiting cases are shown in table 1. When τ is negligibly small, the Deal–Grove equation is obtained. In the other limit, when $A^2/4B$ is small, we obtain the pure retarded solubility expression:

$$x_0 = B^{1/2}[t + \tau \exp(-t/\tau) - \tau]^{1/2}. \tag{30}$$

Table 1. Limiting cases for equation (29).

$t \gg \tau$	$t \gg A^2/4B$		$x_0 \simeq (Bt)^{1/2}$
$t \ll \tau$	$t \ll A^2/4B$		$x_0 \simeq (2B/A\tau)t^2$
$t \gg \tau$	$t \ll A^2/4B$		$x_0 \simeq (B/A)t$
$t \ll \tau$	$t \gg A^2/4B$	$t \ll (A^2\tau/2B)^{1/2}$	$x_0 \simeq (2B/\tau)^{1/2}t$
$t \ll \tau$	$t \gg A^2/4B$	$t \gg (A^2\tau/2B)^{1/2}$	$x_0 \simeq (B/2\tau)^{1/2}t$

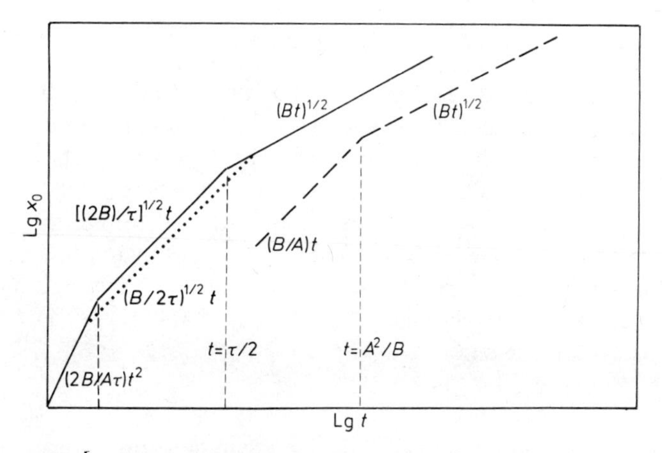

Figure 3. Schematic representation of various growth rates. The lines are extrapolated in the transition regions to demonstrate the rate laws. The situation with $\tau = 0$ is represented by the broken curve. The axes are not coincident.

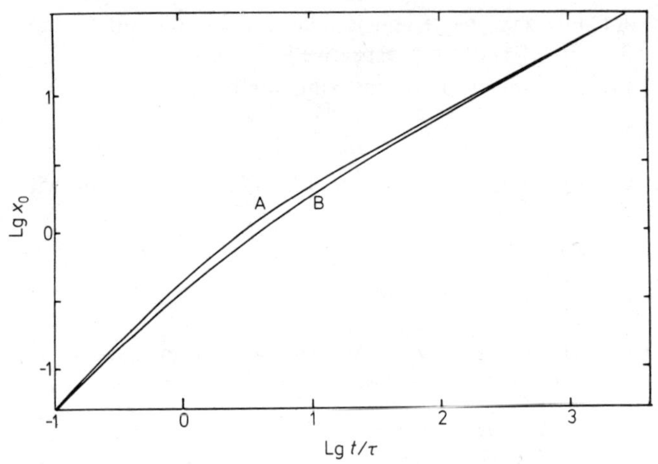

Figure 4. Linear–parabolic expressions with $A^2/4B = \tau$. Curve A, $x_0^2 = k\tau[(t/\tau) + \exp(-t/\tau) - 1]$; curve B, $x_0^2 = k\tau[1 + (t/\tau)]^{1/2} - 1$, (Deal–Grove). The growth rates in the transition region have, theoretically, a maximum difference in x_0 of only 22%.

The form of equation (29) can be examined by plotting the limiting cases on a lg–lg scale. This is shown in figure 3. The full curves correspond to $A^2/4B \ll \tau$, i.e. the dissolution of water in silica inhibits the growth rate. The dotted lines correspond to $A^2/4B \gg \tau$, i.e. the boundary reactions are rate-determining.

Both linear–parabolic expressions are plotted in figure 4 with $A^2/4B = \tau$. The maximum difference is only 22%. The experimental determinations are, however, precise enough to be able to discriminate between both expressions.

6. Oxidation in O_2–H_2O ambients

When silicon is oxidised in oxygen, trace amounts of water influence the oxidation rate. The parabolic rate constant increases significantly when the partial pressure of water is raised from less than 1 to 25 ppm (Irene and Ghez 1977).

The catalytic role of water in the oxidation process can be understood if we assume sequential diffusion of an oxygen atom, first as part of an oxygen molecule, then as part of a water molecule or rather as part of $(H_3O^+ + OH^-)$. If we increase the partial pressure of water in the ambient, at a certain point, the permeability will become greater than the permeability† of oxygen. Now water will be consumed at the silicon interface and converted to hydrogen. The hydrogen diffuses back to the ambient where it will be oxidised by the oxygen to form water. This causes the water concentration to rise, the transport of water to the Si interface will increase, together with the flux of hydrogen towards the oxygen. The situation stabilises when all the oxygen is consumed by hydrogen before it can reach the Si interface and then the fluxes of water species and oxygen species must be equal under the steady-state condition:

$$J_{(H_2O)_2} = J_{O_2} = \text{constant}. \tag{31}$$

This is shown schematically in figure 5. On the right-hand side of the crossover point, there is no concentration gradient of water because each water molecule consumed is

† The transport of oxygen and water is expressed here in terms of the permeability $P = DC$:

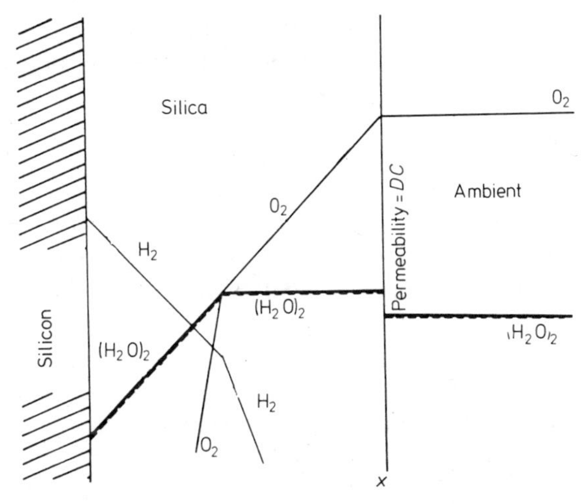

Figure 5. Schematic representation of the permeability profiles of O_2 and $(H_2O)_2$. The transport of oxygen atoms may be regarded as sequential. Water species transport oxygen atoms on the left-hand side where the water permeability exceeds the (decreasing) oxygen permeability.

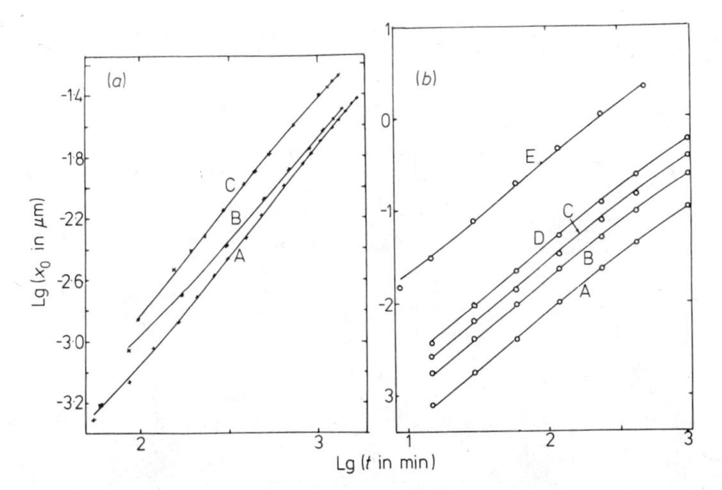

Figure 6. (*a*) Data of Irene and Dong (1978), fitted to equation (33), to demonstrate that the initial growth may be regarded as part of the growth process in mixed ambients. Deviations are within experimental error. Curves A, B, C: standard, B and P doped Si respectively; (*b*) Data of Deal *et al* (1978) fitted to equation (33). The deviations of these data from Deal–Grove theory are much larger. Curves A, B, C, D, E represent 0, 2, 5, 10% and pyrogenically formed water respectively.

replenished by oxidation. No oxygen molecules react with the silicon directly if the water pressure in the ambient is large enough.

It can be shown that the series transport of oxygen and water may be represented by an expression where the parabolic rate constants are added (Wolters 1979):

$$x_0^2 = (k_{H_2O} + k_{O_2})\, t. \tag{32}$$

This expression is only valid for pure parabolic growth. When the solubility of water changes with time it may be shown that we can introduce the time function T from expression (21) (Wolters 1979). The linear–parabolic expression for growth then becomes

$$x_0^2 = (k_{H_2O} + k_{O_2})t + k_{H_2O}\tau[\exp(-t/\tau) - 1]. \quad (33)$$

It may be seen by expansion of the exponential term that at very small values of t/τ there must be an 'initial' parabolic growth, that exceeds the linear process, when the silicon is oxidised by oxygen directly. The period of 'initial' growth is when

$$t < (k_{O_2}/k_{H_2O})\tau. \quad (34)$$

Expression (33) was fitted to the experimental data of Irene and Dong (1978) and Deal *et al* (1978). Values of k_{O_2}, k_{H_2O} and τ were calculated by a non-linear regression method. The results are shown in figure 6(*a*) and (*b*), the deviations being within experimental error.

7. Discussion

When the data of Pliskin (1966) for oxide growth in pure water ambients are fitted on the curves corresponding to $\tau \ll A^2/4B$ and $\tau \gg A^2/4B$ it appears that the Deal–Grove theory gives the best curve-fitting. The transport is thus inhibited at one of the

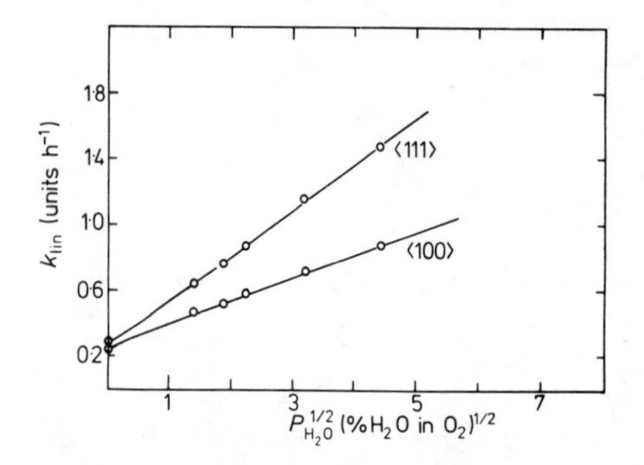

Figure 7. Data of Deal *et al* (1978) for k_{lin} plotted against $P^{1/2}_{H_2O}$.

boundaries. This also follows from the fact that the characteristic time k_{par}/k_{lin}^2 ($=A^2/4B$) varies with the inverse of the water pressure (Pliskin 1966).

For water–oxygen ambients it can be made clear that $\tau \gg A^2/4B$ and thus the solution reaction of water in silica is the rate-determining step in the transport of oxidant. This follows from the fact that k_{par}/k_{lin}^2 ($= \tau$) is almost independent of the partial water pressure (Wolters 1979).

In the Deal–Grove theory k_{lin} must vary linearly with partial water pressure. It is shown in figure 7 that there is obviously a linear relationship between k_{lin} and $P^{1/2}_{H_2O}$. This relationship follows from equation (33) by ignoring the first 'initial growth' term and expanding the second term. For $t \ll \tau$ we can omit the higher-order terms

$$x_0^2 \simeq k_{H_2O}\tau[(1/2!)(t/\tau)^2] \simeq k_{lin}^2\tau \quad (35)$$

and thus

$$k_{\text{lin}} \simeq k_{\text{H}_2\text{O}}^{1/2} \simeq P_{\text{H}_2\text{O}}^{1/2}. \tag{36}$$

The catalytic action of water on growth in an oxygen ambient and the occurrence of the 'initial growth', together with the perfect fit of equation (33) to the experimental points in figure 6(*a*) and (*b*) are strong indications that equation (33) gives a more distinctive description of the growth curves than the well-known Deal–Grove curves.

8. Conclusions

It has been shown in §2 that the solution of water can be regarded as a two-stage process. The complex solubility behaviour can be explained, at least in part. Inspection of the diffusion profiles of hydrogen in §3 leads to the conclusion that water is transported through silica in bimolecular form, probably as H_3O^+ and OH^- in ambipolar diffusion. In §4 it is shown that the model given in §§2 and 3 can explain the linear relationship between k_{par} and $P_{\text{H}_2\text{O}}$ and that this relationship is not a consequence of Henry's law.

In §5 an expression is derived for oxide growth when the diffusion coefficient changes slowly with time. In the equation obtained, the particular case of 'Deal–Grove' behaviour is implicated.

In §6 the 'catalytic' role of water has been explained in terms of sequential transport of an oxygen atom by an oxygen molecule and water species (H_3O^+ and OH^-). This mechanism can provide an explanation for the 'initial growth' process and for the variation of k_{lin} with the square root of the partial water pressure. The expression derived fits the experimental points perfectly.

Acknowledgments

The author acknowledges the critical reading of this manuscript by P Hart and discussions with D Breed, L Heyne, J Snel, C van Opdorp and J Verwey.

References

Breed D J and Doremus R H 1976 *J. Phys. Chem.* **80** 2471–3
Burkhardt P J 1967 *J. Electrochem. Soc.* **114** 196–200
Crank J 1975 *The Mathematics of Diffusion* (Oxford: Clarendon)
Deal B E and Grove A S 1965 *J. Appl. Phys.* **36** 3770–8
Deal B E, Hess D W, Plummer J D and Ho C P 1978 *J. Electrochem. Soc.* **125** 339–346
Drury T and Roberts J P 1963 *Phys. Chem. Glasses* **4** 79–90
Irene E A and Dong D W 1978 *J. Electrochem. Soc.* **125** 1146–51
Irene E A and Ghez R 1977 *J. Electrochem. Soc.* **124** 1757–61
Pliskin W A 1966 *IBM J. Res. Dev.* **10** 198–206
Revesz A G and Evans R J 1969 *J. Phys. Chem. Solids* **30** 551–64
Roberts G J and Roberts J P 1964 *Phys. Chem. Glasses* **5** 26–32
Walrafen G E 1975 *J. Chem. Phys.* **62** 297–9
Wolters D R 1979 to be published

Electron trapping in SiO_2

D R Young
IBM T J Watson Research Center, Yorktown Heights, New York 10598, USA

Abstract. Electron trapping in SiO_2 has become an important limitation in the design of n-channel MOSFET structures. This consideration is particularly important for short-channel devices. This paper discusses the electron-trapping behaviour of the SiO_2 layer used for MOSFET's. The effects of processing conditions, SiO_2 thickness and temperature of measurement are discussed. Techniques for characterising the electron traps are described that enable the trap cross section, trap density and location of the traps in the SiO_2 to be determined.

1. Introduction

Electrons can acquire sufficient energy in Si semiconductor devices to surmount the Si to SiO_2 interface barrier of 3·2 eV and thus flow through the SiO_2. This can occur as a result of carriers flowing from source to drain in MOSFET's (Abbas and Dockerty 1975, Troutman 1978) or it can occur from reversed-biased p–n junctions (Bosselaar 1973, Pepper 1973, Verwey and De Maagt 1974). Some of the electrons that flow through the SiO_2 are trapped and build up space charges that change the device characteristics. This becomes increasingly important for short-channel devices, since the electrons lose less energy due to collisions in the Si and therefore a larger percentage can surmount the barrier and be trapped in the SiO_2. In addition to these detrimental effects, electron traps can be used to improve the electrical breakdown characteristics of SiO_2 (DiMaria *et al* 1977a, b) and can be used to store charge in charge-storage devices (Kahng *et al* 1974, Yoshimo *et al* 1977, Williams 1965). These practical considerations suggest the desirability of obtaining a better understanding of the factors controlling the electron-trapping characteristics of SiO_2 and of developing methods for enhancing the trapping when this is desired.

2. Experimental procedures

Various methods have been used to induce electron current flow which is required for trap characterisation, into the SiO_2. These methods include photoionisation (Williams 1965), p–n junction devices biased into avalanche (Verwey and Heringa 1977), MOSFET structures with an underlying forward-biased epitaxial layer (Verwey 1973), MOSFET structures with optical generation of minority carriers in Si (Ning and Yu 1974) and RF avalanche techniques using MOS capacitor structures (Nicollian *et al* 1969). The last technique has been chosen for most of this work because of the simplicity of the devices required (MOS capacitors) and the uniformity of the injected current density. The charge build-up arising from trapped charge can be monitored by changes in device character-

0305-2346/80/0050-0028$02.00

istics, by the use of photo I–V measurements (DiMaria 1976) and by C–V measurements (Kerr *et al* 1964).

The voltage shift (ΔV_{FB}) measured by the C–V technique is given by

$$\Delta V_{FB} = qN\bar{x}/\epsilon_{ox} \tag{1}$$

where q is the electronic charge, N is the total number of filled traps in the oxide per unit area and $\bar{x}$ is the centroid of the charge measured with respect to the gate electrode. The essential point is that this measurement depends on $N\bar{x}$ and N and $\bar{x}$ cannot be measured independently. The photo I–V technique enables two independent measurements to be made, the voltage shift for electron emission from the silicon and the voltage shift for electron emission from the metal gate electrode. The sign of the applied voltage is used to determine the emitting electrode. Positive gate voltage is used for silicon emission and negative gate voltage for gate electrode emission. These two independent measurements make it possible to separate N and $\bar{x}$.

The relationships are given by

$$\bar{x}/D_{ox} = [1 - (\Delta V_G^-/\Delta V_G^+]^{-1} \tag{2}$$

$$N = (\epsilon_{ox}/D_{ox})(\Delta V_G^- - \Delta V_G^+) \tag{3}$$

where D_{ox} is the oxide thickness and ΔV_G^+, ΔV_G^- are the voltage shifts (for the same current) for the gate voltage positive and negative respectively. It should be noted that ΔV_G^+ and ΔV_G^- have opposite signs for a given charge: a negative oxide charge gives a positive ΔV_G^+ and a negative ΔV_G^-. If the charge is uniformly distributed throughout the oxide, then $\Delta V_G^+ = -\Delta V_G^-$ and $\bar{x}/D_{ox} = 0{\cdot}5$. In this case the actual charge density/unit area is a factor of two larger than the effective charge $N\bar{x}/D_{ox}$ determined from C–V measurements. Another important difference between these measurements arises because the photo I–V technique is insensitive to charges at or very close (< 30 Å) to the interfaces, whereas charge at the Si–SiO_2 interface does affect the C–V measurement. As a result, differences between ΔV_{FB} and ΔV_G^+ can be used as a measure of charge very close to this interface.

The avalanche injection technique can be used to inject electron currents covering a wide range (10^{-12}–10^{-6} Å). Internal photoemission is only useful for the smaller currents. Small currents are used for traps with large cross sections and large currents are used for traps with small cross sections as observed in dry, well annealed oxides. The procedure that we use to characterise traps is as follows: For a single trap the flat-band voltage shift (ΔV_{FB}) is given by

$$\Delta V_{FB} = \Delta V_0(1 - C^{-t/\tau}) \tag{4}$$

where ΔV_0 is the flat-band voltage shift when all the traps are saturated and τ, the time constant, is given by

$$\tau = q/I\sigma. \tag{5}$$

The current density is I and σ is the trap cross section. The saturated flat-band voltage shift is

$$\Delta V_0 = \frac{\bar{x}}{\epsilon_{ox}} \int_0^{D_{ox}} N_V \, dx \tag{6}$$

where N_V is the volume density of the traps. From equation (4) we can obtain

$$\ln [d(\Delta V_{FB})/dt] = \ln (\Delta V_0/\tau)(t/\tau). \qquad (7)$$

A plot of ln $d(\Delta V_{FB})/dt$ against t should yield a straight line of slope $-1/\tau$ and intercept $\ln(\Delta V_0/\tau)$. This plot makes it possible to see whether there is only one trap. If several traps are involved but have cross sections separated by at least a factor of two, then we see a break in the curve which means that equation (4) needs to be replaced by a sum of several exponentials. The traps with the largest cross section are saturated first and then the next smaller traps become important, etc. In practice, we first fit the data for the longest times to an exponential (this would characterise traps with the smallest cross section). We subtract this exponential from the original data and then fit the next largest trap. With our automatic apparatus, 500–1000 measurements can be made in one run, and it has been found possible to characterise three different traps by this procedure. Our computer programs assist in this characterisation. To cover a wide range of trap cross sections, it is necessary to change the current; this can be done on a run-to-run basis or it can be done during a particular run. It is convenient to scale the effective time-scale when the current is changed during a particular run to enable us to use the same programs. An example of this is shown in figure 1. The original data are shown on this scaled time-scale and the effect of subtracting the exponentials corresponding to the various traps evaluated is also shown.

A schematic diagram of the automatic avalanche injection apparatus is shown in figure 2. In the injection mode a RF waveform (500 kHz square wave here) was applied to the sample. The amplitude of this square wave was adjusted automatically to maintain a constant, preset, average current through the sample as monitored by the electrometer. In more recent versions of this apparatus, a 50 kHz saw-tooth waveform was used; this was required to allow the same apparatus to be used for hole injection. Periodically, the avalanche injection process was interrupted to measure the flat-band voltage shift automatically as an indication of the charge build-up in the insulator. The data were stored and analysed in a computer.

The samples were MOS capacitors using 0·1–0·2 Ω cm p-type silicon, 200–1500 Å of dry SiO_2, and evaporated Al electrodes. Various heat treatments were used as described in subsequent sections. Cleaning procedures were similar to those used for contemporary

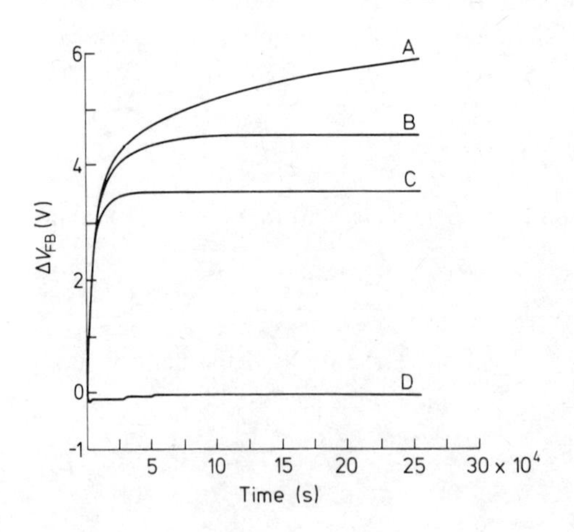

Figure 1. ΔV_{FB} plotted against time. Time-scale scaled after 2000 s by a factor of six to correct for the increase in current from 5×10^{-8} to 3×10^{-7} A that occurred at that time. The actual time of the experiment was 50 000 s, but the effective time was 300 000 s. Curve A, original data; curve B, without $4{\cdot}7 \times 10^{-26}$ cm² trap; curve C, without $5{\cdot}0 \times 10^{-19}$ cm² trap; curve D, without $2{\cdot}9 \times 10^{-18}$ cm² trap.

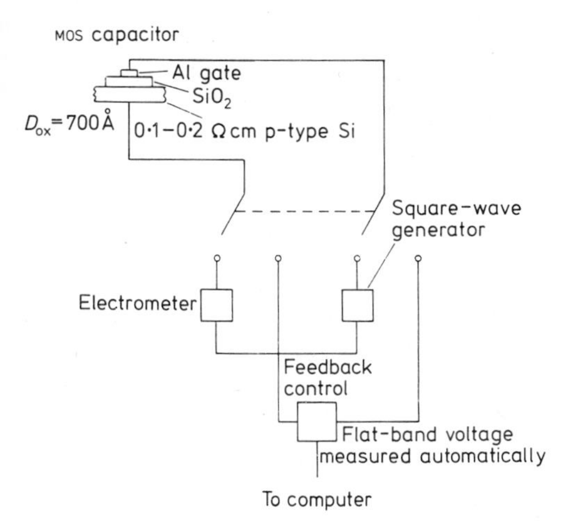

Figure 2. Schematic drawing of the apparatus.

high-quality MOSFET processing and the Al was applied as soon as possible after high-temperature processing (1000 °C) to minimise contamination of the unprotected SiO_2. The Al was used not only as an electrode but also for protection of the SiO_2. The Al thickness was nominally 3000 Å; however, this had to be reduced to ~100 Å to permit photo I–V measurements. Avalanche injection can be used with the thin or the thick Al electrodes.

3. Description of previous results

Nicollian *et al* (1971) studied electron trapping in SiO_2 as a result of the diffusion of 'water' into the SiO_2. They used MOS capacitors and the avalanche injection technique (Nicollian *et al* 1969). This work clearly demonstrated the importance of 'water' in SiO_2 in determining the electron-trapping characteristics. Their results indicated a square-root dependence on the water pressure which they suggested indicates that the water breaks up into two fragments in diffusing through the SiO_2. By using etch-back techniques, they were able to measure the concentration profile of the induced traps and to determine a capture cross section of $1{\cdot}5 \times 10^{-17}$ cm^2. Their model suggests that the capture of an electron on a water-related site results in the formation of atomic hydrogen which is released and diffuses out of the samples.

Ushirokawa *et al* (1973) measured a trap cross section of $\sim 2 \times 10^{-18}$ cm^2 that they ascribed to water-related centres. An extensive study of trapping in steam-grown oxides has been carried out by DiMaria *et al* (1975). The centres they studied are believed to be due to complexes of Na, grown into the SiO_2 during the growth process and water-related species. The traps have a depth of 2·4 eV as measured by photoemission techniques. The capture cross section for these traps was $\sim 10^{-14}$ cm^2.

Gdula (1976) studied the effect of processing on electron trapping in SiO_2. The oxides he studied were grown in dry O_2, O_2 + HCl, O_2 + H_2O steam and CVD SiO_2. He observed a variation in trapping efficiency of three orders of magnitude depending on forming and annealing conditions. The dry O_2 process gave the lowest trapping efficiency and the CVD SiO_2 gave the largest trapping efficiency. Gdula also observed an anomalous effect that subsequent work has shown to be caused by the build-up of positive charge at the

$Si-SiO_2$ interface (Young *et al* 1979, Weinberg *et al* 1979). This occurred as a result of the passage of an electron current through the SiO_2 and it has been shown that the addition of H_2O to the SiO_2 accelerates this effect (Weinberg *et al* 1979).

Ning *et al* (1975) studied traps with a capture cross section of 3×10^{-13} cm^2 that are related to the built-in oxide charge. This was altered by using various post-metallisation treatments in forming gas.

Impurities added to the SiO_2 enhance the trapping rate. Examples of this are the work on implanted Au (Chen *et al* 1972), deposited W (Kahng *et al* 1974, Young *et al* 1977), implanted Al (Young *et al* 1978, DiMaria *et al* 1978) and implanted As and P (DiMaria *et al* 1977a, b). It is interesting to note that electrons cannot be trapped on Na^+ ions diffused into SiO_2 at room temperature but can be trapped at 77 K with a very small trap cross section of 10^{-19}–10^{-20} cm^2 (DiMaria *et al* 1976).

For a comprehensive review of the nature of traps in SiO_2, readers are referred to the recent review by DiMaria (1978).

4. Recent results

Measurements at room temperature and at 77 K have yielded the results shown in figure 3. The 'wet' samples had ~10^4 ppm of H_2O added to the O_2 during oxide growth at 1000 °C. The samples were immediately removed from the oxidation furnace on completion of the growth . In this figure, PMA refers to post-metallisation heat treatment at 400 °C for 20 min in forming gas and NPMA refers to the absence of this treatment. These results clearly indicate the enhanced trapping at 77 K and also show a reduction in trapping if the dry PMA process is used. The shape of the curves at room temperature is affected to some extent by the positive charge effect observed by Gdula (1976). Subsequent work has shown a further reduction in trapping if the samples are heated at 1000 °C in a N_2 or As environment after the oxidation (hereafter called POA). Measurements of thickness dependence show a strong dependence on this parameter for data obtained at 295 K (see figure 4). A (thickness)^2 dependence is observed (as shown in figure 5), which suggests a trap density that is uniform throughout the SiO_2 and has

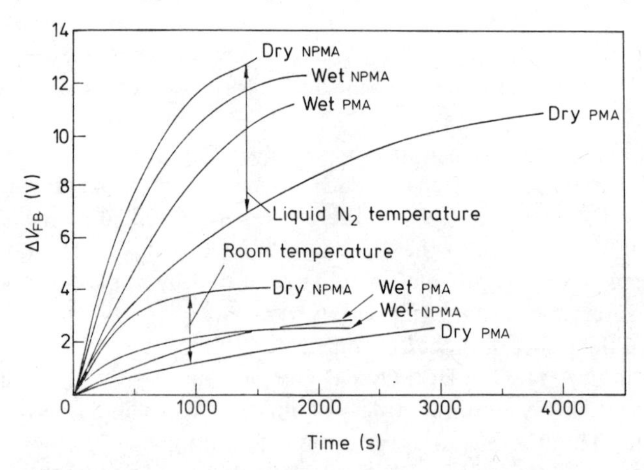

Figure 3. ΔV_{FB} plotted against time for measurements at room temperature and 77 K. Current = 3×10^{-7} A; current density = $5{\cdot}78 \times 10^{-5}$ A cm^{-2}; SiO_2 thickness = $5{\cdot}25 \times 10^{-6}$ cm.

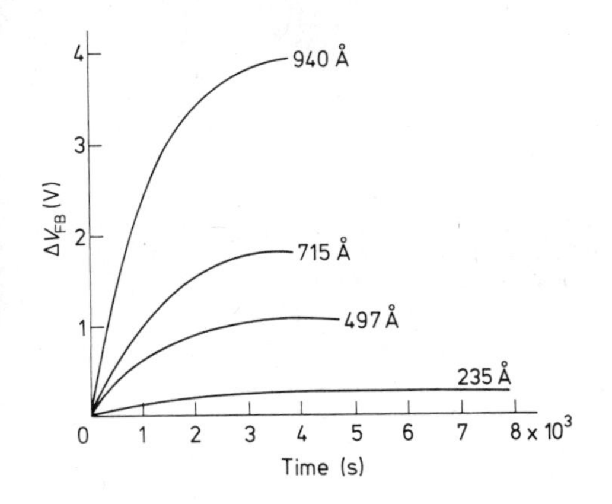

Figure 4. ΔV_{FB} plotted against time for several oxide thicknesses. POA: 1 h, 1000 °C, N_2; PMA: 20 min, 1000 °C, N_2; $I = 5{\cdot}78 \times 10^{-5}$ A cm^{-2}. Measurements at 300 K.

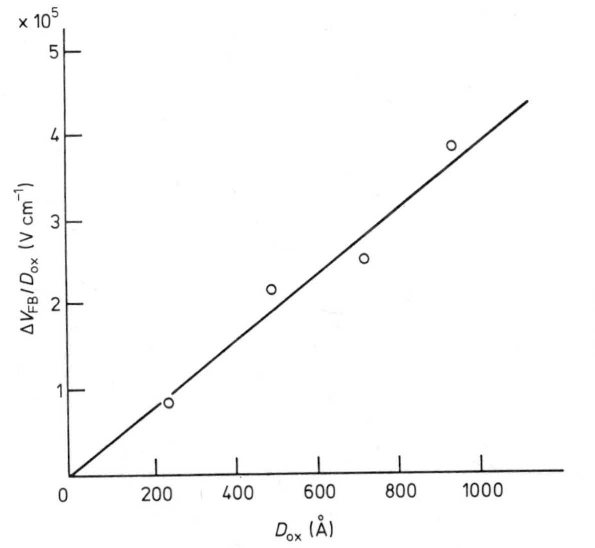

Figure 5. (Average ΔV_{FB})/D_{ox} (taken to 3000 s) from results of figure 4 function of the oxide thickness. $I = 5{\cdot}78 \times 10^{-5}$ A cm^{-2}.

a constant volume density for the various samples. This observation is consistent with photo *I–V* studies that indicate that the centroid of the trapped charge is close to the centre of the oxide but shifted slightly toward the Al–SiO_2 interface (Young *et al* 1979). Measurements at 77 K (see figure 6) result in a thickness dependence that is given by

$$\Delta V_{FB} \sim (D_{ox} - 9 \times 10^{-7}) \tag{8}$$

where D_{ox} is the oxide thickness in cm. This suggests a centroid for the trapped charge that is 90 Å from the Si–SiO_2 interface. Photo *I–V* measurements have also confirmed this observation. This enhanced trapping at 77 K was observed by DiMaria *et al* (1976). Ning (1978) has studied the re-emission of electrons trapped at 77 K as a function of temperature and observed a trap depth 0·3 eV below the conduction band edge of SiO_2. The difference in the dominant traps at room temperature and at 77 K is further demonstrated by a difference in response to POA as shown in figure 7 (Young *et al* 1979). The

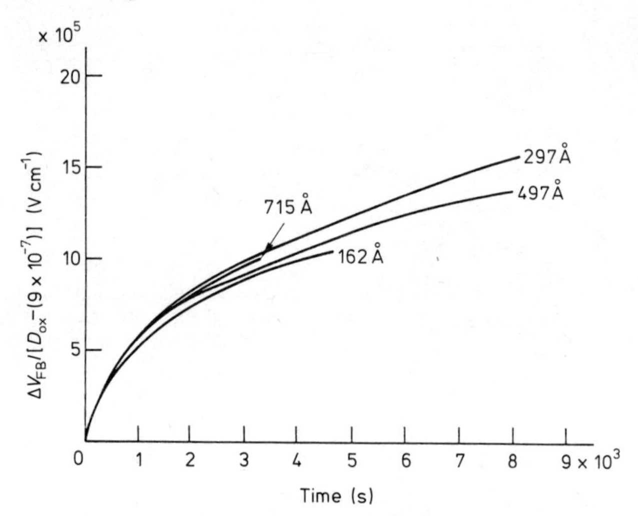

Figure 6. $\Delta V_{FB}/(D_{ox} - 9 \times 10^{-7})$ plotted against time. POA: 1 h, 1000 °C, N_2; PMA: 20 min, 400 °C, N_2; $I = 5{\cdot}78 \times 10^{-5}$ A cm^{-2}. Measurements at 77 K.

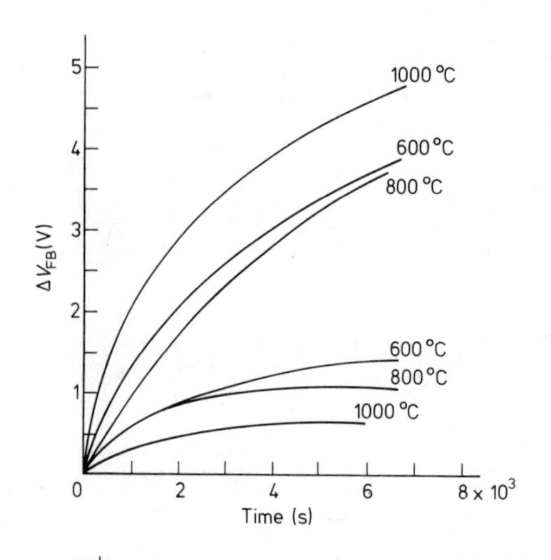

Figure 7. Effect of post-oxidation annealing temperature on flat-band voltage shifts for measurements at 295 K (lower curves) and 77 K (upper curves). D_{ox} = 500 Å; current = 3×10^{-7} A.

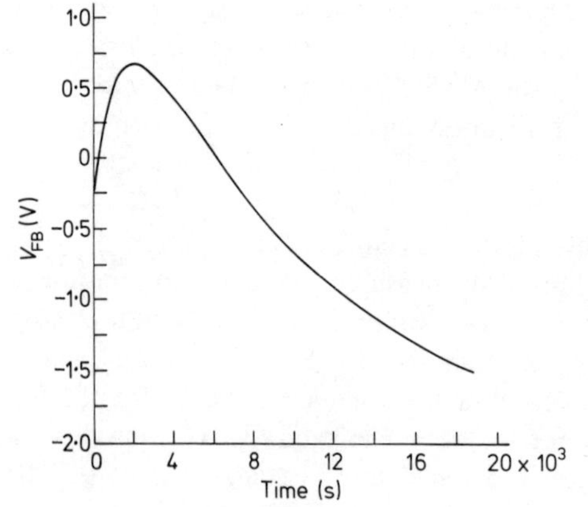

Figure 8. ΔV_{FB} plotted against time showing the anomalous turn-around effect. POA: 30 min, 1000 °C; PMA: 20 min, 400 °C, N_2; $I = 3 \times 10^{-7}$ A; D_{ox} = 500 Å; measurements at 295 K.

room-temperature trapping is reduced by POA treatment at 1000 °C, compared to lower temperatures. Measurements at 77 K indicate an optimum temperature of 800 °C for POA.

Passage of a relatively large current (current density $5{\cdot}9 \times 10^{-5}$ A cm^{-2}) results in the build-up of positive charge which partially compensates the negative charge. This effect is shown in figure 8 and will be referred to as the turn-around effect. The work of Gdula (1976) also showed this effect. Photo *I–V* measurements (Young *et al* 1979) have shown that this effect results from the build-up of positive charge at the Si–SiO_2 interface. It is suggested that this results from the passage of the current through the sample, as shown by figure 9, where ΔV_{FB} is plotted as a function of the total charge density flowing through the oxides for applied currents of 3×10^{-7} A and 3×10^{-8} A. It can be seen that the results are similar, showing that the turn-around effect depends on (current × time) or charge. The positive charge results from the generation of donor states in the oxide close to the Si–SiO_2 interface. These states can be charged and discharged by the application of moderate electric fields after the avalanche injection is terminated, as indicated by figure 10. This charging and discharging effect is enhanced by an increase in temperature to 100 °C, as shown by figure 11. The voltage required to discharge these states is positive, which is the sign of the voltage required for avalanche injection. This suggests the possibility of measuring the trapping at elevated temperatures to avoid complications in the analysis of the results arising from the positive charge effect. Figure 12 shows that this is indeed the case and measurements are now made at 100–120 °C to avoid this effect.

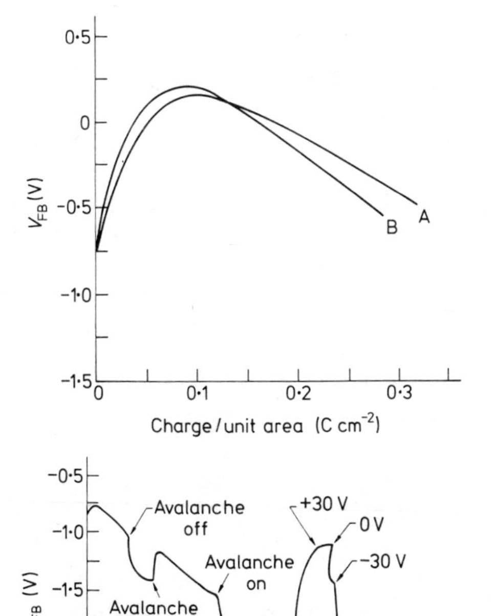

Figure 9. ΔV_{FB} plotted against charge/unit area for samples with applied currents of 3×10^{-7} A (curve A) and 3×10^{-8} A (curve B).

Figure 10. Effect of electric field on V_{FB} after avalanche injection. $I = 5{\cdot}78 \times 10^{-5}$ A cm^{-2}.

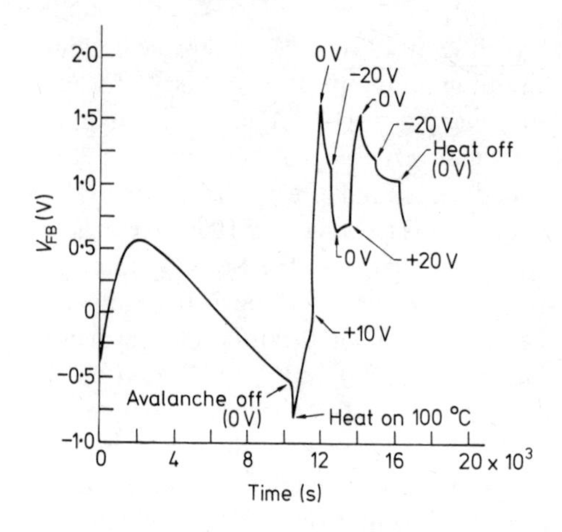

Figure 11. Increased effect of electric field on V_{FB} after avalanche injection due to elevated temperature. 500 Å SiO_2, dry process; POA: 30 min, 1000 °C, N_2; PMA: 20 min, 400 °C, forming gas; $I = 3{\cdot}7 \times 10^{-7}$ A.

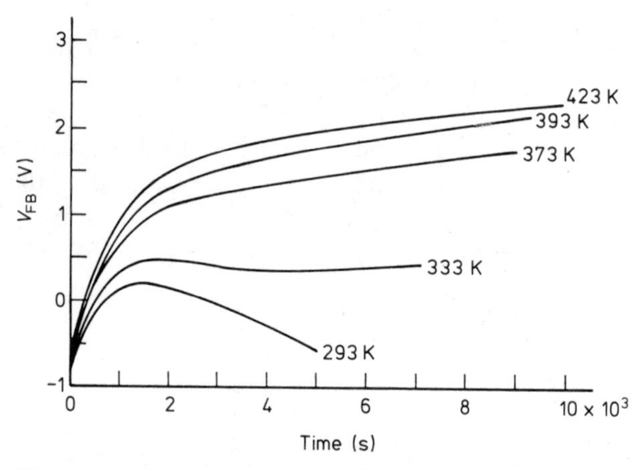

Figure 12. Elimination of 'turn-around effect' at elevated temperatures. $I = 3{\cdot}7 \times 10^{-7}$ A.

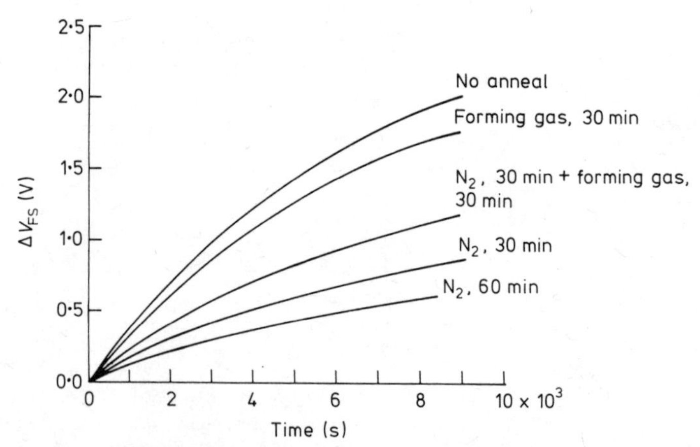

Figure 13. Comparison of various POA treatments. Measurements at 20 °C; $D_{ox} = 5 \times 10^{-6}$ cm; $I = 9{\cdot}64 \times 10^{-6}$ A cm^{-2}.

Table 1. Trap characterisation.

Conditions	σ	N_{eff}	N_V	
Forming gas, 30 min, PMA	$2{\cdot}93 \times 10^{-18}$	$1{\cdot}47 \times 10^{12}$	$5{\cdot}9 \times 10^{17}$	(C)
	$5{\cdot}02 \times 10^{-19}$	$4{\cdot}12 \times 10^{11}$	$1{\cdot}6 \times 10^{17}$	(B)
	$4{\cdot}69 \times 10^{-20}$	$1{\cdot}08 \times 10^{12}$	$4{\cdot}3 \times 10^{17}$	(A)
Forming gas, 60 min, NPMA	$8{\cdot}8 \times 10^{-18}$	$1{\cdot}9 \times 10^{12}$	$7{\cdot}6 \times 10^{17}$	(D)
	$3{\cdot}9 \times 10^{-18}$	$3{\cdot}7 \times 10^{11}$	$1{\cdot}5 \times 10^{17}$	(C)
	$8{\cdot}8 \times 10^{-19}$	$4{\cdot}6 \times 10^{11}$	$1{\cdot}8 \times 10^{17}$	(B)
N_2, 30 min, PMA	$2{\cdot}9 \times 10^{-18}$	$6{\cdot}7 \times 10^{11}$	$2{\cdot}7 \times 10^{17}$	(C)
	$4{\cdot}4 \times 10^{-19}$	$2{\cdot}2 \times 10^{11}$	$8{\cdot}8 \times 10^{16}$	(B)
	$9{\cdot}9 \times 10^{-20}$	$6{\cdot}5 \times 10^{11}$	$2{\cdot}6 \times 10^{17}$	(A)
N_2, 60 min, PMA	$1{\cdot}9 \times 10^{-18}$	$5{\cdot}0 \times 10^{11}$	$2{\cdot}0 \times 10^{17}$	(C)
	$4{\cdot}9 \times 10^{-19}$	$1{\cdot}2 \times 10^{11}$	$4{\cdot}8 \times 10^{16}$	(B)
	$8{\cdot}7 \times 10^{-20}$	$6{\cdot}3 \times 10^{11}$	$2{\cdot}5 \times 10^{17}$	(A)
N_2, 30 min, NPMA	$7{\cdot}2 \times 10^{-18}$	$1{\cdot}1 \times 10^{12}$	$4{\cdot}4 \times 10^{17}$	(D)
	$5{\cdot}3 \times 10^{-19}$	$1{\cdot}8 \times 10^{11}$	$7{\cdot}2 \times 10^{16}$	(B)
	$8{\cdot}6 \times 10^{-20}$	$5{\cdot}8 \times 10^{11}$	$2{\cdot}3 \times 10^{17}$	(A)

The trap cross sections σ are in cm^2, N_{eff} is the effective areal density in cm^{-2}, N_V is the volume density in cm^{-3} calculated from N_{eff} using a centroid correction of two (assuming the traps are uniformly distributed throughout the SiO_2) and dividing by the thickness. The time (min) refers to the time for heat treatment at 1000 °C. Traps with the same letters are thought to be identical.

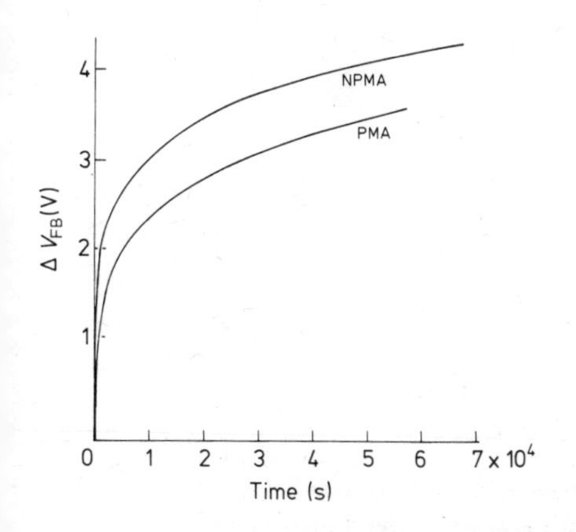

Figure 14. Effect of PMA on electron trapping. Measurements at 120 °C; $I = 5{\cdot}78 \times 10^{-5}$ A cm^{-2}.

Recent work has been done on the effect of POA and PMA treatments on the trapping characteristics; measurements were made at 120 °C. These results are summarised in table 1. The use of forming gas for POA certainly enhances the trapping, compared with results with N_2 or Ar. Figure 13 shows the effect of various POA treatments on the trapping characteristics and the reduction resulting from the POA treatment in N_2 for 60 min can be seen. These comparisons are correct, but the shapes of these curves are slightly modified by the 'turn-around effect' since the measurements were made at 20 °C.

It is interesting to note that a POA treatment in forming gas that contains H_2 does not decrease the trapping rate significantly. The PMA treatment does, however, significantly decrease trapping as shown in figure 14. The effect of this treatment is to reduce the cross section of the dominant trap from $\sim 8 \times 10^{-18}$ cm^2 to $\sim 3 \times 10^{-18}$ cm^2. The two smallest traps measured ($\sim 9 \times 10^{-20}$ cm^2 and $\sim 4 \times 10^{-19}$ cm^2) are independent of processing conditions; larger traps (2×10^{-18} cm^2 and 8×10^{-18} cm^2) are, however, strongly dependent on these conditions.

5. Summary

Electron trapping in SiO_2 has important technological implications. MOS capacitors can be used to evaluate trapping and photo I–V and C–V measurements on these devices permit accurate evaluation of the trapping characteristics. These studies indicate the importance of processing conditions and measuring temperatures in determining the trap characteristics. The generation of donor states close to the Si–SiO_2 interface has been observed to result from the passage of current introduced by avalanche injection from Si into the SiO_2. A method has been described to avoid complications due to this effect in interpreting the 'bulk' electron-trapping results.

Acknowledgments

This work was supported by the Defense Advanced Research Agency (RADC) and monitored by the Deputy for Electronic Technology, under Contract No F-19628-78-C-0225. The author has benefited from numerous discussions with D J DiMaria, Z Weinberg, E A Irene, J M Aitken, R F DeKeersmaecker, A B Fowler and M I Nathan. Technical assistance was rendered by J A Calise and C M Serrano. Some of the samples were made in the IBM Research Center Silicon Process Facility.

References

Abbas S A and Dockerty R C 1975 *Appl. Phys. Lett.* **27** 147
Bosselaar C A 1973 *Solid St. Electron.* **16** 648
Chen L I, Pickar K A and Sze S M 1972 *Solid St. Electron.* **15** 979
DiMaria D J 1976 *J. Appl. Phys.* **47** 4073
DiMaria D J 1978 in *Proceedings of the International Topical Conference on the Physics of* SiO_2 *and Its Interfaces* ed S T Pantelides (New York: Pergamon Press) p160
DiMaria D J, Aitken J M and Young D R 1976 *J. Appl. Phys.* **47** 2740
DiMaria D J, Feigl F J and Butler S R 1975 *Phys. Rev.* **B11** 5023
DiMaria D J, Young D R, DeKeersmaecker R F, Hunter W R and Serrano C M 1977b *The Electrochemical Society Fall Meeting, Atlanta, Georgia, Abstract No* 212 (unpublished)
DiMaria D J, Young D R, Hunter W R and Serrano C M 1978 *IBM J. Res. Dev.* **22**
DiMaria D J, Young D R and Ormond D W 1977a *Appl. Phys. Lett.* **31** 680
Gdula R A 1976 *J. Electrochem. Soc.* **123** 42
Kahng D, Sundberg W J, Boulin D M and Ligenza J R 1974 *Bell Syst. Tech. J.* **53** 1723
Kerr D R, Logan J S, Burkhardt P J and Pliskin W A 1964 *IBM J. Res. Dev.* **8** 376
Nicollian E H, Berglund C N, Schmidt P F and Andrews J M 1971 *J. Appl. Phys.* **42** 5654
Nicollian E H, Goetzberger A and Berglund C N 1969 *Appl. Phys. Lett.* **15** 174
Ning T H 1978 *J. Appl. Phys.* **49** 5997
Ning T H, Osburn C M and Yu H N 1975 *Appl. Phys. Lett.* **26** 248

Ning T H and Yu H N 1974 *J. Appl. Phys.* **45** 5373
Pepper M 1973 *J. Phys. D: Appl. Phys.* **6** 2124
Troutman R R 1978 *Solid St. Electron.* **21** 283
Ushirokawa A, Suzuki E and Warashina M 1973 *Japan. J. Appl. Phys.* **12** 398
Verwey J F 1973 *J. Appl. Phys.* **44** 2681
Verwey J F and DeMaagt B J 1974 *Solid St. Electron.* **17** 963
Verwey J F and Heringa A 1977 *IEEE Trans. Electron Devices* **ED-24** 519
Weinberg Z A, Young D R, DiMaria D J and Rubloff G W 1979 to be published
Williams R 1965 *Phys. Rev.* **140** A569
Yoshimo H, Kiuchi K and Yashiro T 1977 *Japan. J. Appl. Phys.* **16** 441
Young D R, DiMaria D J and Bojarczuk N A 1977 *J. Appl. Phys.* **48** 3425
Young D R, DiMaria D J, Hunter W R and Serrano C M 1978 *IBM J. Res. Dev.* **22**
Young D R, Irene E A, DiMaria D J, DeKeersmaecker R F and Massoud H Z 1979 *J. Appl. Phys.* to be published

Electron trapping and detrapping characteristics of arsenic-implanted SiO_2 layers

R F DeKeersmaecker† and D J DiMaria
IBM Thomas J Watson Research Center, Yorktown Heights, New York 10598, USA

Abstract. The electron trapping and detrapping properties of As^+-implanted thermally grown SiO_2 layers incorporated into MOS structures are described. The samples were charged by avalanche injection from the silicon substrate or internal photoemission from either interface. The charge state of the MOS devices was studied using capacitance versus voltage ($C-V$) and photocurrent versus voltage (photo $I-V$) measurements. After annealing at high temperatures, indications were found for a correlation between the trapping phenomenon and the nature of the implanted ions. The possibility of removing the trapped charge both by optical and thermal excitation is reported.

1. Introduction

The intensive use of ion implantation in the fabrication of modern integrated circuits has encouraged a study of its influence upon sensitive device regions such as the SiO_2–layer of metal–silicon dioxide–silicon (MOS)-type structures. Furthermore, energy levels in the band gap of SiO_2 have attracted fundamental interest (Feigl *et al* 1976) and ion implantation has emerged as an elegant means of introducing the impurities that will generate these levels.

We will review some of the properties of arsenic-implanted SiO_2 layers incorporated into MOS structures. A more detailed account of the experiments will be published elsewhere (DeKeersmaecker and DiMaria 1979).

2. Experimental details

2.1. Sample fabrication

The samples used were 0·1–0·2 Ω cm p-type ⟨100⟩ silicon wafers. After thermal oxidation in a nominally dry oxygen ambient at 1000 °C to a thickness ranging from 490 to 1440 Å, the wafers were implanted with As^+ at room temperature with fluences ranging from 3×10^{12} to 1×10^{14} cm^{-2} and ion energies between 10 and 100 keV. All wafers were then cleaned and annealed in nitrogen at 1000 °C for 30 min unless otherwise specified. A shadow mask was used to deposit thin (120–150 Å) circular aluminium electrodes with an area of $5{\cdot}2 \times 10^{-3}$ cm^2 to form MOS capacitors. The semitransparent metal electrodes were used to permit generation of photocurrents from both the silicon and aluminium contacts. Finally, after deposition of the metal, all MOS devices were annealed at 400 °C in forming gas for 20 min.

† Present address: Katholieke Universiteit Leuven, ESAT Laboratorium, Kardinaal Mercierlaan 94, B-3030 Heverlee, Belgium.

0305-2346/80/0050-0040$01.00

2.2. Charging

The avalanche injection technique (Nicollian *et al* 1969) was used in most cases to inject electrons into the oxide of the MOS structures. In our set-up (Young *et al* 1977), the time-averaged injection current can be kept constant. Injection is periodically interrupted in order to record the flat-band voltage reading.

From the evolution in time of the flat-band voltage shift ΔV_{FB}, the capture cross section σ_c and the effective trap density N_{eff} can be determined according to (Aitken and Young 1974)

$$\Delta V_{FB}(t) = eN_{eff}(L/\epsilon)\,[1 - \exp(-\sigma_c\, jt/e)] \tag{1}$$

where e is the electron charge, L is the oxide thickness, ϵ is the low-frequency oxide permittivity and j is the injection current density.

The effective trap density is the trap density if all traps were located at the Si–SiO_2 interface and is related to the integrated trap density N_T by

$$N_{eff} = (\bar{x}/L)\, N_T \tag{2}$$

where $\bar{x}$ is the charge distribution centroid, measured from the Al–SiO_2 interface. The integrated trap density is given by

$$N_T = \int_0^L N_t(x)\, dx \tag{3}$$

where N_t is the local trap density in the oxide.

If more than one trapping centre is present in the oxide (each centre is characterised by its own capture cross section $\sigma_{c,i}$ and effective trap density $N_{eff,i}$), $\Delta V_{FB}(t)$ will be given by the summation of the individual components of the form of equation (1). Since the charging time constant of one trap

$$\tau = e/(\sigma_c j) \tag{4}$$

is also determined by the current density, the latter will be a useful parameter for the study of various capture cross sections.

The distribution centroid of bulk oxide charge can be determined from photocurrent versus voltage characteristics (photo I–V) for the MOS devices. For this purpose, monochromatic light was used with 5 (4·5) eV photons to excite electrons over the SiO_2–Si (Al) energy barrier. The voltage shifts obtained from such characteristics after charging the oxide traps depend on the charge density Q and its centroid $\bar{x}$, as seen from the following equations (DiMaria 1976):

$$\Delta V_g^+ = -\bar{x}Q/\epsilon \tag{5}$$

$$\Delta V_g^- = (L - \bar{x})\, Q/\epsilon \tag{6}$$

where the superscripts + and − refer to the gate polarity. Photo I–V measurements have been used routinely in our experiments in order to determine $\bar{x}$ according to the relation

$$\bar{x} = L\,[1 - (\Delta V_g^- / \Delta V_g^+)]^{-1}. \tag{7}$$

Complementary 1 MHz capacitance versus voltage (C–V) measurements were performed in order to monitor perturbations in the charge distribution.

2.3. Discharging

Optical detrapping of the trapped electrons was studied as a function of photon energy. For that purpose the flat-band voltage was tracked accurately after illumination for a given time at zero gate bias. The internal electric fields in the oxide layer (caused by the trapped charge) were sufficiently large to sweep out the excited electrons. On the other hand, the total change in the flat-band voltage shift during a discharging experiment was small compared with the initial flat-band voltage shift, in order to keep the electric field configuration in the oxide constant.

From such detrapping measurements, together with measurements of the incident photon flux and a calculated convolution of the light standing wave pattern with the spatial trap distribution, an effective photoionisation cross section can be determined (DeKeersmaecker *et al* 1978).

Thermal emptying of the trapping sites was studied both by isothermal measurements of the change in the flat-band voltage shift and by thermally stimulated current (TSC) measurements (Hickmott 1975).

3. Results

3.1. Charging studies

The analysis described in the previous section was used to reduce the avalanche injection data shown in figure 1. It can be seen that the flat-band voltage shift depends linearly

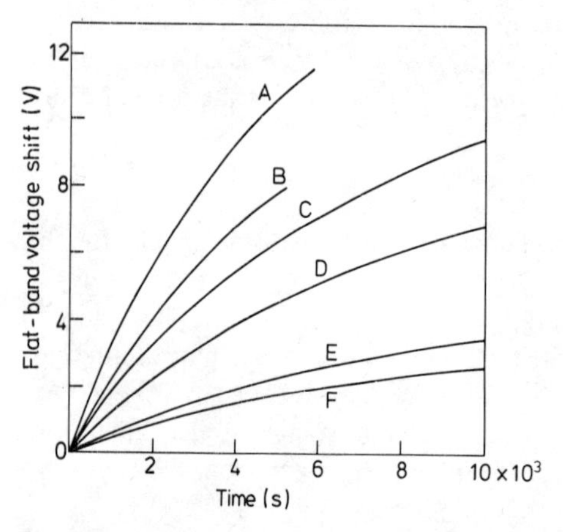

Figure 1. Flat-band voltage shift as a function of time during avalanche injection at a current $I = 9 \times 10^{-11}$ A for implantation energies of 100, 80, 60, 40, 20 and 10 keV (curves A, B, C, D, E, F respectively). All samples ($L \approx 1400$ Å) were implanted with 1×10^{13} As^+ cm^{-2} and annealed in N_2 for 30 min at 1000°C.

upon ion energy. Since the charge centroid, as determined from photo *I–V* measurements (see figure 3) varies linearly with ion energy, it is concluded that after annealing at high temperatures, the integrated oxide charge and therefore the integrated trap density is independent of ion energy for an As^+ fluence of 1×10^{13} cm^{-2}.

Results for the various capture cross sections and the corresponding integrated trap densities are listed in table 1. The data were collected from charging experiments with samples implanted at various ion energies. The origin of the different cross sections is not clear at present. The centre with the highest cross section may correspond to a positively charged centre, the capture cross section of which has been lowered by the electric

Table 1. Capture cross sections and integrated trap densities for the various centres resolved from avalanche injection data. The samples (oxide thickness $L \approx 1400$ Å) were implanted with 1×10^{13} As^+ cm^{-2} at various energies and annealed in N_2 at 1000 °C for 30 min.

σ_c (cm^2)	N_T (cm^{-2})
$(0{\cdot}4-1{\cdot}6) \times 10^{-14}$	$(0{\cdot}6-7{\cdot}4) \times 10^{11}$
$(0{\cdot}9-1{\cdot}8) \times 10^{-15}$	$(1{\cdot}8-5{\cdot}6) \times 10^{12}$
$(1{\cdot}5-5{\cdot}4) \times 10^{-16}$	$(2{\cdot}5-5{\cdot}7) \times 10^{12}$
$(0{\cdot}9-4) \times 10^{-17}$	$(\sim 5-7) \times 10^{12}$

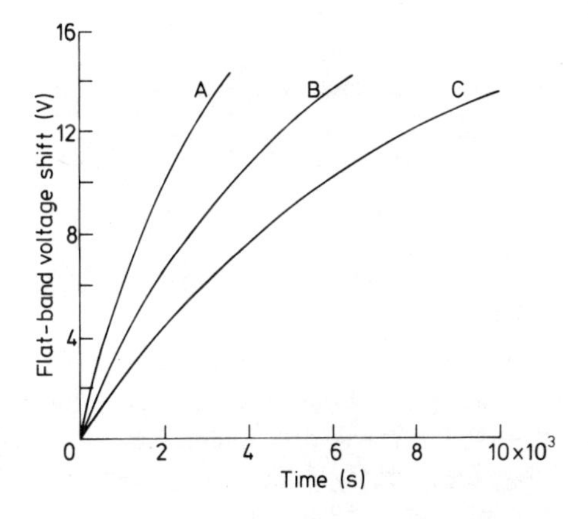

Figure 2. Flat-band voltage shift plotted against time during avalanche injection at a current $I = 9 \times 10^{-11}$ A. The oxide thickness was approximately 1360 Å. All samples were annealed in N_2 for 30 min at 600, 800 or 1000 °C (curves A, B, C respectively). Samples were implanted with 1×10^{13} As^+ cm^{-2} at an energy of 80 keV.

field in the oxide during avalanche injection (3–4 MV cm^{-1}). Its highest density is found with the lowest value of its capture cross section; in general, the trap density is rather low and represents only a few hundred millivolt.

There is a spectrum of cross sections in the range 10^{-15}–10^{-16} cm^2 which dominates the trapping behaviour. The sum of the trap densities over this spectrum is equal to (0·7–1) times the ion fluence. A trapping centre with $\sigma_c = (1-3) \times 10^{-17}$ cm^2 was also found in some control samples which were processed in the same way except for the ion implantation. Its integrated density is increased by approximately a factor of 10 after implantation.

Figure 2 shows the flat-band voltage shift plotted against time during avalanche injection for three samples annealed for 30 min in N_2 at 600, 800 and 1000 °C. The general trend is for higher annealing temperatures to reduce the electron trapping rate. Attempts to investigate the completeness of the annealing after the treatment at 1000 °C were hampered by some slight irreproducibilities in the trapping behaviour. These observations led us to conclude that poor control of some processing details (e.g. implantation current density, heating or cooling rate during annealing) might have an impact upon the trapping efficiency.

Figure 3 compares the charge distribution centroids (from photo I–V data) with theoretical (LSS theory) and experimental (SIMS and He^+ backscattering) determinations

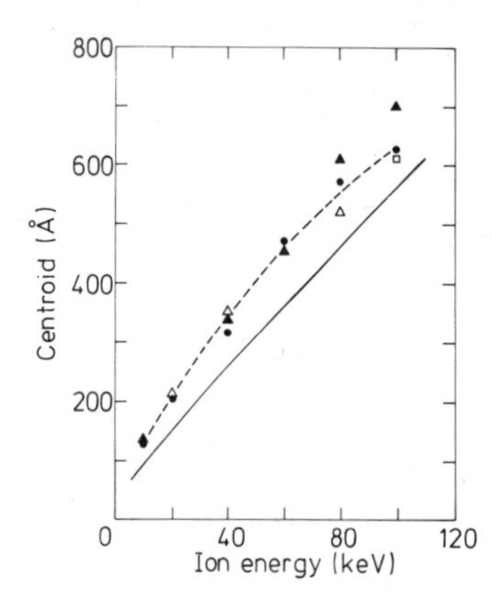

Figure 3. Experimental charge distribution centroid (from photo I–V data, ●), experimental ion distribution centroid (from SIMS, ▲, △, and He^+ backscattering, □) and theoretical ion distribution centroid (LSS theory according to Lindhard *et al* 1963, —) as a function of ion energy. The wafers in our experiments were implanted with 1×10^{13} As^+ cm^{-2} (for photo I–V) or 1×10^{14} As^+ cm^{-2} (for SIMS, ▲) and annealed at 1000 °C in N_2 for 30 min. The SIMS results from Schimko *et al* (1975, △) are for 5×10^{14} As^+ cm^{-2} implants. The He^+ backscattering result (Tsukamoto *et al* 1977) is for a 1×10^{16} As^+ cm^{-2} implant at 100 keV.

of the ion distribution centroids. There is good agreement between the experimental values for charge centroids and ion centroids. From this it is concluded that the charge distribution is closely tracking the distribution of implanted ions. We believe that the high annealing temperatures used in our study reduce the structural damage in the oxide to a level where the nature of the implanted ions and their spatial distribution dominate the electron trapping in a charging experiment.

The charge centroid $\bar{x}$ was measured for an integrated charge density Q that varied over almost two orders of magnitude. Within the small experimental error (20–30 Å), no dependence of $\bar{x}$ upon Q was found. For the smallest values of Q, only the $10^{-14}\,cm^2$ trap and a fraction of the $10^{-15}\,cm^2$ trap had been charged, whereas for the largest values of Q, centres with cross sections in the range 10^{-16}–$10^{-17}\,cm^2$ had been charged to a large extent. Since $\bar{x}$ was constant, it was concluded that the spatial charge distributions and their respective trap distributions have very similar centroids.

It was also found that the charge centroid does not depend upon the ion fluence (in the range 3×10^{12}–$10^{14}\,cm^{-2}$), the electron injection mechanism (avalanche injection or internal photoemission), the oxide thickness (490–1430 Å) or the ambient temperature during the measurement (83–295 K). The charge distribution centroid was also independent of the annealing temperature over the range 600–1100 °C which indicates that the major defects were annealed after treatment at 600 °C, since the defect distribution is known to be shallower than the ion distribution (see e.g. Crowder and Title 1970).

3.2. Discharging studies

Figure 4 shows a set of photo I–V characteristics for negative bias for an implanted sample at various stages of an illumination experiment. A similar progressive reduction in ΔV_g^+ is obtained for positive bias. It is therefore concluded that optical excitation removes trapped negative charge from the bulk of the oxide. Concomitant centroid calculations show that the charge is released in a homogeneous way (with constant centroid), provided small portions of the charge are being removed. When larger portions are removed, optical interference effects become important. The centroid will then ultimately coincide with a minimum in the standing wave pattern of the light (DiMaria *et al* 1978).

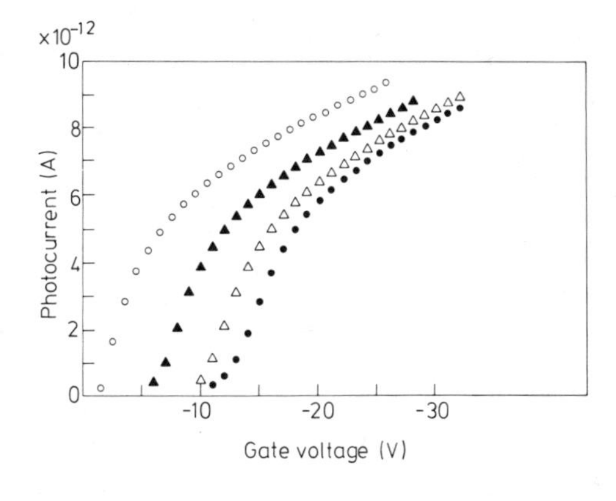

Figure 4. Photocurrent for 4·5 eV light as a function of applied gate voltage for negative bias for a sample with an initial ΔV_{FB} = 7·8 V, at successive stages of a detrapping sequence: ● initial photo I–V data; △ after detrapping, $\hbar\omega$ = 5·28 eV; ▲ after detrapping, $\hbar\omega$ = 4·96 eV; ○ after detrapping with a deuterium lamp. L = 1274 Å; implantation energy 60 keV; fluence 1×10^{13} As^+ cm^{-2}.

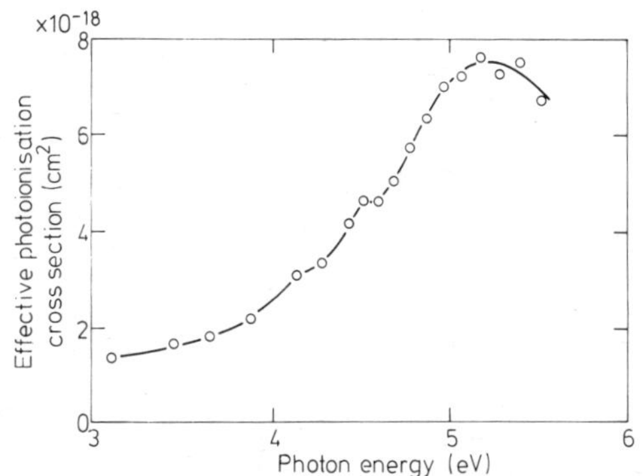

Figure 5. Effective photoionisation cross section as a function of photon energy for a sample (L = 418 Å) implanted with 1×10^{13} As^+ cm^{-2} at 10 keV, initially charged by avalanche injection up to ΔV_{FB}(initial) = 3 V. At every photon energy, the MOS was illuminated for 5 min with V_g = 0 V; ΔV_{FB}(final) = 2·3 V.

The onset of spectrally resolved photodetrapping can be considered as a measure of the optical trap depth (with respect to the edge of the oxide conduction band), whereas the magnitude of the reduction in the flat-band voltage shift under illumination (corrected for the light intensity) is a measure of the photoionisation cross section. It follows from the details of the analysis (DeKeersmaecker and DiMaria 1979) that an effective photoionisation cross section is obtained, which is a convolution of the photoionisation cross section with the optically accessible energy distribution of traps.

The effective photoionisation cross section spectrum is shown in figure 5. A threshold of 3·3 eV is found from the linear portion of the spectrum, which also shows a broad maximum at 5·2 eV. It has been suggested that one of the electron traps observed in As-implanted SiO_2 corresponds to the unaffected As 4p level found in the middle of the SiO_2 band gap when arsenic is incorporated into the SiO_2 network at an oxygen site (DeKeersmaecker *et al* 1978). The experimental value for the photoionisation threshold, 3·3 eV, is in excellent agreement with the prediction from such a model.

For a set of discrete trap energy levels E_{ti}, the effective photoionisation cross section will be given by

$$\sum_i \sigma_{pi}(E_{ti}, \hbar\omega)\, n_{ti}(E_{ti})$$

which is the sum of the contributions σ_{pi} from the individual levels, weighted by the normalised density of trapped electrons per unit energy which is optically accessible.

The optical detrapping threshold gives us information about the most shallow trap energy level; any deeper trap will affect the spectrum beyond that threshold. If, furthermore, such a trap is only filled to a small degree, its contribution to the photoionisation spectrum may easily go undetected.

Thermal removal of the trapped charge has also been observed. The broad experimental TSC spectrum (current versus temperature) could be modelled by assuming a Gaussian energy distribution of trapping centres with a standard deviation of 0·2 eV around a median thermal trap depth of 1·2 eV. In the model, a frequency factor of the order of the atomic vibration frequency (10^{11}–$10^{13}\,s^{-1}$) could be used. The thermal activation energy range appears to be in accordance with expectations from the optical absorption spectrum (Mott and Gurney 1964).

4. Discussion

The number of active electron trapping sites was found to be proportional and (at least for fluences up to $10^{13}\,cm^{-2}$) nearly equal to the implanted arsenic fluences, but independent of the ion energy. Moreover, the location of the charge distribution centroid after charging the electron traps was found to be proportional to the energy of the implanted ions and in excellent agreement with the experimentally determined ion distribution centroid. These observations lend credibility to our assumption that the electron trapping phenomenon is directly related to the implanted ions.

The results of other experiments performed in this laboratory, which will be reported in forthcoming publications, shed more light on the present work. Although the atomic masses of aluminium and phosphorus are nearly identical, it has been observed that these ions exhibit considerably different electron trapping behaviour after high-temperature annealing. It is thus concluded that the nature of the implanted ions is of prime importance in determining the electron trapping characteristics, regardless of the exact physical origin of the trapping sites.

It is possible to conceive a memory device based upon optical injection and trapping for storage and optical detrapping for erasure, as was demonstrated recently by DiMaria *et al* (1978).

Acknowledgments

The authors would like to acknowledge helpful discussions with D R Young and S T Pantelides, sample preparation by E D Alley and the Fabrication Technology group, SIMS measurements by J Webber and B Conlin of the IBM East Fishkill facility, TSC measurements by T W Hickmott and the technical assistance of J A Calise and F L Pesavento. One of the authors (RFD) was supported by an IBM Postdoctoral Fellowship. He is now with the Belgian National Research Foundation (NFWO). This research was supported by the Defense Advanced Research Projects Agency and monitored by the Deputy for Electronic Technology, RADC, under Contract F19628-76-C-0249.

References

Aitken J M and Young D R 1974 *J. Appl. Phys.* 47 1196
Crowder B L and Title R S 1970 *Radiat. Effects* 6 63
DeKeersmaecker R F and DiMaria D J 1979 *J. Appl. Phys.* to be published

DeKeersmaecker R F, DiMaria D J and Pantelides S T 1978 in *The Physics of SiO_2 and its Interfaces* ed S T Pantelides (New York: Pergamon) p189
DiMaria D J 1976 *J. Appl. Phys.* **47** 4073
DiMaria D J, DeKeersmaecker R F and Young D R 1978 *J. Appl. Phys.* **49** 4655
Feigl F J, Butler S R, DiMaria D J and Kapoor V J 1976 in *Thermal and Photostimulated Currents in Insulators* ed D J Smyth (Princeton: The Electrochemical Society) p118
Hickmott T W 1975 *J. Appl. Phys.* **46** 2583
Lindhard J, Scharff M and Schiøtt H E 1963 *K. Dansk. Vidensk. Selsk. Mat. Fys. Meddr* **33** No. 14
Mott N F and Gurney R W 1964 *Electronic Processes in Ionic Crystals* 2nd edn (New York: Dover) p160
Nicollian E H, Goetzberger A and Berglund C N 1969 *Appl. Phys. Lett.* **15** 174
Schimko R, Richter C E, Rogge K, Schwarz G and Trapp M 1975 *Phys. Stat. Solidi* **A28** 87
Tsukamoto K, Akasaka Y and Horie K 1977 *Japan J. Appl. Phys.* **16** 663
Young D R, DiMaria D J and Hunter W R 1977 *J. Electron. Mater.* **6** 569

Ionisation thresholds of electron traps in SiO_2†

D D Rathman, F J Feigl and S R Butler‡
Sherman Fairchild Laboratory, Lehigh University, Bethlehem, Pennsylvania 18015, USA

Abstract. The photoinjection–photodepopulation technique has been used to investigate the kinetics of trap emptying in wet thermal oxide films on Si. The electron trapping centres studied were associated with impurities incorporated in the SiO_2 films. The total negative oxide charge Q_{OT} resulting from electrons captured by these centres was measured by photo I–V characteristic voltage shifts. Components of Q_{OT} associated with three specific trapping centres were separated by determining the characteristic photoionisation cross section σ_p associated with each centre. In the photon energy range 1·5–7 eV, σ_p varied between 10^{-19} and $10^{-16}\,cm^2$. Values of the photoionisation threshold E_T (eV) and maximum areal trap density N_T (cm^{-2}) were: 2·5 eV and $1 \times 10^{12}\,cm^{-2}$ (Coulomb trap); 4·0 eV and $3 \times 10^{12}\,cm^{-2}$ (neutral trap); 5·5 eV and $3 \times 10^{13}\,cm^{-2}$ (neutral trap).

1. Introduction

Trapping of electrons during transport across amorphous SiO_2 films produced by thermal oxidation of crystalline Si was first reported by Williams (1965). At Lehigh University, we have had a continuing interest in this problem for about 10 years and the phenomenon has recently been the subject of intensive study, notably at the IBM Watson Laboratory. Comprehensive reviews have recently been presented by DiMaria (1978) and, at this conference, by Young (1980).

The trapping of injected electrons has recently become a matter of technological interest, because electron injection has become an important phenomenon. If unintentional, as in short-channel devices, the electron trapping results in a build-up of negative oxide charge Q_{OT} (see Deal 1979) which degrades designed device characteristics. If intentional, as in non-volatile memory devices, the build-up of Q_{OT} is itself a designed device (switching) characteristic (DiMaria *et al* 1978).

We present here the results of a comprehensive study of a particular class of electron traps in a particular class of thermal oxide films. We suggest that these traps have several characteristics. Firstly, they are found to some degree in all thermal SiO_2 films. This statement is based in part on our direct experience with films from four major industrial laboratories, as well as our own films. Secondly, the traps are related to impurities, notably, but not exclusively, H and Na (see DiMaria 1978, Butler *et al* 1976). Thirdly, areal densities of negative oxide trapped charge $Q_{OT}/e \leqslant 10^{12}\,cm^{-2}$ are observed; the 'less than' can be 'very much less than', particularly in annealed oxides prepared under extremely dry and clean conditions. Finally, the dominant traps (those with the highest density) are electrically neutral before trapping an electron.

† Research supported by US Navy Office of Naval Research, Electronic and Solid State Science Program.
‡ Supported by National Science Foundation, Division of Materials.

0305-2346/80/0050-0048$01.00

We will demonstrate that the neutral traps can be photoionised by light with photon energies in the range 4–7 eV. Previous studies indicated correctly that photoionisation was not possible for photon energies below 4 eV (DiMaria 1978). The primary quantitative result of this study is a determination of the trapping levels E_T of all electron traps with capture cross sections greater than 10^{-17} cm^2 (the 'particular class' of traps) in wet-grown oxides prepared at Lehigh University.

2. Experimental procedure and details

MOS devices were prepared on (100), 1–5 Ω cm, n-type Si substrates. The SiO_2 layer was formed at 1230 °C in a wet oxygen ambient, obtained by bubbling O_2 through a distilled water boiler operated at 95 °C. The SiO_2 film thickness L was in the range 1–1·5 μm. The metal field-plate was a semitransparent Al layer, 10–15 nm thick, evaporated onto the SiO_2 surface in a technical vacuum at room temperature. Thick Al contact pads were deposited over a small part of the field-plate and on the reverse surface of the Si. The area of the rectangular uncovered portion of the optically semitransparent Al field-plate was 0·2–0·3 cm^2. This was the active device area for the measurements described below.

No post-oxidation or post-metallisation annealing steps were performed before measurements were taken. The oxidation conditions were designed to maximise the incorporation of impurities (notably Na) leached from the fresh glassware of the oxidation system (Butler *et al* 1976). The above statements define the 'particular class' of thermal oxide films to which the experimental results described below apply.

The basic experimental programme is measurement of a sequence of electron photoinjection–trapped electron photodepopulation cycles; each cycle was performed with light of a different wavelength during the photodepopulation step. The equipment used was quite simple: a current meter (Keithley electrometer) and bias supply (Fluke precision voltage source) were connected across the MOS device and various UV light sources were used to illuminate the MOS oxide through the semitransparent Al field-plate. The UV light sources were centred at 185 nm (Hg lamp plus filter), 205 nm (D_2 lamp plus filter), 225 nm (D_2 lamp plus filter), 235 nm (Xe arc plus monochromator), 254 nm (Hg lamp plus filter), 275 nm (Xe arc plus monochromator) and 310 nm (Xe arc plus monochromator). The intensity of the UV sources within the oxide films was estimated by a combination of direct and indirect measurements. This intensity is designated hereafter as a photon flux F_λ or $F_{h\nu}$, with units of photons cm^{-2} s^{-1}. λ is the central wavelength for each source and $h\nu$ the corresponding photon energy.

The photoinjection–photodepopulation cycle is illustrated in figure 1. During injection under positive bias ($V>0$), light of energy $h\nu > E_b$ produces an internal photoemission current I across the oxide. E_b is the potential barrier at the Si–SiO_2 interface. Individual electrons are captured into localised states within the oxide film. This results in the build-up of a negative oxide charge $(L/\bar{x})Q_{OT}$ ($\bar{x}$ is the spatial centroid (first moment) of this charge distribution). Note that the exciting light can simultaneously remove trapped electrons from the oxide film if $h\nu > E_T$, where E_T is the trap depth relative to the SiO_2 conduction band edge. In all our experiments, $h\nu = 6{\cdot}7$ eV during photoinjection.

In zero-bias photodepopulation, with $V = 0$, the electron injection current is suppressed by the large negative oxide space charge $(L/\bar{x})Q_{OT}$. The exciting light can only

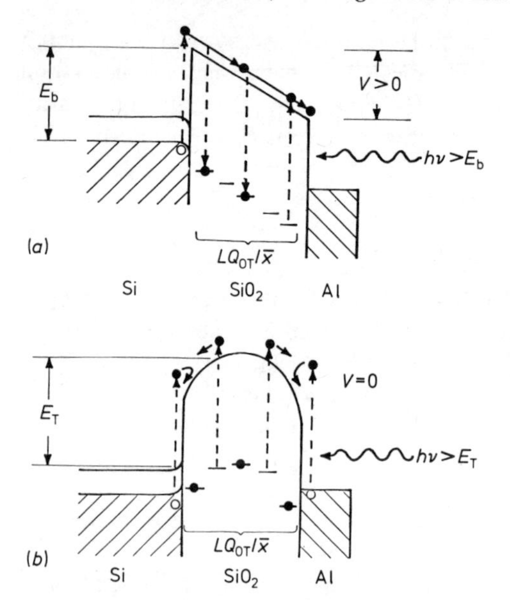

Figure 1. MOS energy-level diagrams illustrating the photoinjection (*a*) and photodepopulation (*b*) steps in the experimental sequence.

remove trapped electrons from the oxide film (and hence reduce the space charge). This will occur only if $h\nu > E_T$.

The complete experimental sequence consists of an injection–depopulation cycle using light of photon energy $h\nu_1$ during the depopulation step. The two-step cycle is repeated for depopulation photon energy $h\nu_2$, then for $h\nu_3$, etc. In each cycle, a common initial value of Q_{OT} is established by the injection step and is reduced by the depopulation step.

$\bar{x}$ and Q_{OT} are measured at several points during each step of each cycle in the total experimental sequence. The technique used for measurement is a modification of the photo $I-V$ technique developed by DiMaria. The photo $I-V$ logic and the zero-bias photodepopulation technique have been described by DiMaria (1978) and Young (1980).

The results of these measurements can be expressed as a centroid-weighted trapped oxide charge Q_{OT} (Deal 1979) or as a photo $I-V$ characteristic voltage $V_\phi = Q_{OT}/C_{OX}$, where C_{OX} is the device oxide capacitance. Note again that the trap depth E_T, which is the desired result of these measurements, is in fact a photoionisation threshold for the electron trapping centre (see figure 1 and discussion).

3. Results

Two photoinjection–photodepopulation cycles are illustrated in figure 2. The photo $I-V$ characteristic voltage V_ϕ is plotted against the total time for which the sample was illuminated, during either photoinjection or photodepopulation steps. Two depopulation steps, for $h\nu = 6{\cdot}7$ and $4{\cdot}9$ eV, are indicated. Note that each depopulation step begins at the same value of V_ϕ (approximately 55 V).

Data from all the depopulation steps of a complete sequence are shown in figure 3. The ordinate is a photo $I-V$ voltage shift, normalised to constant photon flux. ΔV_ϕ is defined by the equation $\Delta V_\phi = V_\phi(t) - V_\phi(t=0)$, where t is the time of illumination

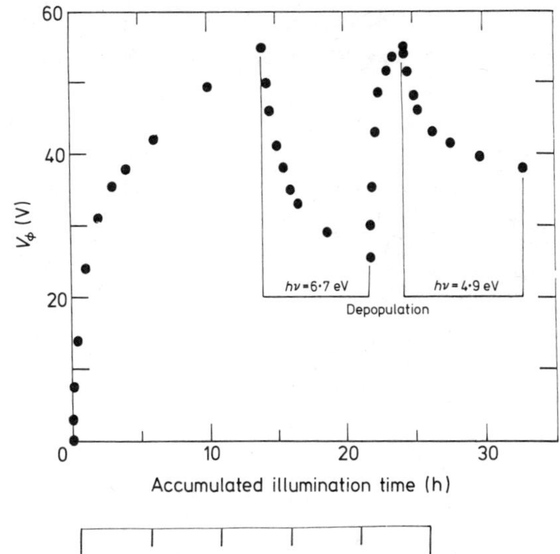

Figure 2. Photo I–V voltage shift V_ϕ plotted against total illumination time over two injection–depopulation cycles of the experimental sequence.

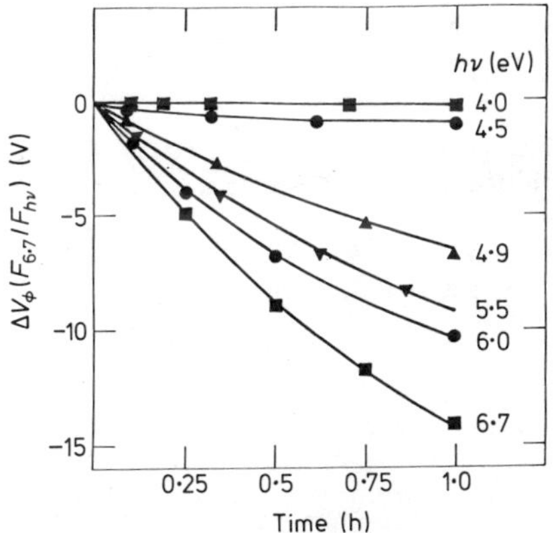

Figure 3. Photo I–V voltage shift ΔV_ϕ, corrected for differences in light intensity, plotted against illumination time for a given photodepopulation step. Six depopulation steps are illustrated. The full curves are analytical fits to the data, described in the text.

during a given depopulation step (the abscissa in figure 3). $V_\phi(t = 0)$ is approximately 55 V, the initial value established by photoinjection.

Finally, figure 4 shows data from one of the photodepopulation steps, with $h\nu = 6{\cdot}7$ eV, over an extended time interval. The data are the same as those displayed in figure 3, except that the ordinate has been converted to an areal density of negative oxide trapped charge, Q_{OT}/e, in units of number of trapped charges cm^{-2} of active device area.

4. Analysis

The data of figures 3 and 4 were interpreted using a multiple-trap model and assuming first-order trapping kinetics for each individual trap. This model is formulated mathematically as follows:

$$\frac{1}{e} Q_{\mathrm{OT}} \equiv N_{\mathrm{OT}}(t) = \sum_{j=1}^{3} N_j(0) \exp(-t/\tau_{\mathrm{p}j}). \tag{1}$$

$N_{\mathrm{OT}}(t)$ is the areal density of total centroid-weighted oxide trapped charge Q_{OT} at time t in a given depopulation step. $N_j(0)$ is the areal density of electron-occupied traps j at

$t = 0$ ($j = 1, 2, 3$). The j trapping levels are each characterised by a trap-emptying rate τ_{pj}^{-1}. The full curves in figures 3 and 4 represent a statistical best fit of equation (1) to the experimental data.

The result of this fitting program is a determination of the best-fit values of the trapping rates τ_{pj} as a function of depopulation light wavelength λ (or photon energy $h\nu$). From this result, we define an effective photoionisation cross section σ_{pj} for trap j as follows:

$$1/\tau_{pj} = F_\lambda \sigma_{pj}. \tag{2}$$

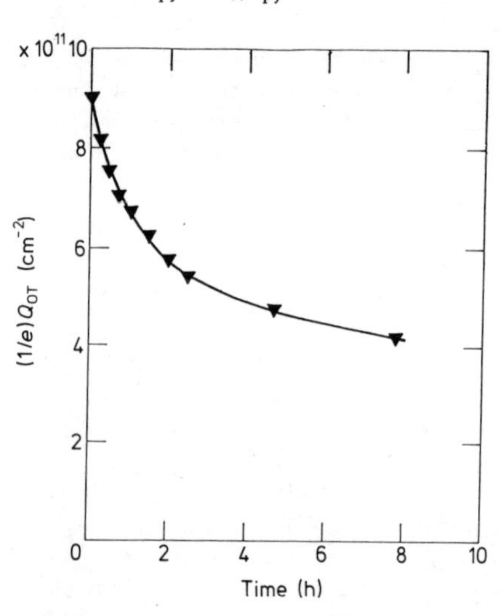

Figure 4. Oxide trapped charge density plotted against illumination time for one photodepopulation step (using light of photon energy $h\nu$ = 6·7 eV). The full curve is an analytical fit to the data.

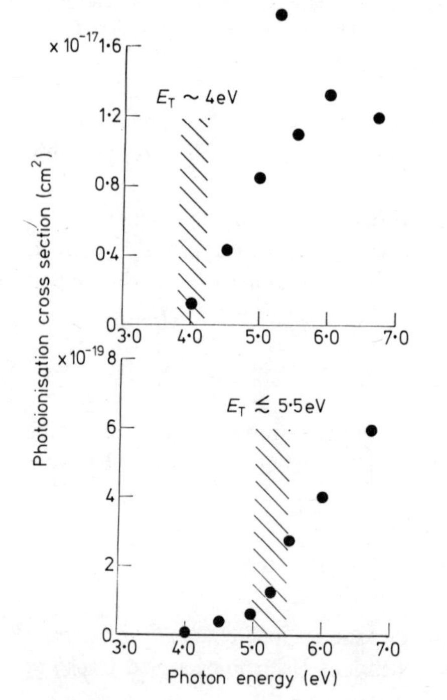

Figure 5. σ_p plotted against $h\nu$ for neutral traps 1 (*a*) and 2 (*b*). The estimated photoionisation thresholds E_T are indicated for each trap.

F_λ is the photon flux for exciting light of wavelength λ. The effective photoionisation cross section thus obtained is plotted as a function of photon energy $h\nu$ in figure 5(*a*) for trap 1 and in figure 5(*b*) for trap 2. Within the error of our overall experimental design and procedure, $\sigma_{p3} < 10^{-20}\,cm^2$ and is effectively zero.

5. Discussion and summary

Detailed interpretation of the results summarised in figure 5 are the object of investigations now in progress at Lehigh University. Recent theoretical studies by Edwards and Fowler (1977) of similar results obtained for shallow acceptors in semiconductors suggest that trap 1 has a photoionisation threshold E_T in the range 3·8–4·2 eV, and that the threshold for trap 2 is at or slightly below 5·5 eV. These estimates are indicated in figure 5. They are consistent with results obtained at Lehigh University for an impurity-related trapping centre with E_T at approximately 2·5 eV (see DiMaria *et al* 1975, Williams 1965).

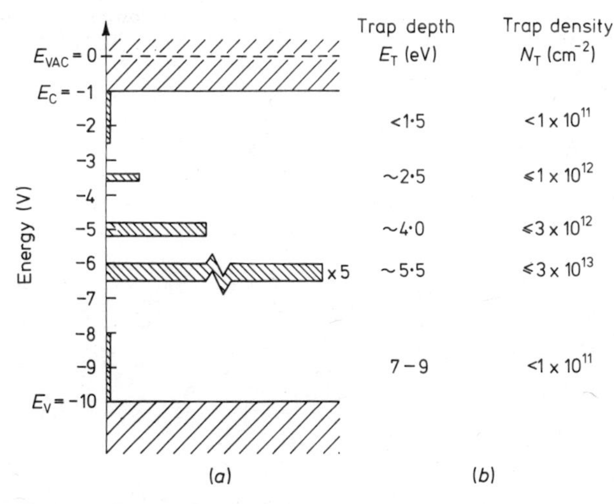

Figure 6. Energy-level diagram for bands (*a*) and localised electron trapping levels (*b*) in amorphous SiO_2.

The trapping centre at 2·5 eV showed Coulombic capture behaviour (DiMaria *et al* 1975), indicating that the trapping centre is electrically positive before electron capture, unlike the neutral traps shown in figure 5. Previously unreported data on the 2·5 eV trap are presented for comparison in figure 6, along with a summary of information on the neutral traps. The trap depths E_T shown in figure 6(*b*) are the photoionisation thresholds determined by the procedures described above. These trap levels are transposed to an energy level diagram for amorphous SiO_2 in figure 6(*a*). E_C is the SiO_2 conduction-band edge, to which the photoionisation thresholds are referenced. E_V in figure 6(*a*) is the SiO_2 valence-band edge.

The trap levels indicated in figure 6 are unoccupied in as-grown MOS oxide layers. They are filled by electron capture during the photoinjection step of an injection–depopulation sequence. Electrons are removed from these levels, via a photoionisation process, during any photodepopulation step.

The band-edge regions in figure 6(*a*) represent either residual or error-limited values of the total negative oxide trapped charge Q_{OT} from our experimental measurement sequence. Thus, the trap densities indicated in the intervals 1–2·5 eV and 8–10 eV in the diagram are estimated upper limits for electron traps.

The trap densities N_T shown in figure 6(*b*) are the total areal densities of the several trapping centres. These are all impurity-related centres and the numbers listed are experimentally determined values for our samples. They can be greatly reduced by processing (Butler *et al* 1976). The trap densities were determined by detailed analysis of the photo-injection step. It is important to note that measured values of components of negative oxide charge Q_{OT} associated with each trap are always of the order of 1–10% of N_T. The experimental and analytical procedures for determination of N_T will be reported by Rathman (1979).

References

Butler S R, Feigl F J, Ota Y and DiMaria D J 1976 in *Thermal and Photostimulated Currents in Insulators* ed D M Smyth (Princeton, New Jersey: The Electrochemical Society) pp149–61

Deal B E 1979 *Rep. IEEE–ECS Subcommittee on Nomenclature* to be published

DiMaria D J 1978 in *The Physics of* SiO_2 *and its Interfaces* ed S T Pantelides (New York: Pergamon) pp160–78

DiMaria D J, DeKeersmaecker R F and Young D R 1978 *J. Appl. Phys.* **49** 4655

DiMaria D J, Feigl F J and Butler S R 1975 *Phys. Rev.* **B11** 5023

Edwards A H and Fowler W B 1977 *Phys. Rev.* **B16** 3613

Rathman D D 1979 *Dissertation* Lehigh University to be submitted for publication

Williams R 1965 *Phys. Rev.* **140** A569

Young D R 1980 *Insulating Films on Semiconductors 1979* (Inst. Phys. Conf. Ser. 50) pp 28–39

Electric conduction and charge distribution in poly-oxide-poly silicon structures

M Conti, G Corda and R Gastaldi
SGS/ATES Componenti Elettronici SpA, Castelletto di Settimo Milanese, Milano, Italy

Abstract. To determine the physical mechanism that controls the high electrical conduction normally found in oxides thermally grown on polysilicon, the I–V characteristics of poly–oxide–poly, metal–oxide–monocrystalline silicon both lightly and heavily doped with phosphorus, poly–oxide–monocrystalline silicon and metal–oxide–texturised silicon devices were tested both at cryogenic and room temperatures. The behaviour of the metal–oxide–texturised silicon samples confirms the important role of the surface asperities that are present at the silicon–oxide interface in the determination of the high conduction properties.

1. Introduction

Silicon dioxide films thermally grown on polysilicon have peculiar conductivity and current ageing properties that are very interesting both from physical and technical points of view. A conduction current of $10^{-7}\,A\,cm^{-2}$ at an apparent electric field as low as $10^6\,V\,cm^{-1}$ has been reported by Anderson and Kerr (1977) for oxide grown at 1000 °C and has also been measured in the present work. These performances are being exploited in non-volatile memory devices which can be written and erased electrically.

With oxides grown on polished monocrystalline silicon, a similar current density requires a substantially higher electric field, about $8 \times 10^6\,V\,cm^{-1}$. Conduction in these oxides was shown to arise from Fowler–Nordheim tunnel emission from silicon to the conduction band of SiO_2 (Lenzlinger and Snow 1969) and a barrier height of 3·25 eV was found, in good agreement with theoretical predictions. When the Fowler–Nordheim conduction concept is applied to oxides grown on polysilicon a much smaller barrier height, in the range 1–1·6 eV, is found (DiMaria and Kerr 1975, Iizuka *et al* 1976).

An important observation reported by Anderson and Kerr (1977) can give an explanation of this discrepancy: the poly–oxide interface is not planar as in polished monocrystalline silicon, but is wrinkled. Therefore at the edge of the asperities a consistent enhancement of the electric field can be expected. This enhancement is considerably larger than that computed for a plane double layer and can explain the high values reported for the conduction. Conduction performances degrade considerably as the charge flows through the oxide and this effect is much larger than in oxides grown on monocrystalline silicon. In the following, this phenomenon is referred to as ageing.

To investigate the conduction and storage mechanisms in oxides grown on polysilicon, many POP (poly–oxide–poly) devices were assembled and tested. Metal–oxide–monocrystalline silicon (MOX) and poly–oxide–monocrystalline silicon (POX) devices were

0305-2346/80/0050-0055$01.00

also assembled, using both low- and high-resistivity n-type silicon. Finally metal–oxide devices processed on texturised surfaces (MOXT) were also tested and showed conduction performances very similar to those of POP devices: this supports unambiguously the asperity conduction mechanism firstly proposed by Anderson and Kerr (1977).

2. Devices

2.1. Poly–oxide–poly devices

A 4000 Å thick layer of polysilicon was deposited by CVD at 600 °C on a 1100 Å thick layer of thermal oxide grown on (100)-oriented n-type silicon (7–10 Ω cm). The film was doped with n-type material by reaction with $POCl_3$ vapour at 920 °C for 30 min and then oxidised in a dry atmosphere at a temperature in the range 1000–1100 °C to grow a layer of thermal oxide 1000–1100 Å thick. A second poly layer was deposited by CVD at 600 °C and then doped with phosphorus by CVD phosphorus glass at 600 °C. POP capacitors $2{\cdot}36 \times 10^{-3}\,cm^2$ in area were delineated by standard photolithographic technique.

The conduction performances of such POP devices were tested systematically by application of a voltage ramp at a predetermined rate and polarity to the second poly layer. The voltage ramp always started from 0 V, could be stopped at any voltage between 0 and ±90 V and its rate could be varied between 0·2 and 10 V s^{-1}. The current was amplified by a logarithmic amplifier and recorded on a x–y plotter. The displacement current flowing through the POP capacitor was partially compensated by a suitable bridge network.

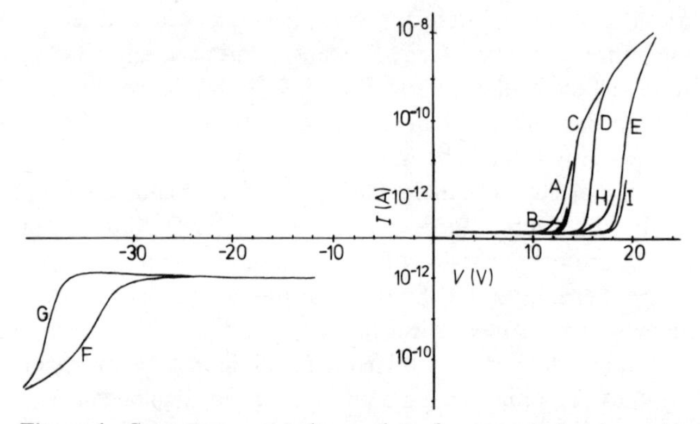

Figure 1. Current versus voltage plots for a typical poly–oxide–poly (**POP**) device for positive and negative voltage ramp excitations. Ageing effects are clearly visible. Curves A–I are discussed in the text.

Figure 1 gives results for a typical POP device in which the oxide was grown in dry oxygen at 1050 °C. Curve A is the I–V plot for the virgin sample with a voltage ramp velocity of 5 V s^{-1}. As a voltage of about 11 V is reached, the conduction current can be clearly distinguished from the background displacement current. The voltage ramp was stopped at 13 V and a new ramp started (curve B). The new I–V plot does not overlap with the previous one but the two tend to merge at high voltages. When, however, the previous maximum voltage is not surpassed a very limited shift in the I–V plot can be detected (see curve C).

Curves D and E show similar plots when additional voltage ramps are subsequently applied to the sample. The threshold voltage V_{th} is defined as the voltage that corresponds to passage of a current $I = 5 \times 10^{-12}$ A ($J = 4 \times 10^{-9}$ A cm^{-2}). In figure 1, as the sample is aged, V_{th} increases from the 'virgin' value of 12·5 V to 19 V. In other samples, an even larger ageing effect can be observed.

Current flow in POP capacitors has in the past been attributed to Fowler–Nordheim tunnel emission in which all the surface is implicitly assumed to be active. Effective barrier heights in the range 1–1·6 eV were calculated (DiMaria and Kerr 1975, Iizuka *et al* 1976), but the procedure seems to be physically meaningless since the current density J that corresponds to the mean electric field $E = V/X_0$ (X_0 is the oxide thickness) is several orders of magnitude larger than that measured. Barrier heights measured from photoelectric emission experiments confirmed a value in the range of 4·3 ± 0·1 eV, as expected for both silicon single-crystal and silicon poly material (DiMaria and Kerr 1975).

The results of conduction experiments performed on the same sample with a negative voltage ramp are also shown in figure 1 and a similar drift in the I–V characteristics can be observed (curves F and G). The initial threshold voltage is however larger, typically 32 V as compared with 12·5 V.

The effect of reversing the polarity of the voltage ramp is stressed in curves H and I. For curve H, the current is found to be considerably larger than that for curve E. This can be interpreted by assuming that electrons previously trapped in the oxide tunnel back to the polysilicon during the negative cycles indicated by curves F and G, which partially restores the original conduction properties. This restoring effect is quickly annealed as a new ramp is applied to the sample (curve I).

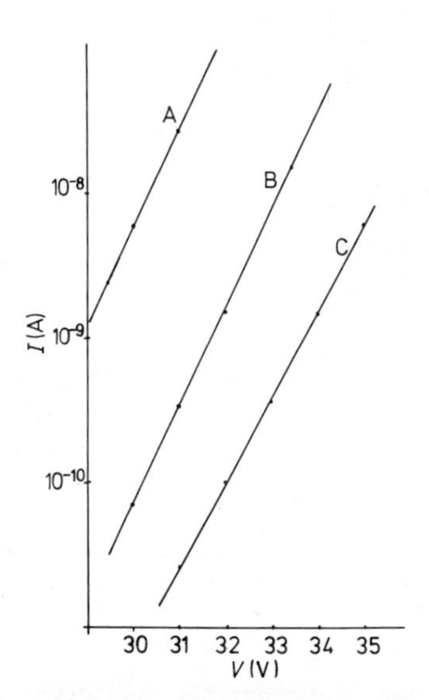

Figure 2. Current versus voltage plots for oxides grown at 1000 °C (curve A), 1050 °C (curve B) and 1100 °C (curve C). The oxide thickness for all three samples is 1100 Å. The device area is $2{\cdot}5 \times 10^{-5}$ cm^2.

The role of oxidation temperature upon conduction was investigated and is shown in figure 2. The dependence of the conduction current on voltage in samples oxidised at 1000, 1050 and 1100 °C is shown for an oxide thickness of 1100 Å. As reported by Anderson and Kerr (1977), the higher is the oxidation temperature the lower is the conduction current, or conversely, the higher is the threshold voltage.

SEM micrographs of the first poly–oxide interface are shown in figure 3. The sample oxidised at 1000 °C shows a dense family of asperities that protrude toward the upper region of the device. These are, however, less pronounced in samples oxidised at 1100 °C.

To investigate the physical characteristics of the traps that are responsible for the ageing effects, conduction measurements under ramp excitation were also performed at cryogenic temperatures. Figure 4 shows *I–V* plots recorded on a sample cooled to 100 K; the behaviour is very similar to that reported previously. When the temperature rises to 300 K, a small decrease in V_{th} can be seen (broken curve). This can be attributed to the thermal re-emission of electrons trapped in SiO_2 (Ning 1978).

Photo depopulation experiments were also carried out. Aged samples were subjected to an intense light flux of selected wavelengths in the range 0·5–0·2 μm. In these samples the second poly layer was replaced by a semitransparent aluminium electrode 100 Å thick. Preliminary measurements showed that important annealing effects can be observed.

The performances of POP devices described above were compared with the conduction performances of similar devices grown on monocrystalline silicon. Three classes of devices were investigated: Metal–oxide–silicon (MOX), poly–oxide–silicon (POX) and metal–oxide–texturised silicon (MOXT).

2.2. *Metal–oxide–silicon devices*

MOX capacitors were obtained by growing a film of oxide 1000 Å thick at 1050 °C on (100)-oriented n-type monocrystalline silicon in an atmosphere of dry oxygen. An

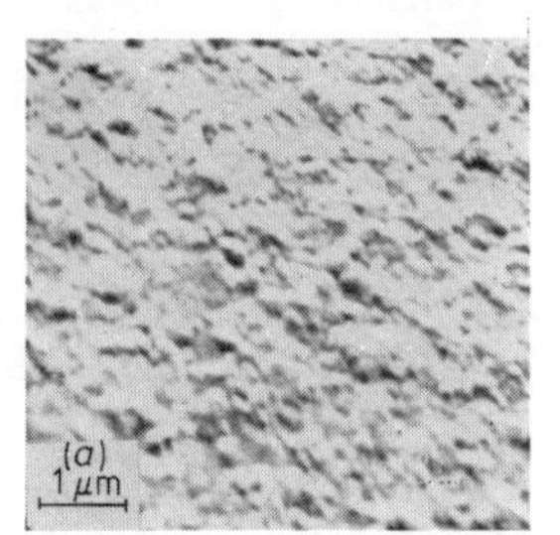

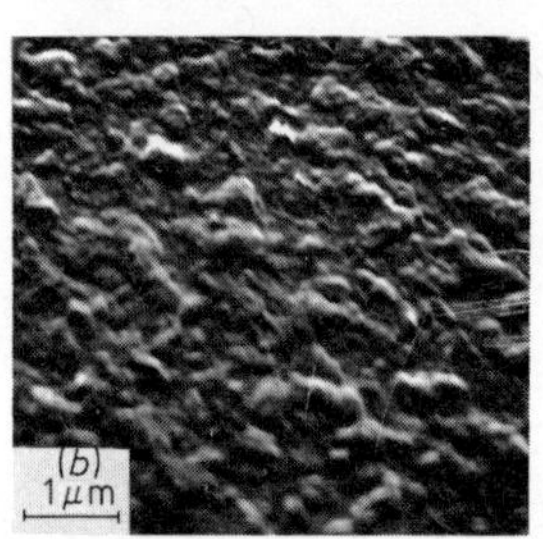

Figure 3. SEM micrographs of the poly–oxide interface for oxide grown at 1000 °C (*a*) and 1100 °C (*b*). 66° tilt.

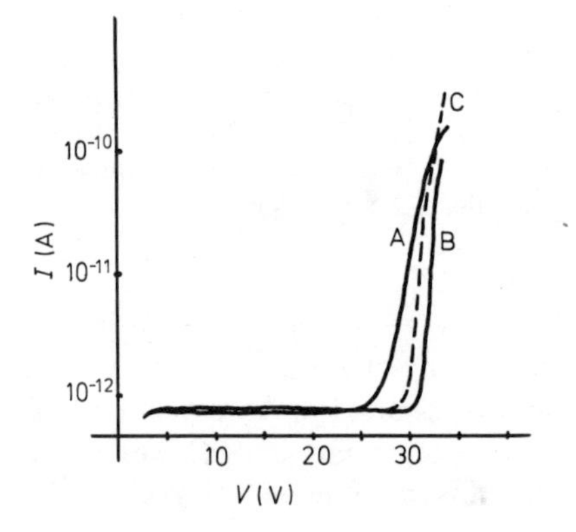

Figure 4. Temperature effects on conductivity for a POP device. Curves A and B show ageing effects recorded at 100 K. When the temperature was raised to 300 K, curve C was obtained.

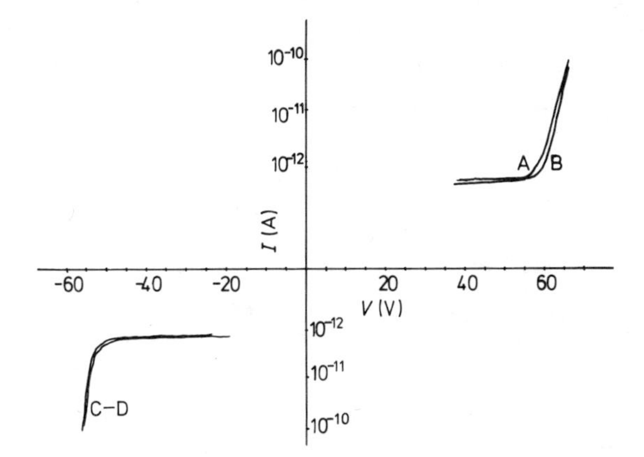

Figure 5. Current versus voltage plots for an aluminium–oxide–monocrystalline silicon device (MOX) under voltage ramp excitation.

aluminium electrode 2 μm thick was deposited by EB evaporation and then alloyed at 500 °C for 20 min. The area of the capacitor was again $2{\cdot}36 \times 10^{-3}$ cm^{-2}.

The conduction properties under positive and negative voltage ramps are in good agreement with those reported by Deal *et al* (1966) and DiMaria and Kerr (1975) for, respectively, electron emission from monocrystalline silicon and aluminium (see figure 5). Very limited ageing effects are present in both materials.

To investigate the influence of monocrystalline silicon heavily doped with phosphorus upon the conduction properties of the oxide, MOX capacitors were assembled by growing, at 1050 °C, a layer of thermal oxide 1130 Å thick on n-type silicon previously doped with phosphorus by reaction with $POCl_3$ vapour at 920 °C. The surface concentration of phosphorus was estimated to be in excess of 10^{20} cm^{-3} and the total dose diffused into the silicon was in excess of 10^{16} cm^{-2}.

Phosphorus is known to accumulate at the Si–SiO_2 interface (Johannessen *et al* 1978) and to give rise to some instabilities in MOS capacitors (Snow and Deal 1966). However, *I–V* plots obtained under positive and negative ramp excitations were identical to those obtained with samples not subjected to phosphorus doping.

2.3. *Poly–oxide–monocrystalline silicon devices*

Poly–oxide–monocrystalline silicon capacitors were obtained by the same procedure as that described above. The metal electrode was replaced by polysilicon deposited by CVD at 600 °C and n-degenerate doping was achieved by a phosphorus-rich oxide layer deposited by CVD at 620 °C. *I–V* plots for positive and negative voltage ramps are shown in figure 6. When the poly electrode was made positive, a low conduction typical of monocrystalline polished silicon was again obtained. However, when the electrode was made negative, some ageing effects were observed.

2.4. *Metal–oxide–texturised silicon devices*

To investigate the influence of surface asperities on oxide conduction some (100)-oriented 1–2 Ω cm n-type polished silicon wafers were preferentially etched by an

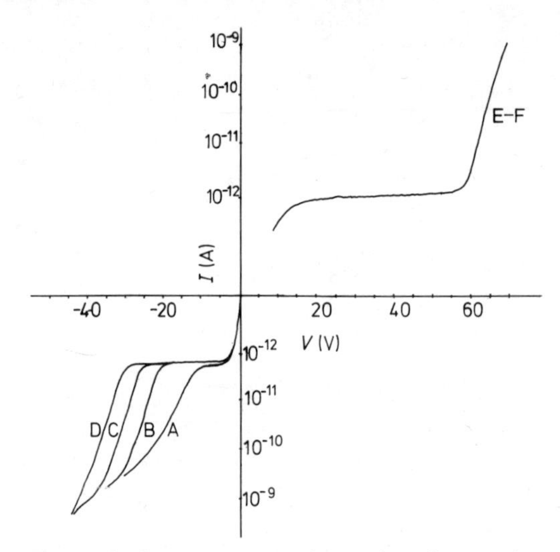

Figure 6. Current versus voltage plots for a poly–oxide–monocrystalline silicon device (POX) under voltage ramp excitation. Notice that ageing effects are only present for a negative bias.

aqueous solution of potassium hydroxide. The wafers were then oxidised in dry oxygen at 1050 °C to form a layer of oxide 1000 Å thick. A 2 μm thick aluminium electrode was then evaporated and alloyed in N_2 at 500 °C for 20 min.

I–V plots of this type of capacitor are shown in figure 7: both the low threshold voltages and the strong ageing effects normally encountered in POP devices are present. SEM micrographs of the SiO_2–Si interface etched in buffered HF to strip off the aluminium and oxide layer are shown in figure 8. A dense network of asperities spaced at approximately 0·1 μm is clearly visible.

Additional evidence that in these devices the conduction current is highly localised can now also be described. The *C–V* plot of a MOXT device was recorded before and after ageing. In spite of the large shift in V_{th} of approximately 18 V, the *C–V* plot remained unchanged, with a detectable drift of less than 0·05 V. This is a further indication that the conduction and the related charge trapping involves only a very limited fraction of the total device area.

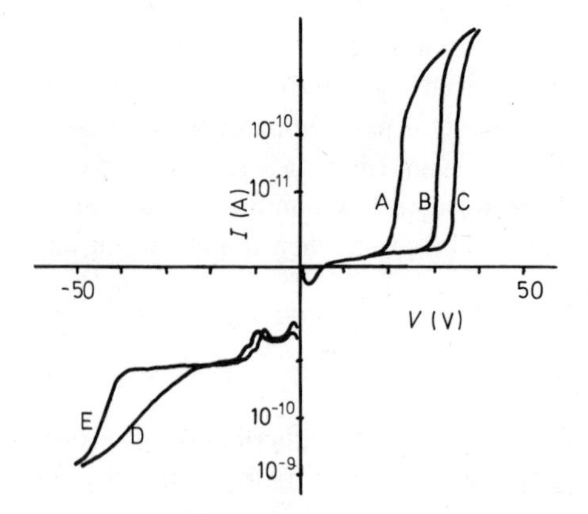

Figure 7. Current versus voltage plots for an aluminium–oxide–texturised monocrystalline silicon device (MOXT) under positive and negative voltage ramps.

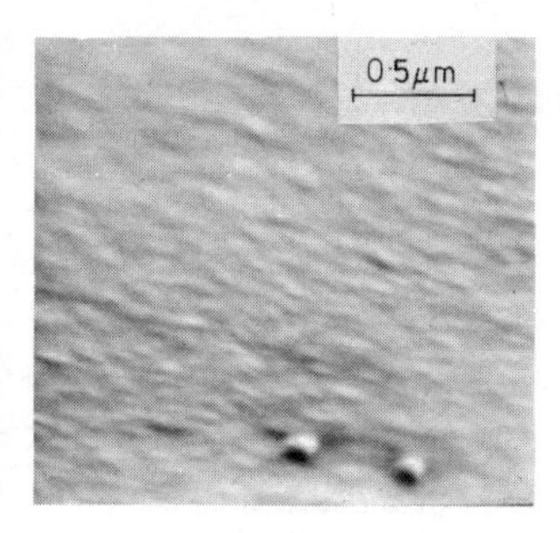

Figure 8. SEM micrograph of the silicon–oxide interface in a MOXT device. Surface asperities are caused by preferential etching of the previously polished (100) face.

3. Conclusions

It seems very likely that the high conduction of oxides grown on polysilicon arises from the large enhancement of the electric field by the asperities on the silicon surface. Important evidence in favour of this mechanism is given by the MOXT experiments in which competing effects such as high phosphorus doping and grain boundary emission are eliminated.

The nature of the electrode is only of minor importance; high conductivity is observed in MOXT devices with both positive and negative voltages. For negative voltages, electrons are emitted by the aluminium gate which is also not flat since it is a replica of the texturised silicon surface.

The conduction mechanism seems to be Fowler–Nordheim tunnel emission of electrons in the conduction band of SiO_2. This is confirmed by the relative insensitivity of the conductivity to temperature. The discrepancy concerning barrier heights could be resolved if the exact electric field profiles were known. This would allow the tunnel equation to be solved for the actual system under consideration.

Ageing effects could in principle be attributed to the intense flux of electrons emitted at the surface asperities and partially trapped in the oxide. The trapped charge density appears to be of the order of a few $10^{12}\,cm^2$, as can be computed from the shift in V_{th} (10 V in a 1000 Å thick capacitor) and is localised near the Si–SiO_2 interface.

Preliminary depopulation experiments suggest that electron traps are situated approximately 2 eV below the conduction band of the oxide.

References

Anderson R M and Kerr D R 1977 *J. Appl. Phys.* **48** 4834
Deal B E, Snow E M and Mead C A 1966 *J. Phys. Chem. Solids* **27** 1873
DiMaria D J and Kerr D R 1975 *Appl. Phys. Lett.* **27** 505
Hickmott T W 1978 *Appl. Phys. Lett.* **49** 3392
Iizuka H, Masuoka F, Sato T and Ishikawa M 1976 *IEEE Trans. Electron. Devices* **ED-23** 379
Lenzlinger M and Snow E H 1969 *J. Appl. Phys.* **40** 278
Johannessen J S, Spicer W E, Gibbons J G and Plummer J D 1978 *J. Appl. Phys.* **49** 4453
Ning T G 1978 *J. Appl. Phys.* **49** 5997
Snow E H and Deal B E 1966 *J. Electrochem. Soc.* **113** 263

Mobility and trapping of ions in SiO_2

J F Verwey
Philips Research Laboratories, Eindhoven, The Netherlands

Abstract. The first sections of this review contain a discussion of the unwanted effects of ions in SiO_2 layers on silicon. These effects are instability in metal–oxide–semiconductor (MOS) transistors, the occurrence of localised electron states at the Si–SiO_2 interface and the wear-out of SiO_2 layers. The instability in MOS transistors is specifically an instability in the threshold voltage. When it is caused by ions, it appears that sodium ions are involved and, to a lesser extent, potassium ions. The number of interface states is also found to be related to the number of mobile ions. Much less is known about the origin of the wear-out mechanism. From the temperature and field dependence it may be concluded that an ionic species is involved, but the actual mechanism is not yet clear.

The trapping and mobility of sodium and potassium in SiO_2 layers is reviewed. Sodium ions are trapped at shallower energy levels than are potassium ions. When the ions are emitted from their traps, the sodium ions have a much higher mobility than potassium ions. Finally, the influence of the addition of HCl during growth is discussed.

1. Introduction

This review starts with a description of the unwanted effects of ions on devices. First of all, the mobile ions in SiO_2 layers cause an instability. It has been found that the presence of sodium ions is the main cause of instability in metal–oxide–semiconductor (MOS) transistors. Apart from device stability, it is also desirable to have a low density of states at the Si–SiO_2 interface. The presence of many interface states gives rise to unwanted effects of enhanced recombination and generation of charge carriers in the silicon and to a lower channel mobility. The relation between ions and interface states is discussed in §3. A third effect is the wear-out of oxide layers, which is a long-term reliability hazard (§4). Some of the unwanted effects, such as instability in MOS transistors, can be ascribed to a particular ion. In other effects, such as wear-out, an ionic species seems to be involved but the actual origin is not yet known.

The trapping and mobility of sodium and potassium are discussed in §§5 and 6 respectively. The influence of HCl addition during SiO_2 growth is reviewed in §7.

2. Instability

Mobile charge in the SiO_2 layer of silicon devices causes instability in the devices. For instance, the threshold or turn-on voltage of a MOS transistor becomes unstable when a bias voltage is applied to the gate electrode. The gate oxide is usually grown at temperatures of about 1000 °C by oxidation of the silicon crystal in oxygen or steam (see Grove 1967, Kooi 1967).

It has been found that mobile charge in the form of ions is always positive. This topic has been reviewed by Eccleston and Pepper (1971) and Deal (1974). The positive charge

0305-2346/80/0050-0062$02.00

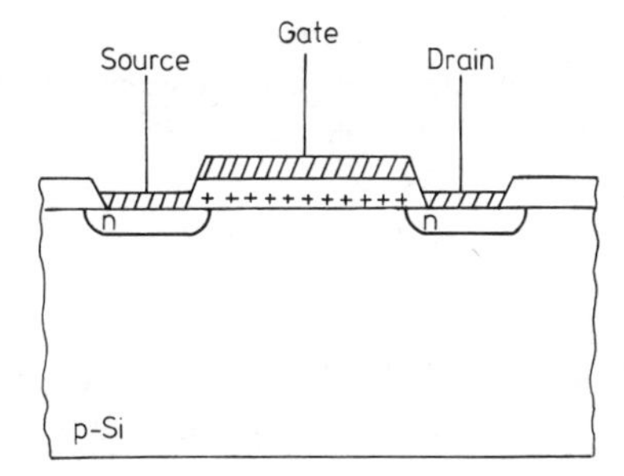

Figure 1. Schematic cross section of a metal–oxide–semiconductor (MOS) transistor with charges in the gate oxide layer.

in the SiO_2 layer of a MOS transistor may be driven to the Si–SiO_2 interface under the influence of a bias on the gate electrode (figure 1). With a high concentration of positive ions this may even happen at room temperature, or at the operating temperature of the MOS transistor, which is somewhat higher. In this way an instability in the threshold voltage V_T occurs.

The instability effect can also be conveniently studied in MOS capacitors as well as in MOS transistors. The flat-band voltage and changes in this parameter are then measured by the capacitance–voltage technique. Changes in the flat-band voltage occur after stressing the capacitor with a voltage bias at temperatures up to 300 °C, the so-called bias–temperature stress. However, changes in the capacitance–voltage curves are not only caused by the transport of mobile ions but also arise from the transfer of charge carriers at the Si–SiO_2 interface or from the creation of interface states (Hofstein 1967a). This may obscure the effect of mobile ions. Other techniques have therefore been used to study these effects. They are discussed in §§5 and 6, because we want to confine this review to the effects of mobile ions.

Two questions to be answered are: why are the mobile ions positive and are there no negatively charged ions present? Some information can be gained from experiments in which there is a layer of phosphosilicate glass (PSG) on top of the SiO_2. These experiments indicate that the mobile ions are predominantly positive. The PSG consists of SiO_2 containing a few per cent P_2O_5 and it is well known that this material removes much of the bias instability from MOS systems. A positive bias on a structure with an outer

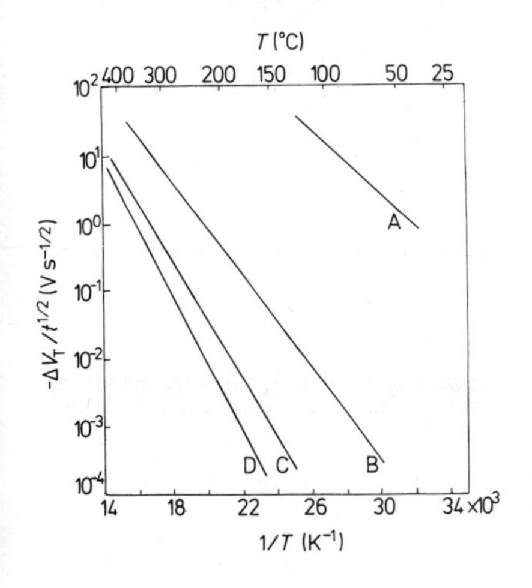

Figure 2. Temperature dependence of the threshold voltage shift ΔV_T in MOS transistors with different PSG layers on the gate oxide. The rate of change $\Delta V_T/t^{1/2}$ is plotted. The sodium concentration was 1×10^{13} cm^{-2}. The PSG parameters are: curve A, undoped SiO_2; curve B, 125 Å, 4% P_2O_5; curve C, 230 Å, 6% P_2O_5; curve D, 190 Å, 8% P_2O_5 (Balk and Eldridge 1969).

layer of PSG has no effect on a contaminated SiO_2 layer, but a negative bias results in a change in the flat-band voltage, indicative of a decrease in the amount of positive charge at the Si–SiO_2 interface. The charge transfer through the SiO_2, which could still be either positive or negative, is, however, eliminated after the negative bias stress, so that a subsequent positive bias stress does not restore the original charge condition at the Si–SiO_2 interface. The PSG layer has apparently trapped the charge firmly. This proves that the mobile charge is predominantly positive (Pepper and Eccleston 1972).

More quantitative data on the passivation induced by the PSG are given in figure 2, taken from a paper by Balk and Eldridge (1969). The figure shows the drift rate of the threshold voltage V_T of field-effect transistors with different PSG layers. The drift rate is plotted as V_T over the square root of time because when the Na^+ drift is small, the charge transported is proportional to the square root of the stressing time. It can be seen that an increase in the amount of P_2O_5 in the PSG layer effectively reduces the drift. However, an upper limit is set on the amount of P_2O_5 because too much of this material causes an instability that arises from polarisation of the PSG (Snow and Deal 1966).

3. Interface states

Positive ions in the SiO_2 layer near the Si–SiO_2 interface may attract electrons in the silicon. These electrons are then no longer able to move freely through the crystal. In this way an electron state is created just below the conduction band in the silicon (Goetzberger *et al* 1968). It has been found (Wang 1977) that the presence of sodium ions gives rise to interface states with energies distributed over the whole gap. More evidence that alkali ions cause interface states is given in the work of Schulz *et al* (1975) and Schulz and Klausmann (1979). They implanted different impurities and found that alkali ions cause a steep increase in the density of interface states near the conduction band, whereas the other elements studied had little effect. Quantitative results to corroborate these data were obtained by J Snel (1979 private communication). He found that the number of interface states in the middle 0·8 eV of the gap shows a linear dependence on the total number of mobile charges (Na^+ and K^+) in the SiO_2. These results are illustrated in figure 3, where the number of interface states N_{ss} is plotted as a function of the flat-band voltage V_{FB} in oxide films contaminated during oxidation

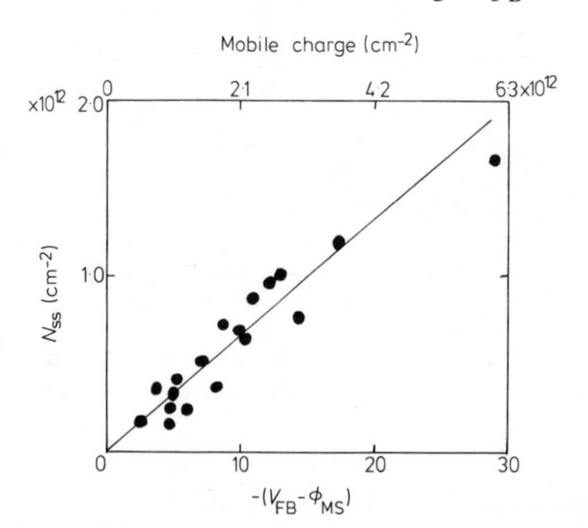

Figure 3. Interface states N_{ss} as a function of the amount of mobile charge in the oxide. The latter parameter was measured via the flat-band voltage V_{FB}.

in the first runs of a high-pressure oxidation system (Kramer *et al* 1978). The flat-band voltage was corrected for the difference in work function ϕ_{MS} and normalised to an oxide thickness of 1000 Å. It was verified by the TSIC technique (§5) that the value of V_{FB} was directly related to the number of ions present in the oxide layer. The latter quantity is therefore put at the top of the figure. The silicon crystal was doped p-type so that $N_A = 3 \times 10^{14}\ cm^{-3}$.

Scattering is another effect that ions may exert on charge carriers in the silicon. It has been clearly demonstrated that electrons in inversion layers are scattered by sodium ions in the SiO_2. Hartstein and Fang (1978) have recently published results relating to this effect.

4. Wear-out

Wear-out is the effect of electrical breakdown of the SiO_2 film during bias–temperature stress. The field applied to the film is lower than the intrinsic breakdown field.

It is thought that an electronic mechanism is responsible for intrinsic breakdown in the range 8–10 MV cm^{-1} (DiStefano and Shatzkes 1976). Electrons are injected into the SiO_2 layer from one of the electrodes (cathode) and acquire sufficient energy to cause impact ionisation with subsequent destruction of the lattice.

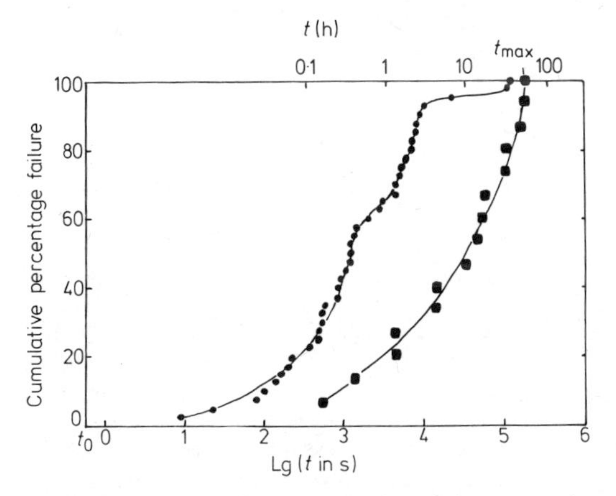

Figure 4. The cumulative percentage failures for a number of poly-Si–SiO_2–Si capacitors (350 Å SiO_2) as a function of stress time measured at 300 °C with positive electrode bias and for applied fields of 5 MV cm^{-1} (■) and 6 MV cm^{-1} (●) (Osburn and Bassous 1975).

In contrast to the intrinsic breakdown mechanism, the mechanism of wear-out is not yet fully understood. Wear-out is an important factor in the lifetime or time-to-failure of integrated circuits, but it should not be confused with the occurrence of zero-hour electrical shorts in the SiO_2. The latter parameter can be translated into the number of defects per unit area (Osburn and Chou 1973) and is important for the yield of integrated circuits. Apparently the defects are formed during growth of the oxide layers (Osburn and Ormond 1972) or during annealing of the aluminium electrode (Osburn and Chou 1973).

The time-of-failure of simple MOS capacitors is strongly dependent on temperature and field. This suggests that an ionic mechanism underlies the wear-out. An example of the type of measurement obtained is given in figure 4 for a number of poly-Si–SiO_2–Si capacitors. Figure 4 is replotted from the results of Osburn and Bassous (1975) and shows the statistical nature of the breakdown events. Rosenberg *et al* (1979) measured wear-out curves starting at times as short as 10^{-6} s. The measuring points were on smoothly increasing curves which suggests that many of the zero-hour shorts are just part of the same wear-out mechanism as the early failures in a wear-out curve that starts at, for example, 1 s.

The effect of sodium on wear-out has been studied by Osburn and Raider (1973) and Osburn and Ormond (1974). The maximum time-to-breakdown (indicated by t_{max} in figure 4) is a function of the concentration of sodium in the SiO_2. However, the influence of the sodium is very pronounced when the ions are present at a relatively high concentration, as can be seen from figure 5 (replotted from figure 4 of the paper by Osburn and Ormond 1974). t_{max} is shown as a function of the field applied to a MOS capacitor consisting of Al on a layer of SiO_2 1000 Å thick. Curve A represents the results for samples with no contamination or containing sodium up to a concentration of 4×10^{11} Na cm^{-2}. It should be noted that this concentration is not very low for modern integrated circuit technology (typically Na $< 5 \times 10^{10} cm^{-2}$). Curve B is for a relatively high concentration of sodium, 8×10^{14} Na cm^{-2}. Here there is a large effect on the maximum time-to-failure t_{max}, but only at relatively high fields. For the strength of field normally encountered during operation of MOS integrated circuits (2 MV cm^{-1} or lower), the influence on the value of t_{max} is small.

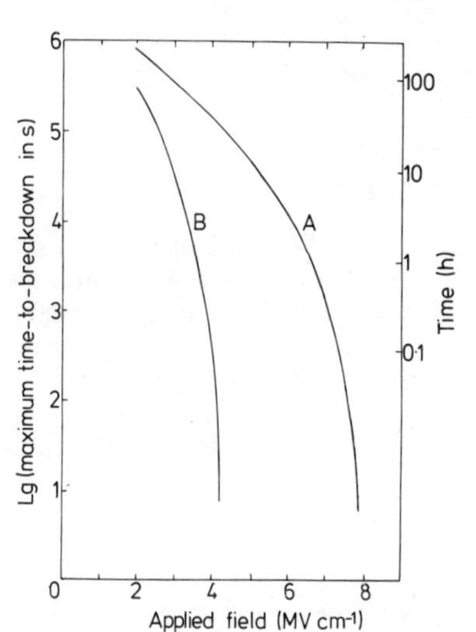

Figure 5. Maximum time-to-breakdown at 300 °C as a function of applied field in oxide layers (1000 Å) contaminated with different amounts of Na. Curve A, Na $\leqslant 4 \times 10^{11} cm^{-2}$; curve B, Na $= 8 \times 10^{14} cm^{-2}$ (Osburn and Ormond 1974). Al electrode, positive bias.

The early failures may also be influenced by sodium. This is easily understood from the movement of sodium ions to the cathodic side of the layer. There they cause a lowering of the barrier followed by increased injection of electrons into the SiO_2, which in turn may initiate breakdown. The phenomenon of enhanced electron injection has been observed by Osburn and Ormond (1974) and is illustrated in figure 6. It can be seen that as a function of time (at 23 °C) the current increases, attains a maximum and then decreases. The decrease in the current when the sample is cooled to liquid nitrogen temperature is hardly detectable when the current is high; the current maximum is therefore not caused by ions which are likely to be frozen in at this temperature. The decrease in current in the curve measured at 23 °C (figure 6) is ascribed to the fact that the sodium ions move too close to the electrode so that they are unable to enhance the field and

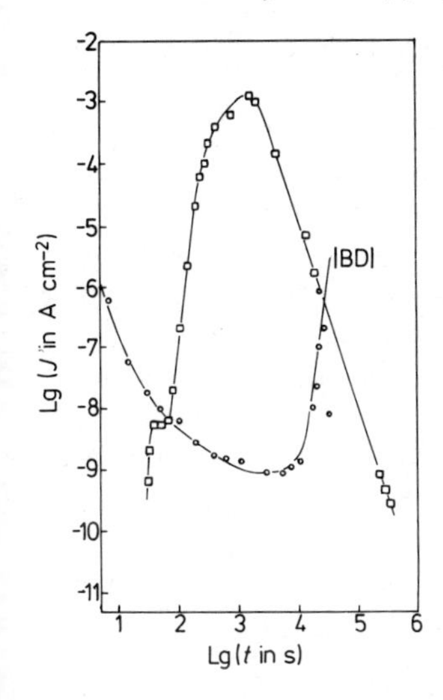

Figure 6. Current through the oxide as a function of time in oxide layers contaminated with Na (1×10^{14} Na^+ cm^{-2}) for an applied field of 4 MV cm^{-1} measured at 23 °C (□) and 200 °C (○) (Osburn and Ormond 1974). BD, breakdown.

reduce the barrier. The maximum moves to the left as a function of increasing temperature. At a temperature of 200 °C (figure 6) the current maximum occurs after a very short time and is therefore seen as a decay in the current as soon as the bias is switched on. In this case a second increase in current is observed which leads to destructive breakdown (BD in figure 6). This increase in the current is not caused by the sodium because it is also observed in non-contaminated samples. However, during the sodium-induced peak, many early failures occur (Raider 1973).

It should be noted that the value of t_{max} is a measure of the wear-out of SiO_2 layers. The conclusion is therefore that in practical situations with Na $< 4 \times 10^{11}$ cm^{-2} and a field lower than 2 MV cm^{-1} the sodium ions are not really involved in the mechanism of wear-out. We wish to emphasise that the actual species involved is of ionic origin because of the temperature and field dependence, but the nature of this ionic species is not yet clear. The anode material is very important in the wear-out and polysilicon is much better than aluminium (Osburn and Bassous 1975). This is important in n-channel MOS integrated circuits where a positive bias is applied to the gate electrodes during operation.

5. Trapping of ions

As indicated in §1 of this paper, some ions are of special interest, namely those that are mobile in SiO_2 and are likely to occur abundantly during device preparation. These ions are Na^+ and K^+. In the SiO_2 layer on Si they are trapped particularly at the interfaces of the SiO_2 film, i.e. at the M—O and at the O—Si interfaces of a MOS structure.

We wish to determine whether this positive charge is caused by sodium and potassium. To answer that question we will first discuss the technique of thermally stimulated ionic current (TSIC; Hickmott 1975). An example taken from the results of Nauta and Hillen (1978) is given in figure 7, where the current through the oxide layer is shown as a function of temperature. The current peak at low temperatures has been ascribed to Na^+ (Hickmott 1975) and the second peak to K^+ (Derbenwick 1977, Nauta and Hillen

1978). Final proof of the origin of the current peaks was found by doping with Na or with K.

As mentioned above, the ions are trapped at one of the interfaces, either the Si–SiO_2 or the SiO_2–Al interface. The proof of this is found in the asymmetry of the current peaks with respect to their position on the temperature axis when recorded with positive or negative bias. Apparently the peak position does not depend on the transport process through the bulk of the SiO_2 but on emission from traps near or at the anode of the structure (Boudry and Stagg 1978).

The ion traps at the interfaces are not at a single energy level but can be described by a distribution of states. When a Gaussian spread in activation energy and an emission time constant of 10^{-12} s is assumed, then an activation energy is found that is consistent with an emission-limited process. This means that the energy depth of the traps in the centre of the distribution (about 1 eV for Na^+) is higher than the activation energy of the transport process (about 0·6 eV, see §6). It should be noted that Li^+ ions are also mobile in SiO_2 films (Snow *et al* 1965) but these have been studied less extensively than Na^+ ions because they are unlikely to occur during technological processing.

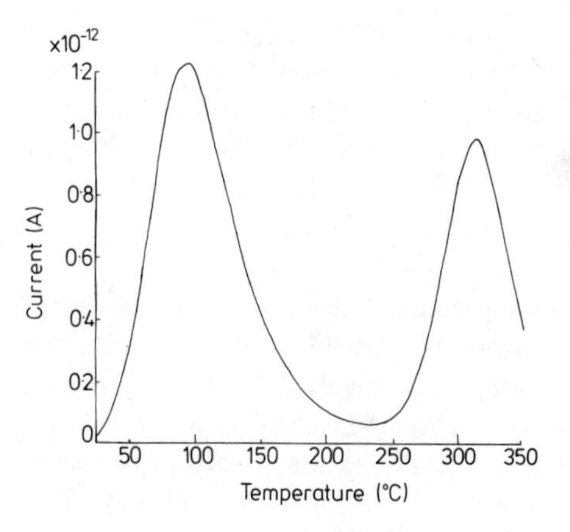

Figure 7. Typical thermally stimulated ionic current (TSIC) curve for an oxide layer containing Na^+ and K^+ ions, measured with a negative bias $V_g = -2·5$ V (Nauta and Hillen 1978).

As yet, TSIC measurements have only been carried out in MOS structures with Al as the metal electrode. H Kraay (1979 unpublished data) obtained results on a structure with polysilicon as the outer electrode. The polysilicon was, as usual, doped with phosphorus. The phosphorus was first deposited, with a subsequent period of drive-in diffusion. After deposition of phosphorus the Na^+ and K^+ current peaks arising from unintentional contamination were clearly visible. After drive-in diffusion, which it was hoped would lower the resistivity of the whole polysilicon layer, the Na^+ current virtually disappeared. Apparently, some PSG had been formed at the polySi–oxide interface after the drive-in diffusion of phosphorus, and the Na^+ ions were trapped. This phenomenon explains why MOS transistors with polysilicon gates generally have good stability even without special PSG stabilisation treatment. It would be interesting to find out why not all of the K^+ ions are gettered during the drive-in diffusion of phosphorus in polysilicon.

Another question concerns H^+ ions in SiO_2. Hofstein (1967b) found some evidence that H^+ ions are mobile in SiO_2, but Burkhardt (1967) observed that hydrogen did not move in a field in steam-grown oxides. Raider and Flitsch (1971) and Raider *et al* (1973) repeated the experiments performed by Hofstein on the doping of oxide layers from ethanol. They found that the instabilities sometimes observed after ethanol treatments were caused solely by unwanted contamination with sodium. Other results (Singh *et al* 1973, Nakayama *et al* 1978, Tangena *et al* 1978) suggest that some effects should be ascribed to the presence of mobile H^+. However, there seems to be no unambiguous proof for any of these data that unwanted sodium contamination is absent. Nevertheless it remains strange that H^+ ions are so immobile in SiO_2 layers on silicon.

6. The mobility of ions

The mobility of ions in SiO_2 was first measured by Hofstein (1967c). He determined the space-charge-limited current of what he considered to be either a proton or a form of sodium migration with a high mobility. He verified that the current through the SiO_2 depended on the square of the voltage applied (Al negative with respect to the silicon). The equation for space-charge-limited currents is (Lampert and Mark 1970)

$$J = (9/8)\epsilon\mu(V^2/L^3) \tag{1}$$

where ϵ is the permittivity of the solid, μ is the mobility of the charge carrier in this solid, V is the applied voltage and L is the thickness. All quantities are known except the mobility and this can be derived from the height of the current. The ion mobility can be written as

$$\mu = \mu_0 \exp(-W/kT). \tag{2}$$

The activation energy found by Hofstein (1967c) was 0·70 eV. The mobility of sodium ions has also been measured by Pepper and Eccleston (1972), Kriegler and Devenyi (1976) and Stagg (1977); the results are collected in table 1 and figure 8. In the second column of table 1 some data are given concerning the method of preparation of the oxide layers investigated. Pepper and Eccleston (1972) determined values of the space-charge-limited current by measuring the time derivative of the charge transported. Kriegler and Devenyi (1976) determined the transit time of the ions similarly by measuring the time derivative of the charge transported. Finally, Stagg (1977) determined the transit time of the ions by measuring the current through the oxide directly as a function of time. Two oxides were investigated, one grown in dry oxygen the other in oxygen with the addition of 10% HCl. These oxides appear to have the same properties with respect to sodium mobility. From table 1 and figure 8 it may be concluded that the different oxides yield about the same results for the sodium mobility; any differences are too small to be of practical interest.

The value of the potassium mobility in SiO_2 layers has also been determined. Stagg (1977) measured the transit time in oxides grown in dry oxygen. Hillen *et al* (1979) measured the oxide current (in SiO_2 grown in dry oxygen) with the aid of the triangular voltage sweep (TVS) method. Values of the potassium mobility are plotted in figure 8. Both determinations are seen to lie on the same straight line and this is also reflected in the numerical values of μ_0 and W. Stagg (1977) and Hillen *et al* (1979) found, respectively, values of $2{\cdot}6 \times 10^{-2}$ and $1{\cdot}16 \times 10^{-2}$ for μ_0 and $1{\cdot}09 \pm 0{\cdot}06$ and $1{\cdot}04 \pm 0{\cdot}03$ for W.

Something more should be said in this review about the TVS method because, along with TSIC studies discussed in §5, it is a very useful tool for ion transport research. In the TVS technique the current through the oxide film is measured as a function of the voltage

Table 1. Sodium ion mobilities.

Authors	Oxide	Temperature (°C)	μ_0 (cm^2 V^{-1} s^{-1})	W (eV)
Hofstein (1967a,b,c)	Dry O_2 1200 °C	40,100	40	0·70
Pepper and Eccleston (1972)	Steam 1150 °C	90–225	3–12	0·62 ± 0·05
Kriegler and Devenyi (1976)	Dry O_2 1150 °C	28–160	0·46	0·63
Stagg (1977)	O_2 1100 °C O_2 + 10% HCl 1150 °C	40–200	1·0	0·66 ± 0·02

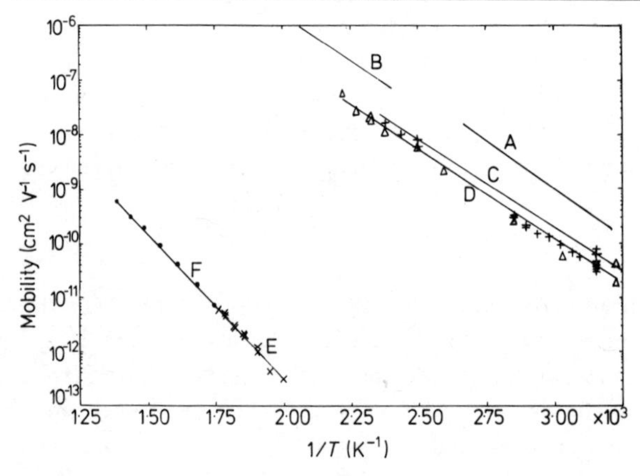

Figure 8. Survey of the mobilities of Na^+ (A, B, C, D; + dry oxide; △ HCl oxide) and K^+ (E, F) ions in SiO_2 layers on silicon. A, Hofstein (1967c); B, Pepper and Eccleston (1972); C, Kriegler and Devenyi (1976); D, Stagg (1977); E, Stagg (1977); F, Hillen *et al* (1979).

in a voltage ramp (Kuhn and Silversmith 1971). The method is generally used to measure the ion current in a quasi-static manner. At any moment of the voltage ramp, the ion distribution is assumed to be in equilibrium with the field applied (Marciniak and Przewlocki 1974, Przewlocki and Marciniak 1975, Hillen 1978). In the TVS current two peaks occur, as in the TSIC curves. One is attributable to Na^+ ions and the other to K^+ ions (Derbenwick 1977). The Na^+ peak is near zero voltage on the MOS structure but the K^+ peak is displaced from zero bias because of the transit time of the K^+ ions (see figure 9). It takes some time for the K^+ ions to be transferred through the SiO_2 film, i.e. the K^+ ions are not measured in a quasistatic condition. In the mobility measurement the peak position is measured as a function of ramp rate (Hillen *et al* 1979).

7. Ions in HCl oxides

A new development in MOS technology is the addition of halogen or halogen compounds. A review of all the effects of halogen compounds in silicon technology has been given by

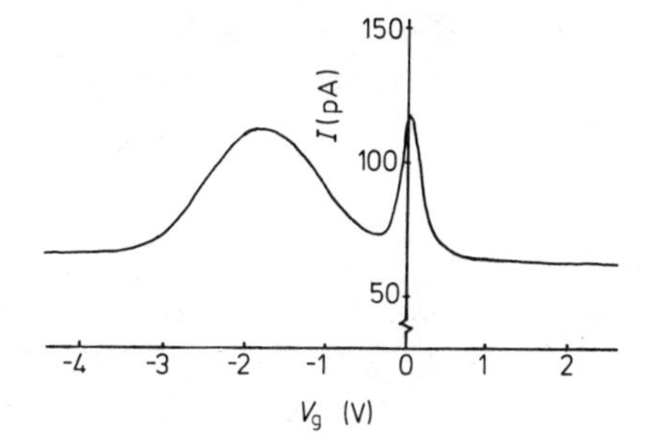

Figure 9. Typical triangular voltage sweep curve in an oxide layer containing Na^+ and K^+ ions measured at 350 °C with a sweep rate $\alpha = -245\ mV\ s^{-1}$ (Hillen *et al* 1979).

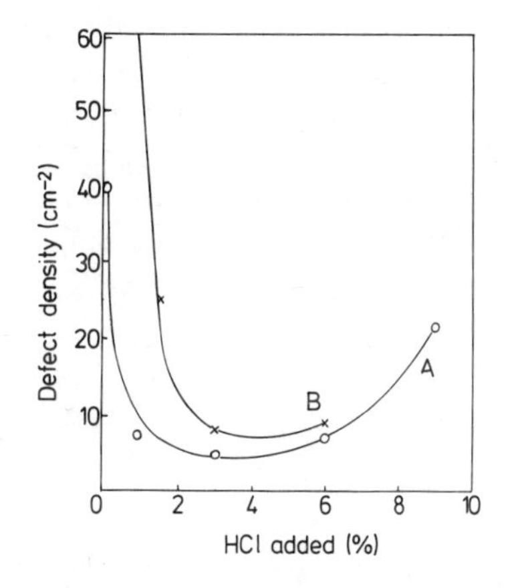

Figure 10. Defect density in SiO_2 as a function of the amount of HCl added during growth. Curve A, HCl added (Osburn and Ormond 1974); curve B, HCl from 1,1,1-trichloroethane (Janssens and Declerck 1978).

Singh and Balk (1978). One effect of the addition of these compounds to the oxygen during the growth of the SiO_2 film is that it improves the film with respect to defect density and wear-out. This is shown for the addition of HCl in figures 10 and 11. Figure 10 is replotted from the data of Osburn and Ormond (1974) and Janssens and Declerck (1978). It shows the defect density in oxide layers grown at 1000 °C in dry oxygen containing various amounts of HCl. It can be seen that the defect density has a minimum when about 3% HCl is present. The defect density decreases with decreasing growth temperature. The density of defects in the oxide layers grown by Janssens and Declerck (1978) without trichloroethane was 140 cm^{-2}. The mechanism by which the density of defects is reduced when HCl is added is not known, and neither is the means by which there is an increase in defects at high HCl concentrations. It is possible that the appearance of a second phase, that grows as increasing amounts of HCl are added (Monkowski *et al* 1978), is responsible for the latter effect. The improvement in the defect density may be due to the removal of an unknown defect-creating impurity from the SiO_2 layer by the HCl. More investigations are needed to study these effects in more detail.

Figure 11 shows the value of t_{max} (see figure 4 in §4) as a function of the percentage of HCl added. The parameter indicated for each curve is the stress field applied in the wear-out test. The addition of, for example, 3% HCl is seen to increase the maximum time-to-breakdown by one to two orders of magnitude. Furthermore, it should be noted that the defect density shown in figure 10 and the wear-out time in figure 11 exhibit different behaviour with respect to the amount of HCl added. The value of t_{max} does not have a maximum at 3% HCl but continues to increase, although at a much lower rate, as more HCl is added. The absence of a maximum in figure 11 suggests that two degradation mechanisms are operative. It should be noted that the improvement of wear-out time with HCl addition is also present in structures which have polysilicon as one of the SiO_2 electrodes (Osburn and Bassous 1975).

The presence of HCl in the oxidising ambient has a remarkable effect on the mobile ions, especially Na^+. The chlorine from the HCl, or other halogen compounds such as trichloroethane and trichloroethylene, is incorporated into the SiO_2 film. Results on oxides made with the addition of trichloroethane have been reported by Janssens and Declerck (1978) and by Linssen and Peek (1978). Kriegler (1972) found that Na^+ ions driven to the Si–SiO_2 interface in a TVS experiment at 250 °C were not re-emitted from that interface when the bias polarity was reversed. In bias–temperature stress measurements it has also been found (Rohatgi *et al* 1977) that a certain amount of the sodium becomes inactive and has no effect on the flat-band voltage. It appears that this passivation effect is proportional to the amount of chlorine incorporated (Rohatgi *et al* 1979a). It has been suggested (Rohatgi *et al* 1979b) that the neutralisation of Na^+ is a simple capture process by neutral Cl species at the Si–SiO_2 interface.

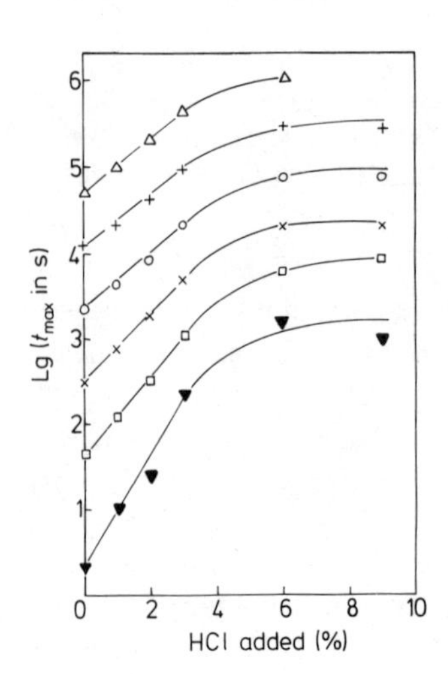

Figure 11. The maximum time-to-breakdown (wear-out time t_{max}) at 300°C for oxide layers (350 Å thick formed at 1000 °C) as a function of the amount of HCl added during growth for applied fields of 2, 3, 4, 5, 6 and 7 MV cm^{-1} (Δ, +, o, ×, □, ▲ respectively). Al electrode, positive bias.

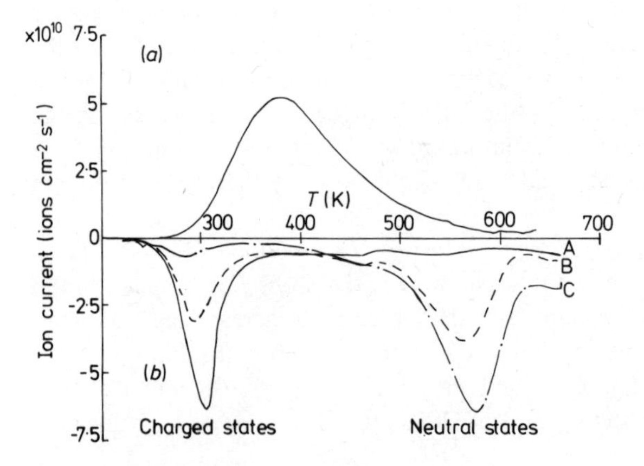

Figure 12. TSIC curves for HCl oxide (6% HCl, $Na^+ = 9 \times 10^{12}$ ions cm^{-2}, $Cl = 4{\cdot}9 \times 10^{15}$ atoms cm^{-2}). (*a*) Positive bias curve; (*b*) negative bias curves recorded after application of a positive bias at 120 °C for various times to obtain different amounts of Na^+ neutralisation: curve A, 0·5 min; curve B, 20 min; curve C, 228 min (Stagg and Boudry 1978).

Additional insight into the processes taking place during sodium transport and trapping is given by the TSIC curves plotted in figure 12, taken from the results of Stagg and Boudry (1978). The curves show the ion current as a function of temperature for a HCl oxide, grown in O_2 containing 6% HCl at 1150 °C. The positive bias curve (*a*) arises from ions transported from the metal to the silicon side of the SiO_2; the negative bias curve (*b*) is due to the ions transported in the other direction. Curve A was recorded after 0·5 min positive bias at 120 °C, curve B after 20 min and curve C after 228 min. The first peak at 300 K, ascribed to the ions in charged states, decreases with time and the second peak, ascribed to ions in neutral states, increases. These curves show that the neutralised ions are in deep traps at the Si–SiO_2 interfaces. The depth of the traps depends on the concentration of chlorine in the SiO_2 (Stagg 1977). Nevertheless, the neutralised ions are charged again under negative bias at a temperature of about 300 °C and are able to move through the SiO_2 layer. It should be noted that interface states are created during negative bias stress in HCl oxides (Hess 1977). In other words, interface states are created during the removal of Na^+ from that interface. This suggests that several processes of various degrees of complication take place during Na^+ transport in HCl oxides.

Acknowledgments

The author is indebted to J Snel and H Kraay for their contributions to, respectively, §§ 2 and 4, to J Stagg (Philips Research Laboratories, Salfords) and M Hillen (University of Groningen) for figure 8 and to Mrs I Sens for editing the reference list and for many figures.

References

Balk P and Eldridge J M 1969 *Proc. IEEE* **57** 1558–63
Boudry M R and Stagg J P 1978 *J. Appl. Phys.* **50** 942–50
Burkhardt P J 1967 *J. Electrochem. Soc.* **114** 196–201
Deal B E 1974 *J. Electrochem. Soc.* **121** 198–205
Derbenwick G F 1977 *J. Appl. Phys.* **48** 1127–30
DiStefano T H and Shatzkes M 1976 *J. Vac. Sci. Technol.* **13** 50–4
Eccleston W and Pepper M 1971 *Microelectron. Reliability* **10** 325–38
Goetzberger A, Heine V and Nicollian E H 1968 *Appl. Phys. Lett.* **12** 95–7
Grove A S 1967 *Physics and Technology of Semiconductor Devices* (New York: Wiley) pp 22–34
Hartstein A and Fang F F 1978 *Phys. Rev.* **B18** 5502–5
Hess D W 1977 *J. Electrochem. Soc.* **124** 740–3
Hickmott T W 1975 *J. Appl. Phys.* **46** 2583–98
Hillen M W 1978 *The Physics of SiO_2 and its Interfaces* ed S T Pantelides (New York: Pergamon) pp179–83
Hillen M W, Greeuw G and Verwey J F 1979 *J. Appl. Phys.* **50** 4834–7
Hofstein S R 1967a *Solid St. Electron.* **10** 657–70
—— 1967b *IEEE Trans. Electron. Devices* **ED-14** 749–59
—— 1967c *Appl. Phys. Lett.* **10** 291–3
Janssens E J and Declerck G J 1978 *J. Electrochem. Soc.* **125** 1696–703
Kooi E 1967 *The Surface Properties of Oxidized Silicon: Philips Technical Library, Eindhoven* pp34–50
Kramer R P, Snel J and van der Pol A G M 1978 *Solid State Device Technology Symp., Montpellier 1978* unpublished

Kriegler R J 1972 *Appl. Phys. Lett.* **20** 449–51
Kriegler R J and Devenyi T F 1976 *Thin Solid Films* **36** 435–9
Kuhn M and Silversmith D J 1971 *J. Electrochem. Soc.* **118** 966–70
Lampert M A and Mark P 1970 *Current Injection in Solids* (New York: Academic) p45
Linssen A J and Peek H L 1978 *Philips J. Res.* **33** 281–90
Marciniak W and Przewlocki H M 1974 *Phys. Stat. Solidi* **A24** 359–66
Monkowski J, Tressler R E and Stach J 1978 *J. Electrochem. Soc.* **125** 1867–73
Nakayama H, Osada Y and Shindo M 1978 *J. Electrochem. Soc.* **125** 1302–6
Nauta P K and Hillen M W 1978 *J. Appl. Phys.* **49** 2862–5
Osburn C M and Bassous E 1975 *J. Electrochem. Soc.* **122** 89–92
Osburn C M and Chou N J 1973 *J. Electrochem. Soc.* **120** 1379–84
Osburn C M and Ormond D W 1972 *J. Electrochem. Soc.* **119** 597–603
—— 1974 *J. Electrochem. Soc.* **121** 1195–8
Osburn C M and Raider S I 1973 *J. Electrochem. Soc.* **120** 1369–76
Pepper M and Eccleston W 1972 *Phys. Stat. Solidi* **A12** 199–207
Przewlocki H M and Marciniak W 1975 *Phys. Stat. Solidi* **A29** 265–74
Raider S I 1973 *Appl. Phys. Lett.* **23** 34–6
Raider S I and Flitsch R 1971 *J. Electrochem. Soc.* **118** 1011–2
Raider S I, Gregor L V and Flitsch R 1973 *J. Electrochem. Soc.* **120** 425–31
Rohatgi A, Butler S R and Feigl F J 1977 *Appl. Phys. Lett.* **30** 104–6
—— 1979a *J. Electrochem. Soc.* **126** 149–54
Rohatgi A, Butler S R, Feigl F J, Kraner H W and Jones K W 1979b *J. Electrochem. Soc.* **126** 143–9
Rosenberg S J, Crook D L and Euzent B L 1979 *IEEE Trans. Electron. Devices* **ED-26** 48–51
Schulz M and Klausmann E 1979 *Appl. Phys.* **18** 169–75
Schulz M, Klausmann E and Hurrle A 1975 *CRC Crit. Rev. Solid St. Sci.* **5** 319–25
Singh B R and Balk P 1978 *J. Electrochem. Soc.* **125** 453–61
Singh B R, Tyagi B D and Marathe B R 1973 *Phys. Stat. Solidi* **A19** K143–7
Snow E H and Deal B E 1966 *J. Electrochem. Soc.* **113** 263–9
Snow E H, Grove A S, Deal B E and Sah C T 1965 *J. Appl. Phys.* **36** 1664–73
Stagg J P 1977 *Appl. Phys. Lett.* **31** 532–3
Stagg J P and Boudry M R 1978 *Rev. Physique Appl.* **13** 841–3
Tangena A G, de Rooij N F and Middelhoek J 1978 *J. Appl. Phys.* **49** 5576–83
Wang K L 1977 *Proc. 3rd Int. Symp. on Silicon Materials Science and Technology* ed H Huff and E Sirtle (Princeton: Electrochemical Society) pp404–13

Lateral diffusion of Na^+ ions at the Si-SiO_2 interface and Na^+ neutralisation in the presence of chlorine

J P Stagg and M R Boudry†
Philips Research Laboratories, Redhill, Surrey RH1 5HA, UK

Abstract. Direct evidence for lateral migration of Na^+ ions at the Si–SiO_2 interface in MOS capacitors has been obtained from measurements on fine-geometry interdigitated structures formed on dry-grown oxides. An interfacial diffusion coefficient $D_I \approx 0{\cdot}003 \exp(-0{\cdot}8\ eV/kT)$ cm^2 s^{-1} is obtained for $400 \leqslant T \leqslant 500$ K. It is suggested that interfacial diffusion may occur also in HCl oxides and could be the rate-limiting step in the Na^+ neutralisation process. To support this suggestion, measurements of the interfacial neutralisation rate and the geometry of chlorine-rich islands obtained from TEM micrographs are compared.

1. Introduction

In this paper we describe observations of the lateral motion of Na^+ ions at the Si–SiO_2 interface in dry-grown MOS capacitors and relate the results to the rates of Na^+ neutralisation measured in HCl-grown structures. The Na^+ ion neutralisation process, observed in oxides containing of the order of 10^{15} chlorine atoms cm^{-2}, is of technological interest because it lessens the flat-band voltage instability in MOS devices that contain Na^+. The physics of the neutralisation process, in particular the rate-limiting stage, is not yet well understood. Valid models to explain the rate-limiting stage should be consistent with the measured dependence of the neutralisation on time. Previous authors, e.g. Kriegler and Devenyi (1973) and more recently Rohatgi *et al* (1979), have assumed a simple first-order equation to describe the interaction of Na^+ with the chlorine neutralising centres. Our previous measurements of the isothermal neutralisation time dependence (Stagg and Boudry 1978) are, however, inconsistent with a first-order equation characterised by a single time constant. The data on lateral migration of Na^+ at the Si–SiO_2 interface presented here suggest that lateral migration may be the rate-limiting step in the neutralisation when chlorine is present. A comparison is made of the interfacial neutralisation rate and the geometry of chlorine-rich islands at the Si–SiO_2 interface measured from TEM micrographs.

2. Experiments on dry-grown oxides

Oxides were grown in pure oxygen for 30 min at 1150 °C on 0·1–0·2 Ω cm (111) Si substrates and annealed in dry nitrogen for 10 min at the same temperature. The oxide

† Now an independent consultant at 42 Dewhurst Road, London W14 0ES.

0305-2346/80/0050-0075$01.00

thickness was about 1000 Å. The samples were then immersed in dilute NaCl solution of known concentration to introduce about 10^{12} Na^+ ions cm^{-2}.

Interdigitated Al structures were then applied by evaporation, using a resist lift-off technique (similar to that described by Baudet *et al* 1976), which allowed the formation of two intermeshing sets of 50 fingers, each 500 μm long and about 5 μm wide, with a gap of about 0·8 μm between adjacent sets of fingers. The sample slice was mounted in an evacuated chamber and probe contact was applied to pads covering a small area, that were integral with each electrode.

Triangular voltage-sweep (TVS) measurements were made to determine the Na^+ ion content of the oxide below each electrode. If the two Al electrodes are referred to as A and B and all potentials are taken with respect to the substrate, the ion level for electrode A was found from the size of the characteristic ion peak in the current to electrode A as the voltage was ramped negative to −4·5 V after 2 min at +4·5 V bias. The potential of B was kept at −1·5 V during this measurement so that the ions under electrode B remained at the Al–SiO_2 interface throughout the procedure. The process was then repeated with the electrodes interchanged. In each case the number of mobile ions measured was equal to the number of Na^+ ions which had been drifted to the Si–SiO_2 interface during the positive bias stress. Since the number of ions released from the Al interface depends on temperature and time of positive bias stress (Boudry and Stagg 1979), only part of the total oxide Na^+ content was measured. Experiments showed that the observed fraction varied with temperature in approximately the same way for each electrode.

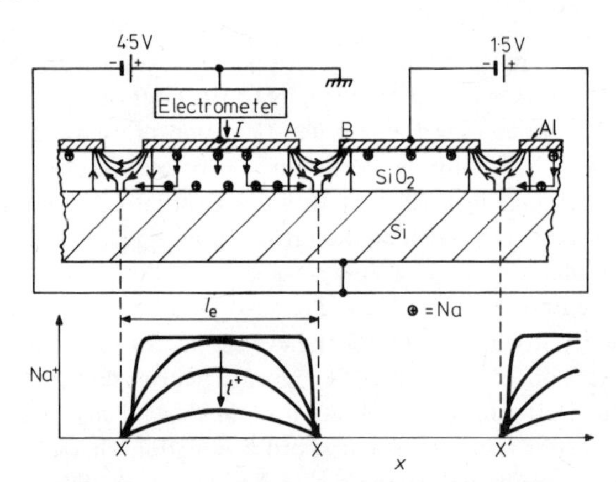

Figure 1. Lateral migration experiment. Note that the oxide thickness is not drawn to scale. The lower part of the diagram shows the lateral distribution of Na^+ at the Si–SiO_2 interface at increasing times t^+.

The experiment to detect the lateral migration of ions at the Si–SiO_2 interface (figure 1) consisted of measuring the number of mobile Na^+ ions under each electrode after increasing times (t^+) of positive bias at a temperature T_1, on electrode A; electrode B was kept at negative bias during this time. Ions released from electrode A move rapidly to the Si–SiO_2 interface directly under A and again are observed by monitoring the electrode current. Any ions which diffuse laterally at the Si–SiO_2 interface to the points X and X′ at which the interfacial electric field reverses, will then be pulled away from the

$Si–SiO_2$ interface to electrode B. The period t^+ was terminated by rapidly sweeping electrode A to negative bias, to remove the remaining ions from the interface.

3. Results

It was found that the number of mobile ions under electrode B increased approximately in proportion to $(t^+)^{1/2}$. The number of mobile ions measured under electrode A decreased or remained approximately constant, depending on the temperature of the TVS measurement. An example of some results that show the increase in the number of mobile ions under electrode B and the decrease under electrode A is given in figure 2. For this experiment (in which the TVS temperature was equal to T_1), the total number of mobile ions measured was constant. The overall effect is interpreted as a transfer of ions by lateral migration at the $Si–SiO_2$ interface. The $(t^+)^{1/2}$ dependence is consistent with a diffusion-type process in which the ions move down to a point of zero concentration near the point of field reversal. The characteristic time of the process depends on the width of the electrode and the diffusion constant. Continual slow detrapping of ions from electrode A tends to replenish the supply of mobile ions under this electrode, which explains the lack of constancy in the total number of mobile ions observed under some conditions.

The migration is reversible in that the ions could be transferred in the opposite direction by reversing the polarities applied to the electrodes. In order to minimise unwanted lateral migration of ions during the TVS measurements, these measurements were usually made at temperatures below T_1.

A simple diffusion model that assumes an initially uniform spatial distribution of ions under electrode A at the $Si–SiO_2$ interface leads to the following relation between the interfacial diffusion constant D_I, the effective electrode width l_e and the fraction of ions transferred after time t^+:

$$\Delta Na/Na_{initial} \simeq (16 D_I t^+/\pi l_e^2)^{1/2} \qquad \Delta Na/Na_{initial} < 0{\cdot}5. \qquad (1)$$

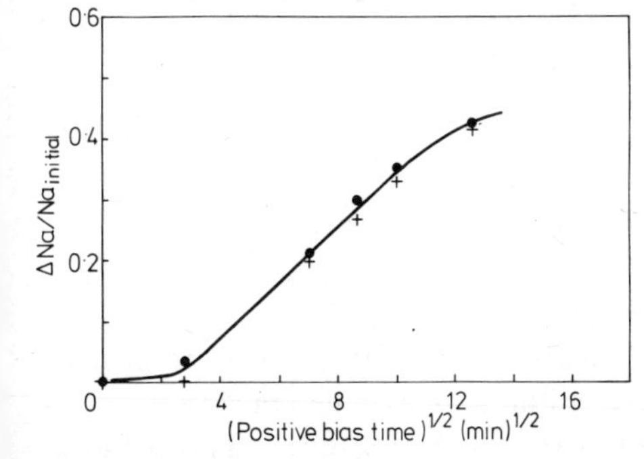

Figure 2. Time dependence of Na^+ ion transfer. ● Decrease for electrode A; + increase for electrode B. Slope = $0{\cdot}045\ min^{-1/2}$; $T_1 = 450$ K; $l_e = 6{\cdot}5\ \mu m$.

Here ΔNa is the number of Na^+ ions transferred and $Na_{initial}$ is the initial number of ions moved to the $Si–SiO_2$ interface under electrode A. The effective electrode width for diffusion l_e is the distance between the field reversal points X and X′ and is thus slightly larger than the actual electrode width. The finite intercept in figure 2 may arise from the smaller initial concentration of ions beneath the edge of electrode A caused by the spreading of the field lines in this region. Values of D_I were calculated from measurements in the temperature range 400–500 K, using equation (1) and are shown as an

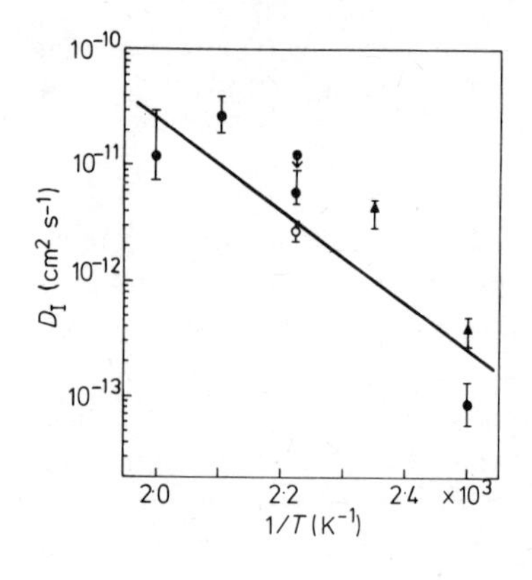

Figure 3. Arrhenius plot of the interfacial diffusion of Na^+ ions. The straight line is a plot of equation (2). TVS measurements were made at 375 K (▲), 400 K (●) and 450 K (○).

Arrhenius plot in figure 3. Although there is a large scatter in the results, the magnitude of D_I is in the range 10^{-13}–10^{-11} $cm^2\ s^{-1}$. This is much smaller than the diffusion coefficient of Na^+ ions in bulk SiO_2 as calculated from the mobility data (Stagg 1977), i.e. $1{\cdot}8 \times 10^{-10}$–$10^{-8}$ $cm^2\ s^{-1}$ for the same temperature range. The straight line in figure 3 is

$$D_I = 0{\cdot}003 \exp(-0{\cdot}8\, eV/kT)\ cm^2\ s^{-1} \tag{2}$$

and describes the results to within a factor of four over the measured temperature range.

4. Errors

The error bars in figure 3 represent errors in assigning slopes to plots of the type shown in figure 2. A further error in these plots, which could account for the scatter in figure 3, arises from the uncertainty in the relationship between the number of ions that actually participate in the lateral migration process and the fraction of these ions measured in the TVS experiments. Experiments suggest that the fraction measured bears constant relationship to the total for each experiment, which allows normalised plots of $\Delta Na/Na_{initial}$ to be constructed. A refinement to the technique would be to measure $Na_{initial}$ directly during the drift period at high field, prior to the lateral diffusion phase.

5. Experiments on HCl-grown oxides

Measurements of the interfacial neutralisation rates for Na^+ in HCl-grown oxides grown on (111) Si substrates showed that the amount of neutralisation was approximately independent of the applied voltage bias. Since the net potential barrier for the neutralisation is lowered by increasing the positive bias, the absence of a strong dependence on bias suggests that the charge exchange is not the rate-limiting process in neutralisation.

In these experiments, the ions were detrapped from the Al electrode and drifted to the Si–SiO_2 interface in a time that was much less than the neutralisation time, by transient application of an electric field larger than that used during the neutralisation period. This effectively separated the time dependence of detrapping, transport and subsequent neutralisation which is observed if this precaution is not taken. The interfacial neutralisation rate was determined by relating the number of charged ions at the interface after time t to the total number of ions initially drifted to the interface. The neutralisation rate was found to be strongly non-exponential. Results for two samples are shown in figure 4.

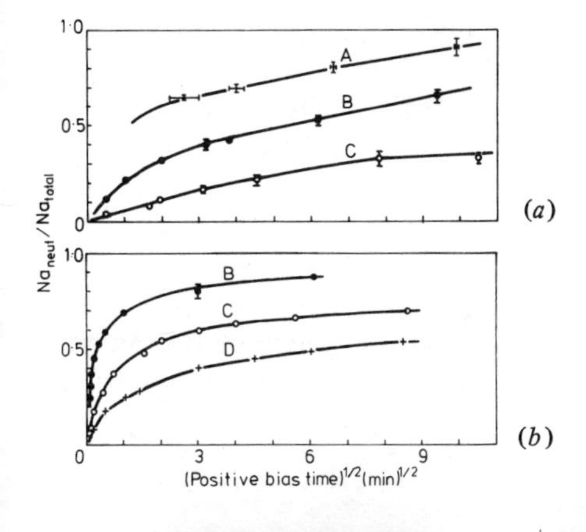

Figure 4. Time dependence of interfacial neutralisation in samples containing $3{\cdot}0 \times 10^{15}$ Cl cm^{-2} (*a*) and $5{\cdot}6 \times 10^{15}$ cm^{-2} (*b*) at temperatures of 500, 450, 393 and 350 K (curves A, B, C, D respectively).

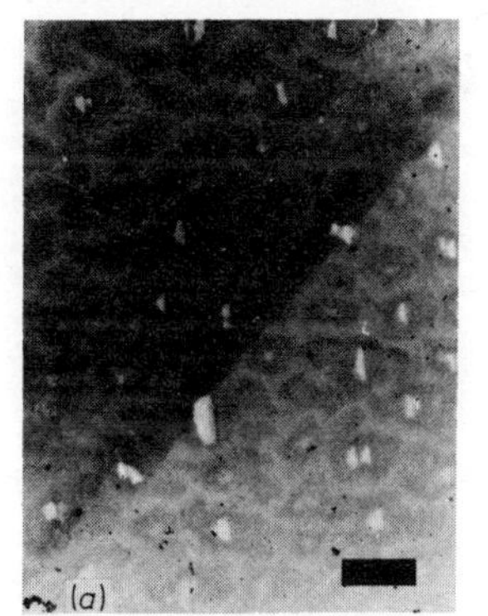

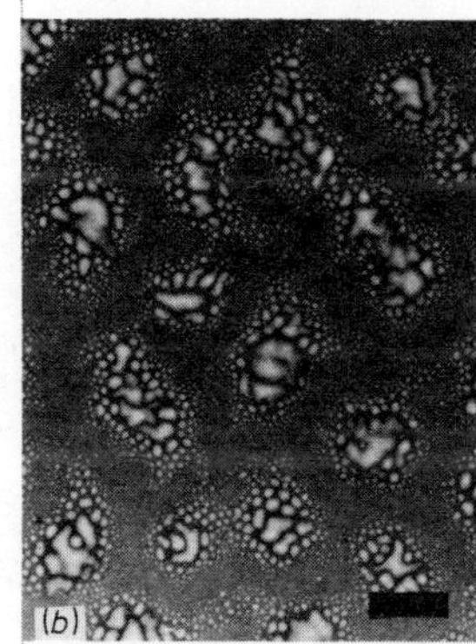

Figure 5. TEM micrographs of planar sections at the Si–SiO_2 interface of the two samples. The marker lengths are 1 μm. (*a*) Sample containing $3{\cdot}0 \times 10^{15}$ Cl cm^{-2}, grown using 11% HCl at 1100 °C for 30 min; (*b*) sample containing $5{\cdot}6 \times 10^{15}$ Cl cm^{-2}, grown using 3·5% HCl at 1200 °C for 30 min.

Table 1. Calculated and measured diffusion lengths.

Oxide chlorine content (10^{15} cm^{-2})	$L = (4D_I\tau/\pi)^{1/2}$ (μm)	L (from TEM) (μm)
3·0	0·6–1·2	$\gtrsim 1$
5·6	0·03–0·06	$\gtrsim 0{\cdot}03$

For small t, the results are reasonably close to a $t^{1/2}$ dependence and are consistent with a diffusion mechanism that involves a spread of characteristic lengths.

The chlorine concentrations in these samples were above the threshold for the formation of chlorine-rich islands as observed previously by Monkowski *et al* (1978); TEM micrographs of the actual samples are shown in figure 5. Table 1 gives a comparison of the diffusion lengths L, calculated from the measured data of figure 4 at short times, with diffusion lengths measured from the micrographs. Given the large experimental errors, the agreement is sufficiently close to propose that interfacial diffusion of Na^+ ions to the chlorine-rich islands is the rate-limiting step in the overall neutralisation process.

6. Conclusions

Direct measurements of lateral migration of Na^+ ions at the Si–SiO_2 interface have been described. We suggest that interfacial diffusion occurs also in HCl-grown oxides and is the rate-limiting stage in the Na^+ ion neutralisation process.

Acknowledgments

We would like to thank Mrs A Gill for making the interdigitated structures, N L Andrew, C K Macvean and Dr P Blood for the chlorine determinations, and Dr J P Gowers and R Gale for the TEM micrographs.

References

Baudet P, Binet M and Boccon-Gibod D 1976 *IEEE Trans. MTT* 372–6
Boudry M R and Stagg J P 1979 *J. Appl. Phys.* **50** 942–50
Kriegler R J and Devenyi T F 1973 *11th IEEE Reliability Physics Symp., Las Vegas* (New York: IEEE) pp153–8
Monkowski J, Tressler R E and Stach J 1978 *J. Electrochem. Soc.* **125** 1867–73
Rohatgi A, Butler S R and Feigl F J 1979 *J. Electrochem. Soc.* **126** 149–54
Stagg J P 1977 *Appl. Phys. Lett.* **31** 532–3
Stagg J P and Boudry M R 1978 *Rev. Physique Appl.* **13** 841–3

Conduction processes in silicon dioxide

K J Dean†, D A Baglee‡, R A Stuart and W Eccleston
Department of Electrical Engineering and Electronics, University of Liverpool, PO Box 147, Liverpool L69 3BX, UK

Abstract. MOS capacitors have been fabricated containing only a single 'defect'. The electrical characteristics of these capacitors suggest that the defect current is space-charge-limited and that a large proportion of the electrons in the oxide are in traps. The results are not consistent with a fixed trap density, which may either vary with the thickness of the layer or with the position in a given layer.

1. Introduction

MOS capacitor structures, produced by thermal oxidation of silicon, pass very low currents at applied fields up to ~5 MV cm^{-1}. More highly conducting regions of oxide are frequently found, however, and these have areas of less than 1 μm^2. They are also characterised by low breakdown fields. The precise location of these areas is difficult to ascertain until breakdown, when the site is destroyed. In order to study the properties of these regions, previous work (Keen 1971, Zakzouk *et al* 1976) has involved the liquid crystal technique which does not destroy the site of the defect. The aim of the work is to study the electronic structure of the oxide and compare it, where possible, with the known properties of the remainder of the film. It is believed that defect-free oxides have electronic conduction controlled by Fowler–Nordheim emission (Lenzlinger and Snow 1969) from the electrodes and the results obtained on such films reflect the properties of these barriers rather than the bulk of the oxides. Defects should not suffer from this difficulty to the same extent, since it is believed that at a defect the condensation of mobile ions at the interface causes enhanced tunnelling, or Schottky emission, of electrons into the oxide.

Previous work using both liquid crystals and MOS capacitor measurements (Zakzouk *et al* 1976) has demonstrated that the defect density depends on:

(i) The total time of application of the field.
(ii) The polarity of the applied voltage. A much higher defect density is seen with the silicon negative rather than positive.
(iii) The strength of the applied field.
(iv) The thickness of the film. A rapid increase in defect density occurs with decreasing thickness. Below 0·1 μm, this increase becomes very rapid with unchlorinated oxides (figure 1).

With the liquid crystal technique, the sample is the lower plate of a parallel-plate capacitor cell. The top plate is glass-coated with a transparent conducting layer on its

† Now at Kings College, University of London, UK.
‡ Now at Colorado State University, USA.

0305-2346/80/0050-0081$01.00

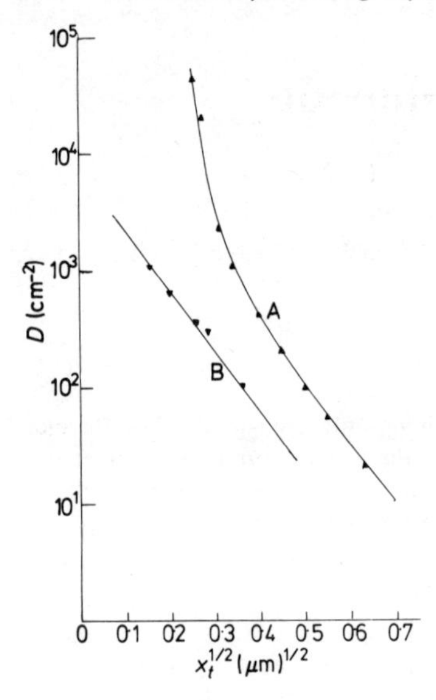

Figure 1. The variation of the defect density D with $x_t^{1/2}$. Curve A, unchlorinated; curve B, chlorinated. Effective applied field 5 MV cm^{-1}.

underside. The negative nematic liquid crystal is sandwiched between these electrodes and on application of a voltage difference across the structures, regions of high conduction become visible as turbulence when viewed with an optical microscope. The liquid crystal material limits the current through the defect to $\sim 10^{-10}$ A, preventing breakdown. On applying a negative voltage to the silicon, the defect density is initially low, but grows with time and reaches a plateau after approximately 30 min. This is believed to be due to the migration of positive ions towards the silicon. Experiments have been performed which show a relationship between the quantity of sodium ions artificially introduced into the oxide and the defect density. In addition, inclusion of chlorine with the gas during the growth of oxide has been shown to reduce the migration of sodium substantially and such oxides have very low defect densities (Baglee *et al* 1977).

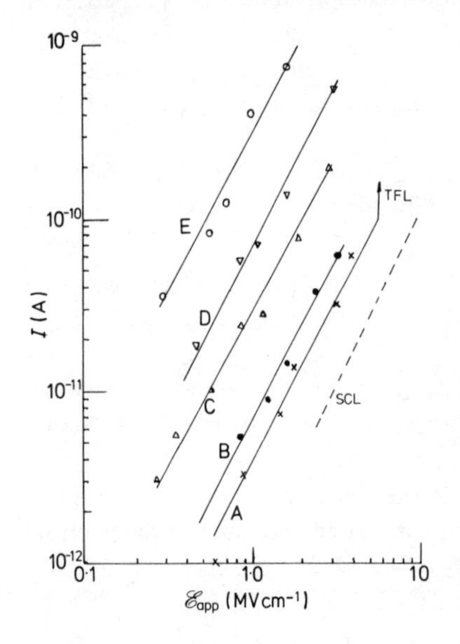

Figure 2. Currents through single-defect capacitors with unchlorinated oxide layers. x_t = 0·45, 0·3, 0·21, 0·15 and 0·12 μm (curves A, B, C, D, E respectively). TFL, trap-filled limit; SCL, space-charge-limited current.

2. Experiments

With a view to understanding the nature of electronic conduction in the oxide, MOS capacitors have been fabricated containing only one defect, for a wide range of oxide thicknesses. To produce such samples, areas of silicon oxide containing only one defect appearing at a relatively high field were found using the liquid crystal technique. The capacitors were then fabricated by producing an aluminium top electrode on these regions. All the oxides studied were made on (100) silicon (n-type). Previous unpublished work has demonstrated qualitatively similar results for (111) and p-type samples. Measurements of this type have given similar results with chlorinated oxides.

Figure 2 shows the variation of current with applied field for samples with differing thicknesses of oxide. This variation of current suggests space-charge-limited behaviour. Figure 3 shows similar results for chlorinated oxides.

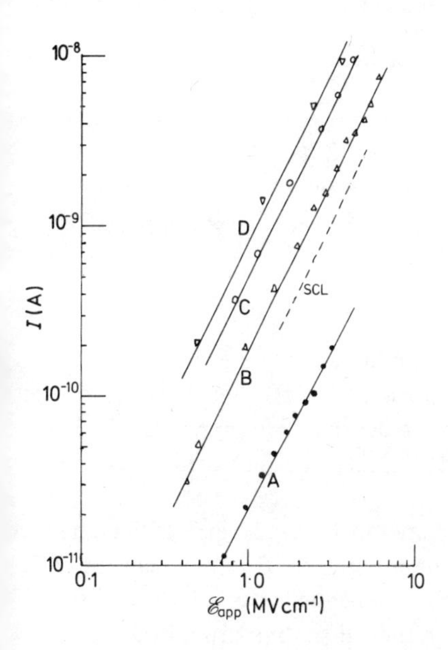

Figure 3. Currents through single-defect capacitors with chlorinated oxide layers. $x_t = 0{\cdot}2$, $0{\cdot}095$, $0{\cdot}055$ and $0{\cdot}039$ μm (curves A, B, C, D respectively). SCL, space-charge-limited current.

Because of the very high density of defects in unchlorinated oxides, for thicknesses below 0·1 μm, single-defect capacitors were impossible to produce. The results with chlorinated oxides go down to thicknesses of 390 Å.

3. Discussion

The density of 'defects' D falls as the oxide thickness increases when the effective applied field $\mathscr{E}_{app}$ is held at a constant value as in figure 1. For a particular thickness, the density of defects can be increased by increasing the applied field. Condensation of sodium ions at particular points on the surface caused enhanced tunnelling, or Schottky emission, of electrons into the oxide. The magnitude of this electron current depends on the number and position of the sodium ions, so that each defect will have a specific value of $\mathscr{E}_{app}$ at which sufficient current is produced to give turbulence with the liquid crystal. We propose the following sequence of events:

(i) For low applied fields the conduction is barrier-limited.
(ii) An increase in the applied field causes increased electron injection into the oxide.
(iii) Further increases in $\mathscr{E}_{app}$ cause space charge to build up in the oxide. At this point the electron current is limited by the bulk oxide and space-charge-limited currents ensue.
(iv) At still higher fields, turbulence appears in the liquid crystal and the additional voltage is largely dropped across this crystal. This process prevents oxide breakdown.

Process (iii) primarily limits the density of defects seen at a particular thickness of oxide with liquid crystal. Two relationships are possible, depending on whether the velocity of the electrons is saturated, or the mobility is field-independent:

$$J = 2\theta\epsilon\epsilon_0 V_{sat}(\mathscr{E}_{app}/x_t) \qquad \text{velocity constant} \qquad (1)$$

$$J = {}^9\!/_8\,\theta\epsilon\epsilon_0\mu(\mathscr{E}^2_{app}/x_t) \qquad \text{mobility constant.} \qquad (2)$$

In equations (1) and (2), J is the current density, ϵ is the dielectric constant, ϵ_0 is the permittivity of free space, V_{sat} is the saturation drift velocity, μ is the electron mobility and x_t is the oxide thickness. θ is the ratio of the free electron concentration to the total electron concentration. Providing the traps are largely empty, θ is given by (Lampert and Mark 1970)

$$\theta = (N_c/N_t)\exp(-\Delta E/kT) \qquad (3)$$

where N_c is the effective density of states at the conduction band edge and N_t is the density of traps. ΔE is the energy of traps. θ is constant for all thicknesses of oxide. Equations (1) and (2) both give a decreasing current density with increasing thickness and are therefore compatible with a decrease in the apparent density of defects D with increasing thickness.

The single-defect current–field measurements, shown in figure 2, agree with equation (2) rather than equation (1), which implies that the mobility is field-independent. The current for a particular applied field shows a strong dependence on thickness, symptomatic of the presence of space-charge-limited rather than barrier-limited currents. It is found, however, that the dependence on thickness is stronger than that predicted by equation (2). We tentatively ascribe this to a variation in θ with thickness of film due to a variation of N_t.

Figure 4 shows the variation of θ with thickness derived from the results of figure 2, for unchlorinated oxides, assuming an electron mobility of 1 $cm^2\,V^{-1}\,s^{-1}$. It can be seen that the ratio of free charge to the total charge concentration increases with decreasing thickness. This suggests that the density of traps falls either as the silicon surface is approached or with decreasing oxide thickness. Another possibility is that with thinner films the 'trap-filled limit' is being reached. With space-charge-limited current the total electronic charge in the oxide is sufficient to shield the interface barrier from the applied field. The total number of electrons in the oxide is, therefore, independent of thickness. With smaller thicknesses the electrons are 'compressed' into a smaller volume and the concentration of electrons is, therefore, increased and θ increases for the condition where the traps contain a significant number of electrons. Equation (3) no longer applies.

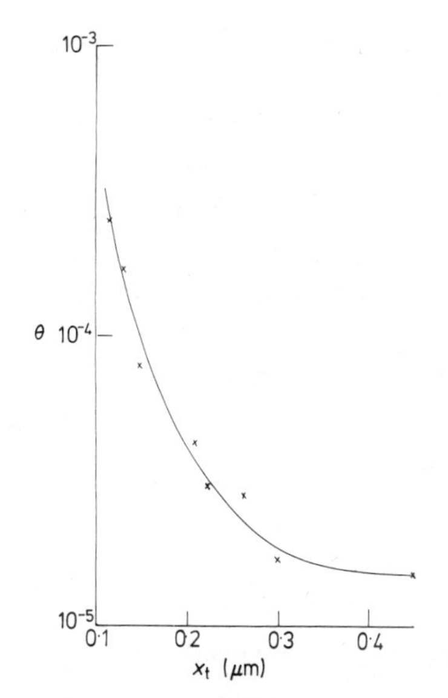

Figure 4. The variation of θ with x_t for unchlorinated oxides.

Application of Fermi–Dirac statistics to the trap and Boltzmann statistics to the conduction band gives

$$\theta \cong [N_c/(N_t - n_t)] \exp [-(\Delta E/kT)] \tag{4}$$

where n_t is the density of trapped electrons. θ is now increased in value and also varies with position. A large increase in current with field is seen and equation (1) is no longer obeyed. For all the experimental points of figure 4 equation (1) is obeyed, so that trap filling cannot, of itself, explain the variation in θ with thickness. For this reason it is proposed that the trap density varies with thickness or with position in a given oxide. Trap filling may also play a part, however, since no current–field plots have been made below 1000 Å because of the difficulty in isolating a single defect.

On one of the curves in figure 2 we show the point at which the current increases rapidly. If we assume that this is due to trap filling, we can obtain a rough estimate of the density of traps by assuming that all the traps are full and solving Poisson's equation. This gives a value of $10^{18}\,cm^{-3}$ for the trap density. Application of equation (3), assuming $N_c = 10^{19}\,cm^{-3}$, gives $\Delta E \geqslant 0{\cdot}3$ eV for $T \geqslant 300$ K. At thicknesses less than 1000 Å unchlorinated oxides show a rapid rise in defect density (figure 2). With the assumption that this is caused by the onset of the trap-filled limit, we again obtain a rough estimate of the trap density of between 10^{17} and $10^{18}\,cm^{-3}$. Nicollian *et al* (1969) have detected electron trapping at a site with an activation energy of 0·35 eV. This 'trap' does not appear to be ionised at high fields and would appear to have a high density at the oxide–metal interface which falls with distance into the oxide. All of these factors are consistent with the present results. The site is believed to arise from the presence of water in the oxidation furnace. A model was proposed by Nicollian *et al*, but the present experiments do not enable us to provide an additional verification of it.

Chlorinated oxides show behaviour that is summarised below:

(i) The currents are still apparently space-charge-limited but are lower than in the corresponding unchlorinated oxide (figure 3).

(ii) θ is found to be lower than with the unchlorinated oxide. It does, however, increase as the oxide thickness is reduced.
(iii) No rapid increase in defect density is seen at smaller thicknesses (figure 1).
(iv) The defect density is always much lower than for the unchlorinated oxide.

On the basis of the present model we assume that there is a higher overall number of electron traps but that the behaviour is essentially similar to that of unchlorinated oxides. The reduced density of defects would appear to be due to the inhibition of sodium condensation at the silicon–silicon dioxide interface by chlorine.

4. Conclusions

The variation in the defect density with thickness of oxide and the dependence of the single-defect current on applied field and oxide thickness all suggest that space-charge-limited currents are present with heavy trapping of electrons. Variation of trap density with thickness or with position in a film is proposed. Chlorinated oxides show a much reduced density of defects, consistent with the presence of a higher density of trapping levels, compared with unchlorinated oxides.

Acknowledgments

The authors wish to express thanks to the SRC for continuing financial support of this work, and to Professor J H Leck for the use of Departmental facilities.

References

Baglee D A, Gill R S, Stuart R A and Eccleston W 1977 *Electron. Lett.* **13** 144
Keen J M 1971 *Electron. Lett.* 7 432
Lampert M A and Mark P 1970 *Current Injection in Solids* (New York: Academic)
Lenzlinger M and Snow E H 1969 *J. Appl. Phys.* **40** 1 278
Nicollian E H, Goetzberger A and Berglund C N 1969 *Appl. Phys. Lett.* **15** 174
Zakzouk A K, Stuart R A and Eccleston W 1976 *J. Electrochem. Soc.* **123** 1551

MOS interface states

M Schulz

Institut für Angewandte Physik, Universität Erlangen-Nürnberg, Glückstrasse 9, D-8520 Erlangen, Federal Republic of Germany

Abstract. Techniques for measuring densities of interface states at the MOS interface and the results obtained are reviewed. Only the conductance technique and CC–DLTS are sensitive enough to measure interface state densities below 10^{10} cm^{-2} eV^{-1} such as are found in MOS structures produced by modern methods. The variation of the capture cross section with the energy position of the interface state is a key feature for characterisation of such a state. In MOS structures on n-type silicon the capture cross section decreases rapidly towards the edge of the conduction band. Its magnitude is dependent on wafer orientation and processing technology. A new model is suggested, based on tunnelling to a discrete level in SiO_2. It is postulated that this level is bent down because of round-off of the interface.

1. Introduction

This review deals specifically with electronic states that occur at the interface between silicon and its native oxide SiO_2. This interface is of great technological importance for device fabrication and has therefore received much attention (see Goetzberger *et al* 1976, Cheng 1977).

Interface states are trapping centres that are located very close to the interface so that rapid interchange of charge is possible with the mobile charge carriers in silicon. The important feature of the MOS interface is its low density of interface states. Densities close to or less than 10^{10} cm^{-2} eV^{-1} can now be achieved in silicon technology. These low densities are possible because electronic surface states in the forbidden gap of silicon are removed by oxidation and are shifted deep into the valence and conduction bands (Ynduráin and Rubio 1971). The observed residual low density of interface states has been achieved only by empirical methods of oxidation and annealing technology without any recourse to scientific models. The lower limit has obviously not yet been reached and its value is not known.

No definite theoretical model is accepted at present that can describe all the observed properties. A review of the current state of understanding is given in this paper. Measurement techniques in use and the results obtained are discussed in the light of a new model which may explain the origin of MOS interface states near the conduction band.

2. Measurement of MOS interface states

2.1. *Quasistatic* C–V *technique*

Quasistatic capacitance–voltage (C–V) measurements on MOS capacitor structures are at present the standard method in laboratory practice for determining the distribution of

0305-2346/80/0050-0087$02.00

the density of interface states in the forbidden gap of silicon. The method is usually automated and the high-frequency (1 MHz) curve is recorded with a commercial bridge. The low-frequency static curve is obtained by recording the displacement current for a slow ($10\,\mathrm{mV\,s^{-1}}$) voltage ramp. A typical set of curves, as recorded by an automated system, for a high-quality MOS capacitor on n-type silicon, is shown in figure 1. Determination of the density of interface states, which is extracted from the small separation of the two traces (Goetzberger *et al* 1976), is only possible with a large error for the samples in use at present. In figure 1 the density of interface states is less than $10^{10}\,\mathrm{cm^{-2}\,eV^{-1}}$. The quasistatic technique can therefore only be used to check the quality of the interface, for example the flat-band voltage and the upper limit of the density of states. More involved measurement techniques are necessary to measure low densities of interface states and the capture cross section. Electrical measurement techniques which have a high sensitivity are the 'conductance technique' (Nicollian and Goetzberger 1967) and the recently developed constant-capacitance deep-level-transient

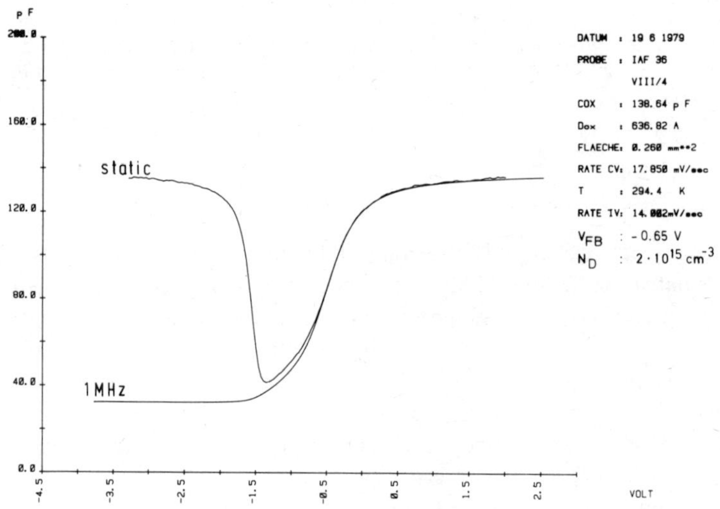

Figure 1. Capacitance–voltage (C–V) traces taken for a typical example of a MOS structure on n-type Si produced at present. The plot was obtained by a fully automated system. The density of interface states is less than $10^{10}\,\mathrm{cm^{-2}\,eV^{-1}}$ near the mid-gap.

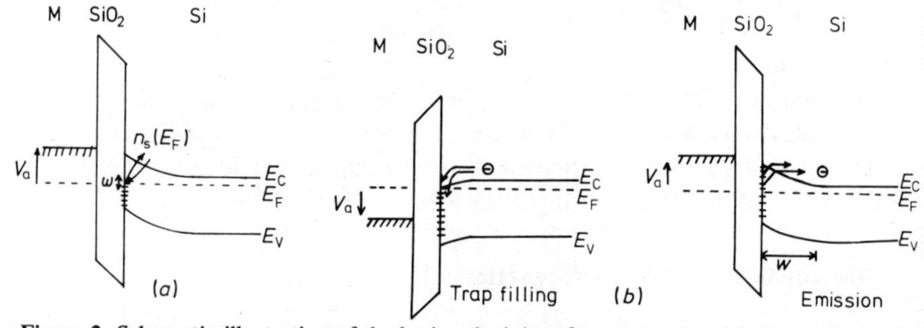

Figure 2. Schematic illustration of the basic principles of measurements by the conductance technique (*a*) and CC–DLTS (*b*). (*a*) $V = V_0 + \delta V(\omega), N_{ss}(E_F) = \sim G_p/\omega, \omega_{max} \approx 2\sigma_n v_{th} n_s(E_F)$; (*b*) $V = V_{acc}$ (trap filling), C_{depl} = constant, $N(\tau) = \sim \delta V(\tau)$, $1/\tau = \sigma_n v_{th} N_C \exp(-E_s/kT)$ (emission).

spectroscopy (CC–DLTS; Johnson *et al* 1978, Schulz and Klausmann 1979). The basic principles of both techniques are illustrated schematically in figure 2.

2.2. Conductance technique

The Fermi level, which may be positioned in the forbidden gap of silicon at the interface by adjusting the applied voltage, is used in the conductance technique to probe the energy position and density of interface states. The conductance signal G_p of the MOS capacitor is proportional to the density of the interface states at the Fermi level. The maximum in the variation of G_p/ω with frequency is approximately at the trapping frequency at the Fermi level $\omega_{max} = \omega_t = 2\sigma_n v_{th} n_s$ where σ_n is the electron capture cross section, v_{th} is the thermal velocity and n_s is the electron density at the interface. The measurement is usually quite elaborate because a wide range of frequencies must be investigated for each position of the Fermi level. The sample temperature has to be reduced in order to increase the trapping time for interface states close to the band edges.

2.3. CC–DLTS

The pure emission time constant

$$1/\tau_e = \sigma_n v_{th} N_c \exp[-(E_{is}/kT)]$$

for an interface state at energy E_{is} is determined in CC–DLTS. N_C is the effective density of states in the conduction band. As indicated in figure 2, the MOS capacitor is pulsed into accumulation to fill all the interface states. The re-emission of the charge is then sensed in the depletion region with the Fermi level near the midgap. The bias voltage in depletion is adjusted to maintain a constant capacitance. The voltage signal is then a superposition of many exponential transients caused by the continuum of interface states. The exponential of a given time constant τ attributable to a level well above Fermi level is selected by the DLTS technique. A spectrum of the density of interface states is obtained in the temperature scan of the DLTS signal/T if the exponential law for the emission time constant given above is valid.

The densities of interface states are therefore analysed by the spectrum of their emission time constant in CC–DLTS rather than by their energy position as in the conductance technique. Since CC–DLTS does not depend on the position of the Fermi level, the result is independent of fluctuations in the interface potential. Potential fluctuations have to be taken into account in the conductance technique where they smear out the energy resolution.

3. Experimental results

3.1. MOS structures on n-type Si

Typical experimental results obtained by the conductance and static techniques are shown in figure 3. In this case, the density of interface states is fairly high so that comparison with the static technique is possible. The observed curves represent the typical shape of the distribution of interface states. It is now accepted that the distribution is U-shaped with a minimum near the midgap and an increase of the density towards both band edges. The maxima that have previously been observed near the band edges (Gray

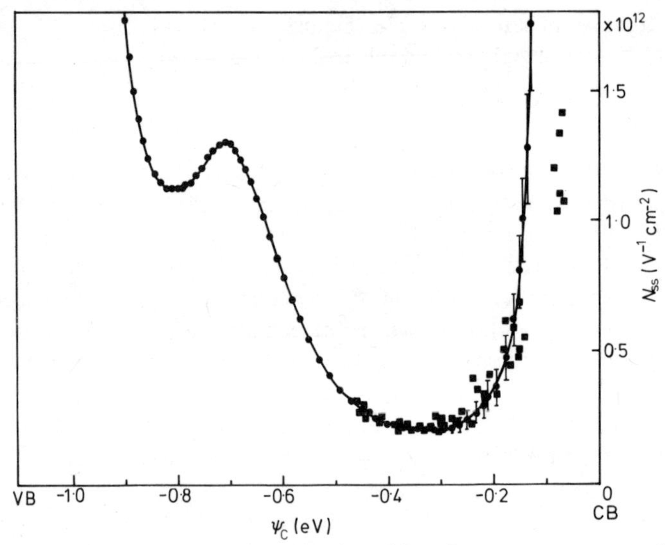

Figure 3. Distribution of the density of interface states as measured by the conductance technique (■) the static technique (●) for a MOS capacitor on n-type silicon (from results of Ziegler and Klausmann 1975).

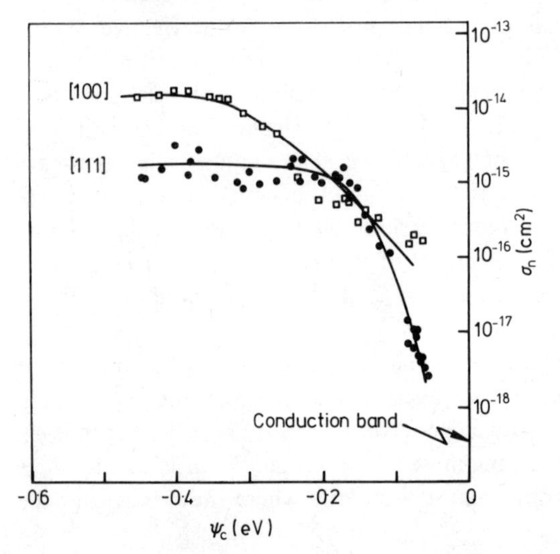

Figure 4. Capture cross sections for interface states as obtained by the conductance technique for a MOS capacitor on n-type silicon (dry oxide, annealed) (from data of Goetzberger *et al* 1976).

and Brown 1966) have been shown to be an artefact of the technique used for measurement (Boudry 1973). The structure that appears 0·75 eV below the conduction band edge has been frequently observed; it can be removed by low-temperature annealing (Johnson *et al* 1978).

Typical cross sections for the capture of electrons from the conduction band are shown in figure 4. The variation of the interface states with energy position shows a strong decrease towards the band edge. The plateau near the midgap is dependent on the orientation of the interface.

Figure 5 shows the frequency dependence of the conductance signal at various temperatures. The width of the maximum is always wider than is expected for an ideal MOS

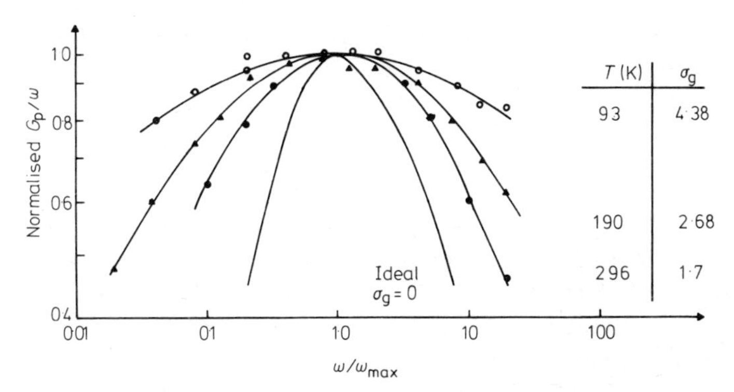

Figure 5. Normalised G_p/ω curves obtained at various temperatures. The variance σ_g of the potential fluctuation is obtained by curve fitting (from Goetzberger *et al* 1976).

interface. The increase in the width is explained by the potential fluctuation at the interface. The variance σ_g for the fluctuation is shown as a parameter.

It is noted here that the widening of the resonance peak is symmetrical at both high and low frequencies. This observation can be used as an argument to rule out tunnelling into SiO_2 states as a possible explanation for the interface states. If the states are spread into the SiO_2, the frequency dispersion is necessarily asymmetric, extending further to low than to high frequencies. Tunnelling to states in SiO_2 leads to long time constants for the capture process. A spread of states into the SiO_2 in excess of 1 Å can be ruled out because of the symmetrical result illustrated in figure 5.

Typical results obtained from CC–DLTS measurements are shown in figure 6. The density of interface states is usually obtained as a function of temperature, which may be converted to an energy distribution if we assume that the trapped charge is re-emitted by thermal emission in proportion to exp $(-E_{is}/kT)$. The energy scale is shown at the top

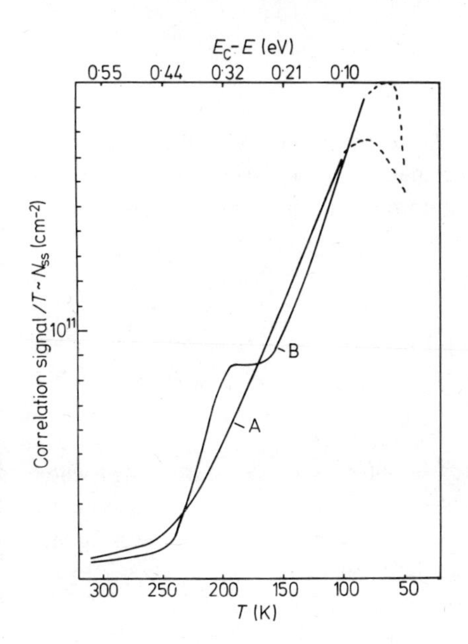

Figure 6. Distribution of the density of interface states as measured by the CC–DLTS technique for two MOS capacitors on n-type silicon. Curve A, clean oxide; curve B, after boron implantation. The spectrum is measured in a temperature scan. The energy scale is shown at the top.

of the figure. The distribution shows the same increase in the density towards the band edge as that shown in figure 3. Near the midgap a low density, less than 10^{10} cm^{-2} eV^{-1}, is still resolved. The weak structure in the implanted sample is caused by radiation damage, which may also be present in the bulk material at the same level.

Results obtained by CC–DLTS measurements for the capture cross section σ_n are shown in figure 7. Evaluation of this cross section is quite complicated; the difficulties have been discussed by Klausmann (1979). The measurements are made as a function of temperature, but the variation of the capture cross section is due to its dependence on energy (Schulz and Johnson 1978). This energy scale is shown at the top of figure 7.

The same decrease in the capture cross section near the band edge is observed as was obtained by conductance measurements (see figure 4). Both techniques therefore produce the same result. Since the evaluation is based on Shockley–Read–Hall statistics, these laws are valid for interface states. It is noted here that the magnitude of the capture cross section is dependent on the processing technology.

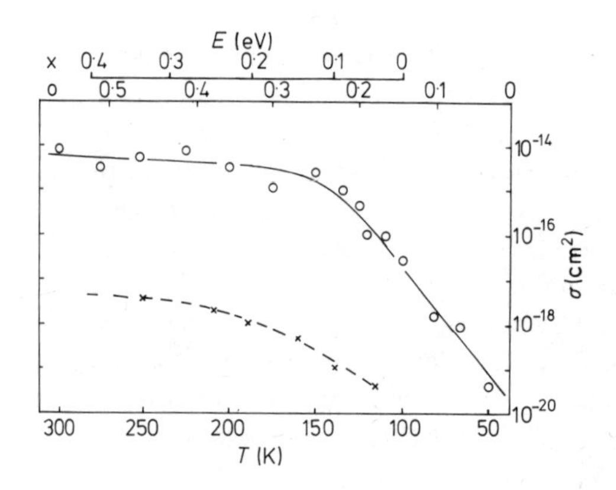

Figure 7. Capture cross sections for interface states as obtained by CC–DLTS for a clean MOS sample on n-type Si, measured with (×) and without (o) annealing. The variation is measured in a temperature scan. The functional dependence, however, is the energy position of the interface states. The energy scale for each curve is given at the top of the figure.

3.2. MOS structures on p-type Si

No conductance measurements are available for the type of MOS structures on p-type silicon that are produced at present. The measurements are impossible because of high potential fluctuations (E Klausmann private communication). Static measurements show a strong increase in the density, similar to the results shown in figure 3.

CC–DLTS measurements show that there is a strong decrease in the density of interface states towards the valence band edge, to values below 10^9 cm^{-2} eV^{-1} (Schulz and Johnson 1977). These results are shown, together with those for MOS structures on n-type material, in figure 8. It is noted that the distribution on p-type silicon looks like a continuation of the distribution of interface states on n-type material. The weak maximum is the residual effect of the peak observed in figure 3 after annealing.

The high densities of interface states in p-type samples near the valence band which are observed by the quasistatic technique are not revealed by CC–DLTS measurements. A possible explanation may be that the emission time constant of these states is outside the range of measurement. This leads to an upper limit for the capture cross section of less than 10^{-20} cm^2.

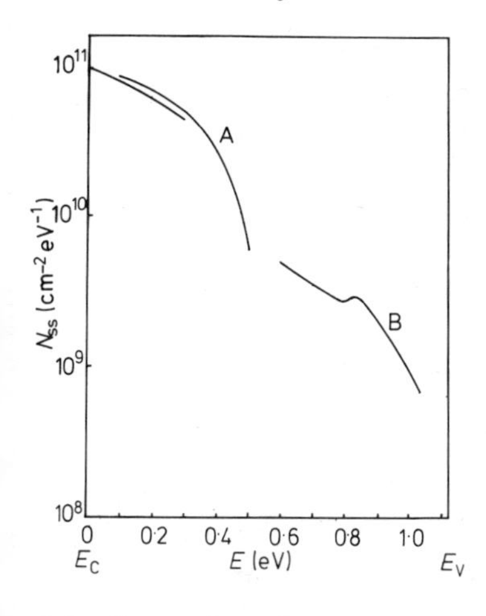

Figure 8. Densities of interface states as measured by CC–DLTS for MOS capacitors on n-type (curve A) and p-type (curve B) silicon. The decay observed for p-type samples seems to be a continuation of the variation observed in n-type material. The strong increase in the density of states shown in figure 3 is not visible.

3.3. Contaminated MOS structures

Since in the past it has been found that the number of interface states could be continuously reduced by improvement in the processing technology, it is possible that impurities have a strong effect on the density of these states. Ion implantation has been used successfully to test the effect of impurities on the density of interface states (Schulz and Klausmann 1979). An impurity can be selected and implanted under very clean conditions. In order to avoid introducing radiation damage, which has a strong effect on the density of interface states, the contaminates were implanted into the bare silicon surface before oxidation. The SiO_2 insulating film was then grown from the implanted layer. During oxide growth the implanted ions are redistributed; it was shown by SIMS measurements (Hurrle and Schulz 1975) that impurities are gettered preferentially at the

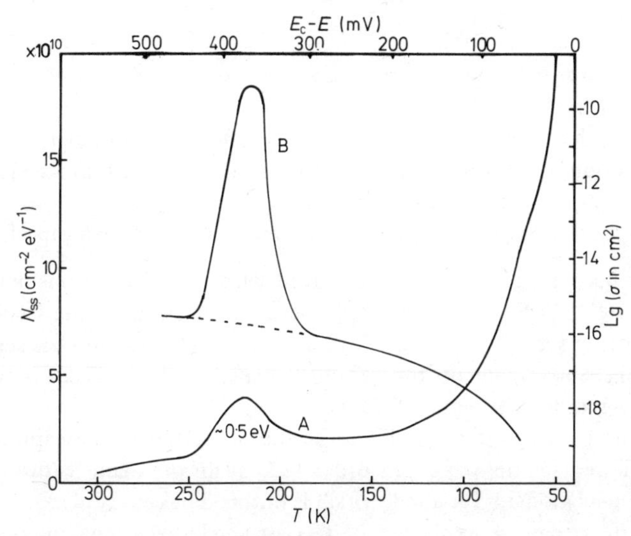

Figure 9. Density of interface states N_{ss} (curve A, left-hand ordinate) and capture cross section σ_n (curve B, right-hand ordinate) as evaluated from CC–DLTS measurements on a Cs-implanted (10^{13} Cs cm^{-2}) MOS capacitor on n-type Si.

interface and the surface. Only the alkali metals seem to affect the distribution of interface states (Schulz and Klausmann 1979). Results obtained by the CC–DLTS technique are shown in figure 9. The distribution of interface states shows the same behaviour as in clean samples. The effect of caesium is to increase the density of interface states close to the band edge. The residual radiation damage is identified as a peak in the distribution of states and in the capture cross section. The capture cross section of states in the continuous distribution decreases towards the band edge in the same manner as in clean samples.

4. Discussion

The data on the properties of interface states collected in this paper suggest that the states in various energy regions are of differing origins. States near the conduction band differ from those states near the valence band that are only visible by the quasistatic technique. The peak shown in figures 3 and 9 is of special origin and may be removed by annealing at low temperatures.

Most of the information presented here is for interface states near the conduction band where CC–DLTS and conductance measurements produce consistent results. From these results, we may draw the following conclusions with respect to models suggested in the literature:

Tunnelling to interface states which are spread out into the SiO_2 (Preier 1967) can be excluded because of the symmetrical variation of the conductance signal with frequency, shown in figure 5.

The 'charge model' (Goetzberger *et al* 1968), where electrons in Si are bound to positive charge centres in SiO_2 in the vicinity of the interface, seems unlikely. For attractive charge centres the probability of capture should increase towards the band edge and not decrease.

The following properties are noted in particular:

There is a definite relation between the energy position of an interface state and its capture cross section, which shows very little spread.

The capture cross section decreases to very low values near the conduction band.

For a given energy position, the probability of capture is dependent on processing technology.

The density of interface states seems to extend right across the forbidden gap of silicon.

Many models have been suggested in the literature which involve various defects in SiO_2 (Cheng 1977). Defects in SiO_2 are very likely because the strain of the lattice mismatch at the interface is absorbed in the first layers of the oxide. The models suggested are based mainly on plausible arguments but very little experimental evidence is available to favour one model over another (Cheng 1977).

A key feature of the interface states seems to be the properties of the capture cross section which are reviewed in this paper. In order to explain the main features of the capture cross section a new model is required, which is outlined below.

All the observed effects can be explained if it is assumed that tunnelling occurs to a discrete defect in SiO_2, the tunnelling depth of which is related to the energy position relative to the edge of the conduction band in silicon. Such a distribution is sketched in

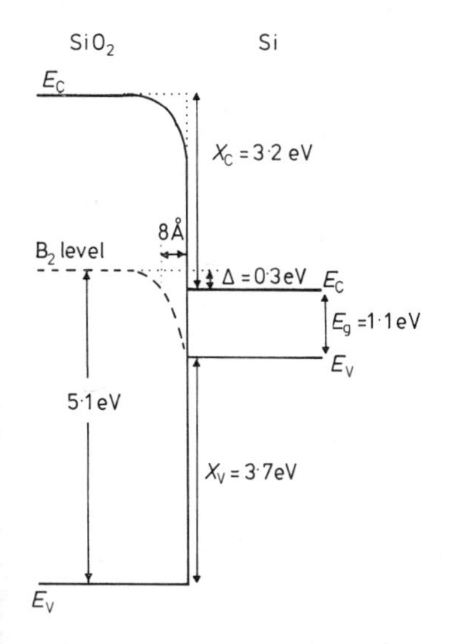

Figure 10. Schematic drawing to illustrate the new model that explains the properties of interface states near the conduction band edge. Tunnelling to a discrete level in SiO_2 is assumed. This level is bent down near the interface. The discrete level postulated to occur 5·1 eV above the valence band in SiO_2 was reported by Hickmott (1972).

figure 10. The decrease in the capture probability near the band edge can then be explained by the increased tunnelling depth. Processing technology may affect the tunnelling depth by tilting the distribution. The distribution of defects with respect to energy is not related to the band gap of silicon and can thus extend across the band gap from the conduction to the valence band.

As indicated in figure 10, this type of tilted distribution may be caused by a discrete level in SiO_2 that is just above the conduction band edge of silicon. This level bends down if there is a band bending or round-off near the interface. A level just above the conduction band edge was reported by Hickmott (1972), 5·1 eV above the SiO_2 valence band which is 0·2–0·3 eV above the Si conduction band minimum where we postulate a level should occur. The round-off of the interface is expected to be strongly affected by the processing and annealing technology so that a dependence of the tunnelling depth and capture cross section on technology is very likely. From the variation in the capture cross section, tunnelling depths ranging from 1 to 8 Å are estimated, which are beyond the resolution (18 Å) of measurements used to determine the width of the interface region (Helms and Johnson 1979).

5. Conclusions

Bending of a discrete energy level in SiO_2 seems to be a very promising explanation for interface states near the conduction band of silicon. It can explain all the observed effects reported for the energy distribution and the capture cross section. The capture cross section seems to be the key feature for deciding between various models. Models based on a variable tunnelling depth to spread out interface states and on electronic states bound to a positive charge can be ruled out by the observed properties of the capture cross section. More experimental data, especially on the optical excitation of shallow interface states and the dependence of the round-off on processing technology are necessary to test the model.

The information available on interface states near the valence band is not sufficient to reach any conclusion with respect to a model.

Acknowledgments

Many fruitful discussions with E Klausmann, IAF, Freiburg and N M Johnson, Xerox, Palo Alto are gratefully acknowledged. Measurements on clean MOS structures were partially performed at Xerox PARC. The work was partially supported by the US-ARO and the German Ministry of Science and Technology.

References

Boudry M R 1973 *Appl. Phys. Lett.* **22** 530
Cheng Y C 1977 *Prog. Surface Sci.* **8** 181
Goetzberger A, Heine V and Nicollian E H 1968 *Appl. Phys. Lett.* **12** 95
Goetzberger A, Klausmann E and Schulz M 1976 *CRC Crit. Rev. Solid St. Sci.* **6** 1
Gray P and Brown D 1966 *Appl. Phys. Lett.* **8** 31
Helms R and Johnson N M 1979 to be published
Hickmott T 1972 *J. Vac. Sci. Technol.* **9** 311
Hurrle A and Schulz M 1975 *CRC Crit. Rev. Solid St. Sci.* **5** 319
Johnson N M, Bartelink D J and Schulz M 1978 *The Physics of SiO_2 and its Interfaces. Proc. Int. Topical Conf., Yorktown Heights* ed S T Pantelides (New York: Pergamon) pp421–7
Klausmann E 1979 *Insulating Films on Semiconductors 1979* (Inst. Phys. Conf. Ser. No. 50) pp97–106
Nicollian E H and Goetzberger A 1967 *Bell. Syst. Tech. J.* **46** 1055–133
Preier M 1967 *Appl. Phys. Lett.* **10** 362
Schulz M and Johnson N M 1977 *Appl. Phys. Lett.* **31** 622
—— 1978 *Solid St. Commun.* **26** (erratum)
Schulz M and Klausmann E 1979 *Appl. Phys.* **18** 169
Yndurain F and Rubio J 1971 *Phys. Rev. Lett.* **26** 138
Ziegler K and Klausmann E 1975 *Appl. Phys. Lett.* **26** 400

The evaluation of transient capacitance measurements on MOS interfaces

E Klausmann
Fraunhofer-Institut für Angewandte Festkörperphysik, Eckerstrasse 4, D-7800 Freiburg, West Germany

Abstract. A new procedure is developed for evaluating the density of interface states and capture cross sections of MOS interfaces by means of the CC–DLTS method. The procedure used hitherto was based on rather restrictive assumptions. The solution cannot be determined unambiguously with only two sets (correlation signals) of measured data. Additional data are required, either from CC–DLTS measurements or the conductance method, to establish the uniqueness of the solution. An experimental comparison of the CC–DLTS and conductance methods is presented.

1. Introduction

The conduction method is generally considered to be the standard procedure for analysis of the properties of the interface states of MOS structures (Nicollian and Goetzberger 1967, Deuling *et al* 1972, Goetzberger *et al* 1976). This method has been used to obtain the most reliable results on the energy distribution of the densities of interface states and capture cross sections over a wide range of energies. The biggest drawback is, however, the great amount of time needed for the measurements and the necessity for the measuring equipment to be permanently attended when in operation. Some progress towards obtaining conductance measurements more rapidly has recently been achieved by a lock-in amplifier technique (Boudry 1978).

A strong competitor to the conduction method has recently emerged in the CC–DLTS technique (constant-capacitance deep-level-transient spectroscopy; Johnson *et al* 1978, Schulz and Klausmann 1979). The most striking advantages are the short measuring time (about ten times faster than the conventional conductance method) and the semi-automatic operation that only needs occasional supervision.

The competitive aspect should not, however, be stressed too much. As pointed out in the papers cited, the two methods also complement each other. For instance, results obtained by the CC–DLTS method are hardly affected by fluctuations in the surface potential and measurements made at low temperatures are more easily interpreted. Information about the statistical distribution of the oxide charges is only provided in full detail by the conductance method. However, the validity of these statements is not as strict as was claimed previously, a result that is a by-product of this paper.

In an experimental comparison of the two methods, some differences in the results were observed. A closer inspection of the procedure used to evaluate the CC–DLTS method showed that some assumptions did not really meet the actual conditions. The

0305-2346/80/0050-0097$02.00

most severe restriction exists in the presupposition that the capture cross sections should be stepwise constant, i.e. independent of energy over small intervals. If this presupposition is abandoned, it will be found that the mathematical treatment no longer results in a unique solution: the density of interface states and capture cross sections are then represented by a family of curves. Only after additional measurements can the proper result be selected.

2. Experimental details

To verify the applicability of the suggested method, a practical example with an n-type silicon MOS capacitor is given here. The capacitor is on a (111)-oriented wafer with an oxide thickness of about 100 nm and a doping density of about $5 \times 10^{15}\ \mathrm{cm}^{-3}$. A great many conductance measurements have been performed previously with capacitors of the same wafer (Deuling *et al* 1972, Ziegler and Klausmann 1975). The conductance data given here have been taken from these papers. Further details about the wafer and its preparation can be found there. Checks were carried out to assure that the wafer had not changed.

In the papers mentioned above, the conductance and CC–DLTS method and the instrumentation involved has been described comprehensively; therefore only the basic ideas behind the CC–DLTS technique are outlined here.

3. The CC–DLTS method

The CC–DLTS method can be used to analyse the density of interface states, the capture cross sections and the energy levels of these states in MOS capacitors. The method makes use of the time and temperature dependence of the electron (or hole) emission from the interface states into the majority carrier band. Firstly the states are occupied by electrons during a voltage pulse (20 μs) into the accumulation regime. Then with the bias in depletion, the excess charges are thermally emitted. In the CC–DLTS method the bias V_G is controlled by a feedback circuit in such a way as to keep the high-frequency MOS capacitance C constant. Consequently the band bending also stays virtually constant during emission and evaluation is made considerably easier.

The time-dependent bias $V_G(t)$ is then sampled at two different times t_i and t_i^* (= $2t_i$) and the difference $\Delta V_G(t_i) = V_G(t_i) - V_G(t_i^*)$ (correlation signal) is obtained. By varying the temperature, the emission rate is changed and the correlation signal shows a maximum at an optimum temperature for a particular interface state. For states of different energy and different capture cross section, the maximum occurs at different temperatures. The correlation signal $\Delta V_G(t_i, T)$ is recorded as a function of the temperature T with the sampling times $(t_i, 2t_i)$ as parameters.

In the papers cited above it has been shown that the correlation signal is related to the pertinent physical quantities by

$$\Delta V_G(t_i, T) = \frac{A}{C_{ox}} \int_{E_s}^{0} N_{ss}(E) \left[\exp\left(-\frac{t_i}{\tau(E)}\right) - \exp\left(-\frac{2t_i}{\tau(E)}\right) \right] \mathrm{d}E \qquad (1)$$

and

$$\tau^{-1} = \sigma(E)\, v_{th} N_C \exp(E/kT). \qquad (2)$$

Here A is the area of the MOS capacitor, C_{ox} is the oxide capacitance, E is the energy level of the interface state within the band gap (referred to the majority carrier band edge), E_s is the energy of the deepest state that can just be charged and discharged by the bias V_G (see figure 1) and $N_{ss}(E)$ is the density of interface states. $\sigma(E)$ is the capture cross section, $\tau(E)$ is the response time of an interface state, t_i is the sampling time, k is the Boltzmann constant, T is the temperature, $v_{th} = 10^7(T/300)^{1/2}\,cm\,s^{-1}$ is the thermal velocity of electrons and $N_C = (2{\cdot}8 \times 10^{19})(T/300)^{3/2}\,cm^{-3}$ is the effective density of states in the conduction band. The numerical values given here refer to electrons as majority carriers in silicon.

The energy E_s can be obtained by high-frequency capacitance or slow ramp measurements (Kuhn 1970). Equation (1) with $i = 1, 2, \ldots$ (together with equation (2)) composes a system of integral equations with the unknown functions $N_{ss}(E)$ and $\sigma(E)$. All other quantities are known or, at least in principle, are amenable to measurement. In a strict mathematical sense, however, nothing is known about a solution, its existence and uniqueness and how many correlation functions with different t_i must be given for such a solution.

In the following the problem will be described approximately by differential equations. Two correlation signals with different t_i will suffice to achieve solutions for practical purposes. These solutions, when put into equation (1), reproduce the measured correlation signals to an accuracy of 1% or better. The measurement of a third correlation signal will be necessary to determine the final solution unambiguously unless additional data obtained by the conductance method are used.

3.1. *Preparatory mathematical transformations*

It is presupposed that the interface states can only be described by a single pair of functions $N_{ss}(E)$ and $\sigma(E)$. These functions must be continuous and differentiable if required. The following developments do not hold if two (or more) types of interface states with different capture cross sections $\sigma_1(E)$ and $\sigma_2(E)$ are present, as may be expected in contaminated MOS structures.

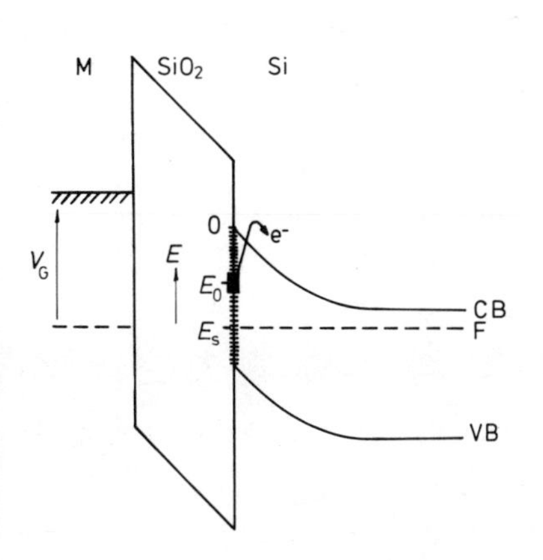

Figure 1. Thermal emission in the energy band model. Electrons are released from occupied interface states above the Fermi level, but only electrons from a small energy interval ΔE in the neighbourhood of the mean energy E_0 contribute to the correlation signal. At sufficiently low temperatures this interval is far above the energy E_s and the correlation signal is therefore independent of this parameter. At higher temperatures and for a sampling time that is not too short, E_0 approaches E_s and at its lower end the interval ΔE is limited by E_s. As a consequence, the correlation signal drops steeply. V_G is the bias voltage, the Fermi level is labelled F and the edges of the conductance and valence bands are marked CB and VB respectively. The energy E_s is given by that point in the band gap at which the Fermi level pierces the interface.

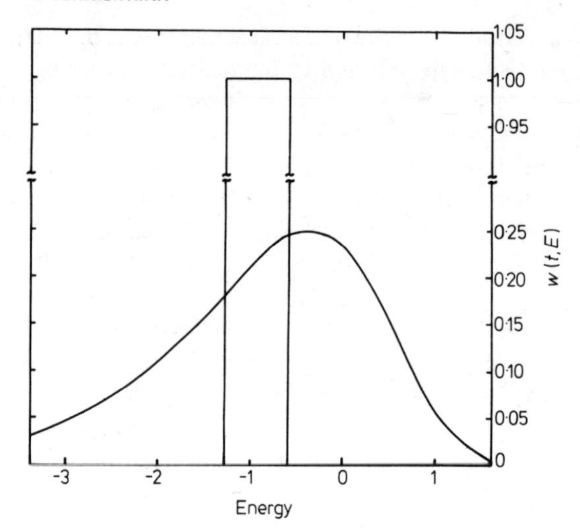

Figure 2. The weighting function $w(t, E)$. It is assumed that $\sigma(E)$ is constant and does not depend on energy. The abscissa represents an energy axis in units of kT/q (it may include an additive constant). If $N_{ss}(E)$ is a smooth function, integral (1) can be approximated by an explicit function (equation 3), which is equivalent to replacing the weighting function by the rectangular function. Its width is given by equation (6), $\Delta E = kT \ln 2$. These formulae can still be retained in the general case, when $\sigma(E)$ actually depends on energy, by adding a correction factor which contains the logarithmic derivative of $\sigma(E)$ (equations 9 and 10).

Let us initially assume that the capture cross sections are constant and independent of energy and that the temperature is sufficiently low. The integral (1) is then independent of E_s, which can be replaced by $-\infty$ (figure 1). The weighting function $w(t, E)$ in the integrand (the expression in square brackets in equation 1) is sharply peaked (figure 2). Its halfwidth amounts to approximately $\pm 1{\cdot}2\ kT$ units. On the condition that the density of interface states $N_{ss}(E)$ does not change very much within the range of halfwidths considered, the integral can be calculated exactly:

$$\Delta V_G(t_i, T) = (kT/q)(A/C_{ox})qN_{ss}(E_0) \ln 2. \tag{3}$$

In this equation $N_{ss}(E_0)$ represents an averaged value of the density of interface states in the vicinity of the maximum of the weighting function.

In calculations used previously in the literature, the abscissa of the maximum of the weighting function has been determined to be E_0. It is better, however, to take for E_0 the abscissa of the centre of gravity of the area under the weighting function curve. Then equation (3) is still exactly valid not only if N_{ss} is constant but also if $N_{ss}(E)$ is a linear function of E:

$$E_0 = -kT \ln (\sigma(E_0)\, v_{th}(T) N_C(T)\, 2{\cdot}52 t_i). \tag{4}$$

This assignment can be interpreted more clearly. Transformation of equation (4) and comparison with equation (2) leads to

$$\tau = 2{\cdot}52 t_i. \tag{5}$$

This means that by the choice of the sampling times (t_i, $2t_i$) just those interface states are selected which possess the response time τ.

Moreover, it can be deduced from equation (3) that the energy interval in which the interface states that make the greatest contribution to the correlation signal are located is given by

$$\Delta E = kT \ln 2. \tag{6}$$

Hence in the regime for which these approximations are valid, the weighting function of the integral (1) is replaced by a rectangular function (also shown in figure 2).

The assumption that the capture cross sections should be constant can easily be dropped. Let us develop $\sigma(E)$ in the vicinity of E_0:

$$\sigma(E) = \sigma(E_0) \exp[\gamma \times (E - E_0)] \tag{7}$$

where γ is the logarithmic derivative of σ at $E = E_0$:

$$\gamma = \mathrm{d} \ln \sigma(E)/\mathrm{d}E. \tag{8}$$

From equation (2) the expression

$$\tau^{-1} = \sigma(E_0)\, v_{\mathrm{th}} N_{\mathrm{C}} \exp\left[\left(\frac{1+\gamma kT}{kT}\right)(E - E_0) + \left(\frac{E_0}{kT}\right)\right]$$

can be obtained.

It can be shown that only the factor by which E is multiplied affects the value of the approximation to integral (1); the additive constant is of no importance. Therefore equations (3) and (6) are still valid if $kT/(1 + \gamma kT)$ is substituted for kT:

$$\Delta V_{\mathrm{G}}(t_i, T) = (A/C_{\mathrm{ox}}) N_{\mathrm{ss}}(E_0)\, \Delta E \tag{9}$$

and

$$\Delta E = \frac{kT \ln 2}{1 + kT(\mathrm{d} \ln \sigma(E)/\mathrm{d}E)|_{E=E_0}}. \tag{10}$$

Equation (4) may be retained. In this more generalised representation, E_0 can still be interpreted as the centre of gravity of the weighting function. It must be remembered, however, that these equations are only valid if the integral (1) is independent of E_{s}, i.e. the interval $E_0 \pm \frac{1}{2}\Delta E$ must be far enough away from E_{s}. To meet this condition, ΔV_{G} must be measured at temperatures that are not too high and at sampling times that are not too long. The system of integral equations can now be replaced by a system of differential equations (equation 9 with $i = 1, 2, \ldots$ and equation 4) that can be solved numerically.

3.2. Method of solving the differential equations

This method originates from a procedure suggested by K Nickel (1979 private communication). To begin with, any value $\sigma(E_0) = \sigma_0$ at any energy E_0 is chosen arbitrarily; and the following three steps are then carried out.

Step 1. Temperatures T_i to be associated with the sampling times t_i are calculated in such a way that equation (4) is fulfilled.
Step 2. These T_i are put into the system of equation (9). These equations must be considered to be conditional for the two unknown quantities

$$N_{ss}(E_0) \text{ and } \gamma = \mathrm{d} \ln \sigma(E)/\mathrm{d}E|_{E=E_0}.$$

Two equations and hence two correlation signals are sufficient to calculate the two unknowns.
Step 3. With equations (7) and (8) and $E = E_0 + \delta E$, another capture cross section $\sigma(E)$ can be calculated. The increment δE may be positive or negative but must be sufficiently small. The quantity E will now be denoted E_0.

Steps 1–3 may then be repeated and in this way functions $N_{ss}(E, \sigma_0)$ and $\sigma(E, \sigma_0)$ can be established. The arbitrarily chosen σ_0 serves as an integration constant. The procedure stops by itself when the temperatures T_i are too high or too low and the corresponding correlation signals fall outside the measured range. It must be stopped in cases where the basic equation (9) is no longer valid. This happens when the temperature T_i approaches the region in which the correlation signal is reduced as a consequence of E_s limiting the emission.

Figures 3, 4 and 5 show the results of a practical example. The necessary explanations are found in the figure captions. Most striking is the large spread of the curves that

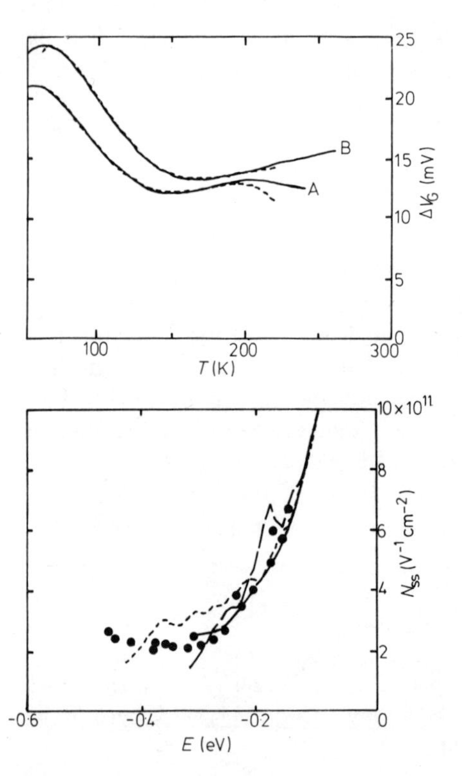

Figure 3. Correlation signals. The full curves are the measured correlation signals for $t_1 = 2{\cdot}5$ ms and $t_2 = 25$ ms. The capacitance, which was kept constant during the emission phase, was $C/C_{ox} = 0{\cdot}537$; this is equivalent to E_s in the midgap. The broken curves were obtained from a check calculation by putting the values of the dotted curves of figures 4 and 5 into the exact equation (1). On average, the calculated and measured values agree to within 0·5% for curve A and 0·8% for curve B. In the method used hitherto, the errors amounted to 100%.

Figure 4. Density of interface states. For the sake of clarity, only three curves out of the family of density of interface state curves are drawn. These correspond to the broken curves in figure 5. The other interface state density curves lie quite close to those displayed in this figure. The points were obtained from conductance method measurements.

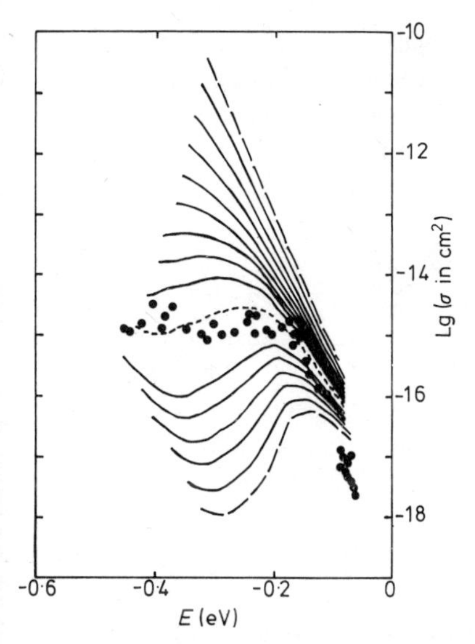

Figure 5. Capture cross sections. The family of curves is limited arbitrarily by the broken curves; the total spread is wider. The dotted curve is the best fit according to the conductance measurements (see captions to figures 3 and 4).

represent the capture cross sections $\sigma(E)$, whereas the corresponding curves for the density of interface states $N_{ss}(E)$ lie comparatively close together. It will be shown below how we may choose the proper functions $N_{ss}(E)$ and $\sigma(E)$ from these families of curves.

3.3. Determination of the integration constant

There are two possible methods:

(i) As has already been shown in figure 5, the capture cross sections can be fitted to values obtained by the conduction method. If in order to expedite the measurements only one or two capture cross sections have been measured by the conductance method, these fitting points should be selected from a suitable energy range; otherwise the errors propagate too much. As can be seen from figure 5 it seems advisable to choose them in the middle of the band gap rather than near the band edge.

(ii) The correlation signal is greatly affected by E_s when E_0 approaches E_s:

$$|E_s| \lesssim |E_0|. \tag{11}$$

The signal then decreases with increasing temperature (see figure 1).

We can use these facts to formulate a method for determination of the integration constant. To this end, we need only measure a third correlation signal with a reduced value of E_s (adjusted to a higher capacitance in the emission phase). The required value of E_s is obtained by HF capacitance or slow ramp measurements. Correlation signals can now be determined according to equation (1) with the solutions $N_{ss}(E, \sigma_0)$, $\sigma(E, \sigma_0)$ for different values of σ. All the quantities required for the computation are known. When evaluating equation (1), attention must be paid to the fact that E_s is temperature-dependent when the capacitance is kept constant. The energy E_s is the sum of the band bending and the Fermi energy:

$$E_s = q\psi_s + kT \ln (N_D/N_C(T)) \tag{12}$$

where N_D is the concentration of dopant. The band bending ψ_s is also dependent on temperature to a small extent, but in the following this dependence will be neglected.

The correlation signals calculated for different values of σ_0 can be distinguished from each other by the steep slope that occurs at different temperatures for each signal. The correct solution can then be found by comparison of the different calculated values with the measured signal.

In figure 6, the two methods for determining the integration constant are cross checked. The full curve represents a correlation signal for the same capacitor as was used for figure 3. The sampling time was chosen as $t_1 = 2{\cdot}5$ ms and the normalised capacitance $C/C_{ox} = 0{\cdot}625$, which is equivalent to $E_s = -0{\cdot}389$ eV at $T = 300$ K. Other correlation signals were computed with different C/C_{ox} (dotted curves); no agreement was apparent and in particular, the measured curve was not nearly as steep as the calculated results.

This discrepancy is caused by not taking fluctuations in the surface potential into consideration. Surface potential fluctuations are a consequence of the statistical distribution of the oxide charges and the charged interface states of the MOS structure (statistical model of Nicollian and Goetzberger 1967). Measurements made by the conductance method verified that a good approximation to the behaviour of these fluctuations is given by a Gaussian distribution.

The influence of fluctuations in the surface potential can be taken into account in the exact equation (1). Good approximate results can also be achieved within the scope of the approximations from which equations (9) and (10) were obtained. Because only electrons from interface states above E_s participate in the emission, the energy interval ΔE given by equation (10) will be reduced if E_0 approaches or even surpasses E_s.

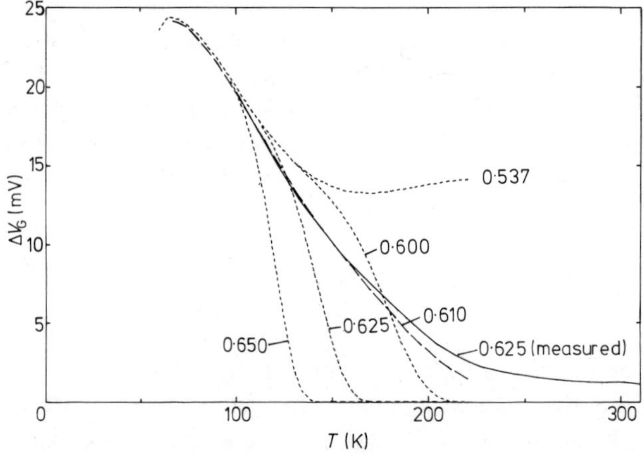

Figure 6. Fluctuations in the surface potential and correlation signals. Correlation signals with a sampling time $t_1 = 2{\cdot}5$ ms are presented. The full curve was measured for the same sample as was used for figure 3; all other curves are calculated. The initial data for the calculations were taken from the dotted curves of figures 4 and 5. The values shown for each curve give the normalised capacitances C/C_{ox} which were kept constant during the emission phase. For the dotted curves the steep slope is seen because the emission is limited by E_s. However, a discrepancy exists between the measurements and the calculations. It can be amended by taking surface potential fluctuations into consideration (broken curve) and in this way rather good agreement is obtained. A small but negligible difference remains in the normalised capacitance values.

Equation (10) must therefore be modified:

$$\Delta E = \int_{E_s}^{0} w(t, E)\, \mathrm{d}E = \frac{kT}{1 + kT(\mathrm{d}\ln \sigma/\mathrm{d}E)|_{E=E_0}} W\left(\frac{E_0 - E_s}{kT}\right). \tag{13}$$

It is almost evident and can easily be shown that this integral W is a function of $(E_0 - E_s)/kT$ only. The function W can be tabulated by numerical integration. From this the generalisation of equation (9) is obtained:

$$\Delta V_G(t_i, T) = \frac{A}{C_{ox}} N_{ss}(E_0) \frac{kT}{1 + kT(\mathrm{d}\ln \sigma/\mathrm{d}E)|_{E=E_0}}$$

$$\times \int_{-\infty}^{+\infty} \frac{\exp\{-[q(\psi_s - \bar{\psi}_s)/kT]^2/2\sigma_g^2\}}{\sigma_g(2\pi)^{1/2}} W\left(\frac{E_0 - E_s}{kT}\right) \frac{q}{kT} \mathrm{d}\psi_s \tag{14}$$

where $\bar{\psi}_s$ is the mean surface potential related to E_s and σ_g is the variance by which the surface potential fluctuations are characterised.

Equation (4) is required when computing E_0 in equation (14). Here the approximation is slightly improved if the factor 2·52 in equation (4) is replaced by 2·06. The factor 2·52 only applies if the weighting function is integrated from $-\infty$ to $+\infty$. Then those values of $N_{ss}(E)$ that lie to the right- and left-hand sides of the centre of gravity of the weighting function and which deviate from $N_{ss}(E_0)$ cancel each other. This property is lost here because the integration starts at a finite lower limit.

When integrating equation (14), attention must be paid to the fact that the quantity E_s itself in the argument of W depends on the integration variable ψ_s (equation 12). The broken curve in figure 6 was computed. According to equation (14) the curve was fitted to the measured values by choosing the capacitance parameter to be $C/C_{ox} = 0{\cdot}61$, which is equivalent to $E_s = -0{\cdot}409$ eV at 300 K. For the variance σ_g a value of 1·8 was taken from conductance measurements and the well-tried approximation for its temperature dependence

$$\sigma_g \propto 1/T$$

was used (Deuling *et al* 1972).

The measured and the calculated curves agree well. In particular, the slope of the correlation signal can be seen to extend over a large range of temperatures when fluctuations in the surface potential are taken into account. The difference between the measured parameter E_s and the value of E_s used for fitting

$$\Delta E_s = 20\,\mathrm{meV}$$

is within experimental error.

4. Summary and conclusions

An improved approximation procedure for evaluating CC–DLTS data has been developed. The new method is more self-consistent than that used hitherto. The restriction that requires the capture cross sections to be independent of energy over small intervals is abandoned, but as a consequence, the solutions for the density of interface states and the capture cross sections are no longer unique. To find the proper solutions, additional measurements, obtained by either the conductance method or by the CC–DLTS technique, must be performed. For the latter method, surface potential fluctuations are of considerable importance.

It is still difficult to give a general estimate of the errors involved in the CC–DLTS method. Densities of interface states can definitely be measured to less than $10^{10} V^{-1} cm^{-2}$, and in the experimental example given here, errors in calculating the capture cross sections amount to about an order of magnitude. An experimental comparison with the conductance method is given; agreement between the two methods is fair.

Acknowledgments

The author is indebted to Professor K Nickel for his valuable advice and to Dr K Eisele for numerous helpful comments. The technical assistance of Mrs A Helde is greatly appreciated. The work was supported by the US Army Research Office.

References

Boudry M R 1978 *J. Phys. E: Sci. Instrum.* **11** 237–47
Deuling H, Klausmann E and Goetzberger A 1972 *Solid St. Electron.* **15** 559–71
Goetzberger A, Klausmann E and Schulz M J 1976 *CRC Crit. Rev. Solid St. Sci.* **6** 1–43
Johnson N M, Bartelink D J and Schulz M 1978 *The Physics of SiO_2 and its Interfaces. Proc. Int. Topical Conf., Yorktown Heights* ed S T Pantelides (New York: Pergamon) pp421–7
Kuhn M 1970 *Solid St. Electron.* **13** 873–85
Nicollian E H and Goetzberger A 1967 *Bell Syst. Tech. J.* **46** 1055–133
Schulz M and Klausmann E 1979 *Appl. Phys.* **18** 169–75
Ziegler K and Klausmann E 1975 *Appl. Phys. Lett.* **26** 400–2

Photoemission from interface states in MOS structures†

D W Greve‡ and W E Dahlke
Sherman Fairchild Laboratory, Lehigh University, Bethlehem, Pennsylvania 18015, USA

Abstract. Interface state photoemission has been observed in MOS structures with 1200 Å thick oxide layers using both photocurrent and photocapacitance measurements. The density of interface states and the photoionisation cross section have been evaluated and the results disagree with those of previous workers.

1. Introduction

Photoionisation of interface states is observed when MOS devices are illuminated with photons of energy $h\nu < E_g$. This effect was first reported by Kamieniecki (1973) and later by Pierret and Roesener (1974, 1976) and Kamieniecki and Nitecki (1978; also Kamieniecki *et al* 1978 unpublished results) using photocapacitance measurements on thick-oxide MOS structures. These authors did not obtain the optical cross section σ_n^0 directly but instead showed that an assumed form could be used to match the experimental results. Direct measurement of the dependence of the photoionization cross section on photon energy is desirable, however, in order to obtain some insight into the physical nature of interface states.

Recently, we reported measurements of photocurrents from interface states in MOS tunnel diodes (Dahlke and Greve 1978). Theoretical work (Dahlke and Greve 1979a) showed that optical cross sections can be easily determined from the temperature dependence of these photocurrents. Here we present the results of similar measurements on devices with thicker oxides. In addition to the optical emission process, we also observe a build-up of the inversion layer because of optical-pair generation.

2. Experimental method

The device used in this study was an n-type MOS device with a 1270 Å thick layer of oxide and Cr metallisation, 5×10^{-3} cm^2 in area. To facilitate measurements, an increased density of interface states was produced by growing most of the oxide at a temperature of 1069 °C and then completing the oxidation at 894 °C. The device was mounted in a variable-temperature cryostat and illuminated through the silicon with photon energy $h\nu < E_g$ from a monochromator. Filters were used to reduce the effect of scattered and higher-order diffracted light from the monochromator. Both the photocapacitance and the photocurrent transients were measured. Photocurrent measurements were performed

† Supported by National Science Foundation grant ENG-7805917.
‡ In partial fulfilment of requirements for the PhD. Support of a Sherman Fairchild Fellowship is gratefully acknowledged.

0305-2346/80/0050-0107$01.00

by chopping the light at 338 Hz and detecting the AC current with a current-sensitive preamplifier followed by a lock-in amplifier. The capacitance was measured at 100 kHz using the same equipment and an AC probe signal of 6 mV RMS.

For both measurements, the interface states were filled by accumulating the device and then the bias was switched to deep depletion. After a delay time $t_d = 11$ s, illumination was begun and the current or capacitance transient plotted on a chart recorder.

3. Qualitative discussion

In this section we discuss qualitatively the optical relaxation from deep depletion to inversion. Immediately after switching to deep depletion, all the states are full and during the delay time t_d, states near the conduction band begin to emit electrons as shown in figure 1(*a*). The trap Fermi level

$$E_{tn} = E_c - kT \ln \nu_n t_d \tag{1}$$

where the occupancy is $1/e$, falls logarithmically with time (Simmons and Wei 1974) with ν_n the attempt-to-escape frequency for electrons. Upon starting illumination at $t = t_d$, electrons are emitted optically (figure 1*b*) and the trap occupancy approaches the light-controlled occupancy

$$f_L = \sigma_p^0/(\sigma_n^0 + \sigma_p^0) \tag{2}$$

determined by the optical cross sections for electron and hole emission σ_n^0 and σ_p^0 respectively. The time required to reach this occupancy is of the order of the optical relaxation time $\tau_L = 1/\phi_0(\sigma_n^0 + \sigma_p^0)$ where ϕ_0 is the photon flux. Once the occupancy f_L is reached,

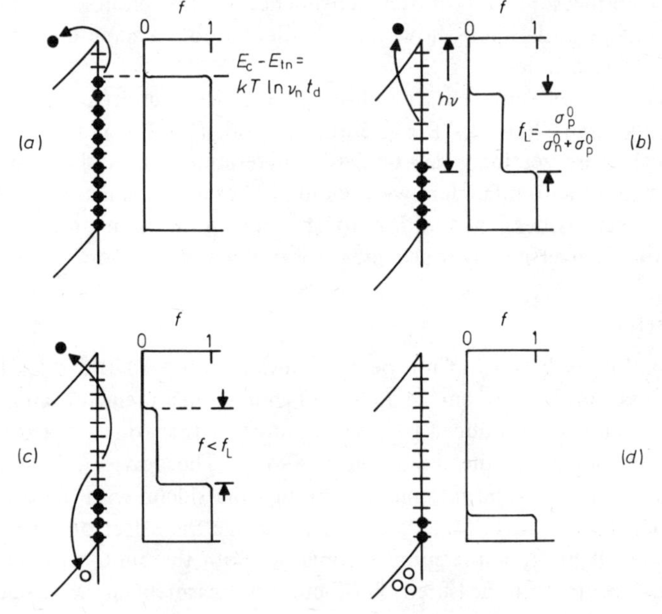

Figure 1. Band diagrams and trap occupancies f at successive stages of the relaxation process: (*a*) thermal electron emission; (*b*) optical electron emission; (*c*) optical pair generation; (*d*) steady state in inversion.

the states near the midgap are only partly filled with electrons and inversion charge is produced by alternate optical emission of holes and electrons as shown in figure 1(*c*). The change in interface charge is slower than when emission dominates, since fewer states are involved and the rate of generation decreases further as the inversion charge builds up and the states become filled with holes. This reduces the rate at which electron emission, the rate-limiting step, occurs. Eventually, the final state of inversion (figure 1*d*) is reached.

4. Measurements

The observed capacitance transient is shown in figure 2. As described above, the initial rapid change in capacitance due to optical emission of electrons is followed by a slower relaxation by optical pair generation. Finally, the capacitance approaches the inversion capacitance observed in the absence of illumination. Decreasing photon energy causes a decreased rate of optical pair generation and consequently the relaxation process takes longer.

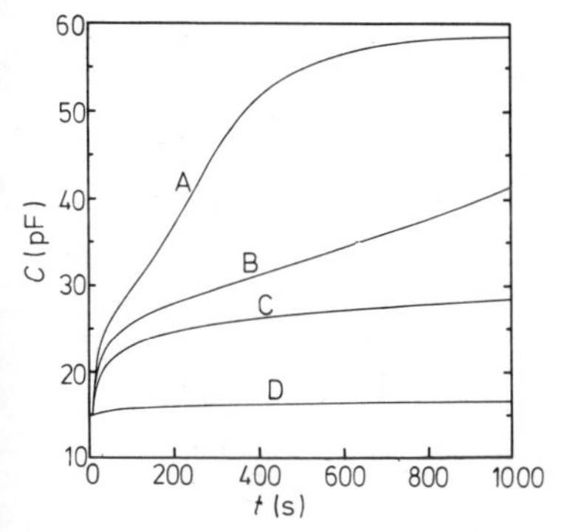

Figure 2. Observed capacitance transient for photon energies of 0·65, 0·62 and 0·59 eV (curves A, B, C respectively) and in the dark (curve D). $T = 83$ K; incident photon flux $\phi_0 \simeq 2 \times 10^{17}$ photons $cm^{-2}\,s^{-1}$.

Referring to the curve for $h\nu = 0{\cdot}65$ eV in figure 2, we will identify the times for which the different band models of figure 1 are valid. Before beginning illumination, we have thermal emission of electrons from the filled states as shown in figure 1(*a*) ($t < 11$ s). After beginning illumination, optical emission of electrons dominates until the occupancy reaches f_L at $t \simeq 30$ s (figure 1*b*). For 30 s $< t <$ 800 s the pair generation process produces holes at a steadily decreasing rate (figure 1*c*). Finally, at $t \simeq 1000$ s, inversion is reached (figure 1*d*).

Figure 3 shows the initial stage of the photocurrent and photocapacitance transients on expanded scales. For the same temperature and photon energy, the same amount of released charge is observed within experimental error for the two measurements. Increasing the deep depletion bias did not increase the amount of released charge so the observed transients cannot be attributed to bulk states.

5. Analysis

As discussed above, the different rates of emission and generation processes permit separation of the two effects. We will use the emission transient to determine the photoionisation cross section σ_n^0 and the density of interface states N_{ss}. Since the hole density

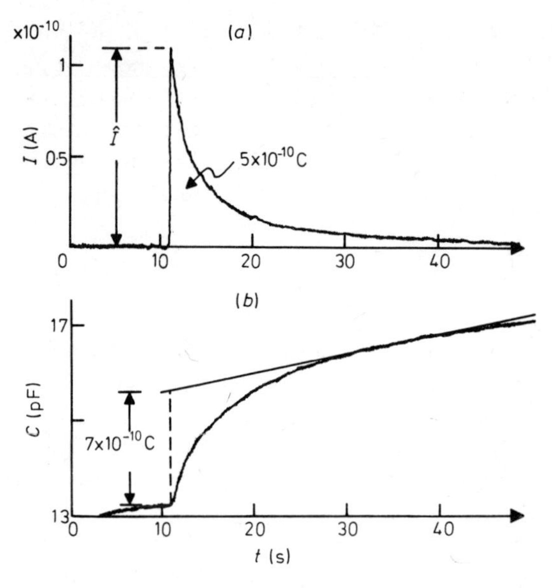

Figure 3. Detail of the initial transient from (*a*) photocurrent studies and (*b*) photocapacitance studies (from figure 2). Measurements were obtained at $T = 83$ K, $h\nu = 0{\cdot}65$ eV; $\phi_0 \simeq 2 \times 10^{17}$ photons cm^{-2} s^{-1}

at the interface is negligible during the emission transient, the results of the theory for the tunnel diode (Dahlke and Greve 1979a) which assumes $p = 0$ can be used without alteration. It can be shown that the total charge released Q_R and the maximum photocurrent $\hat{I}$ are given approximately by the following expressions:

$$\hat{I} = qA\phi_0 \int_{E_v}^{E_{tn}} \sigma_n^0 \; N_{ss} \, dE_t$$
$$Q_R = qA \int_{E_v}^{E_{tn}} \frac{\sigma_n^{0\,2}}{(\sigma_n^0 + \sigma_p^0)^2} N_{ss} \, dE_t \tag{3}$$

where A is the device area and E_{tn} (equation 1) is an explicit function of temperature. At the low temperatures used here, $h\nu < E_{tn} - E_v$ and therefore $\sigma_n^0 \gg \sigma_p^0$ at E_{tn}. Differentiation of equations (3) with respect to temperature then yields

$$N_{ss}\sigma_n^0|_{E_{tn}} = -\frac{1}{k \ln \nu_n t_d} \frac{d\hat{I}/qA\phi_0}{dT}\bigg|_{t_d}$$
$$N_{ss}|_{E_{tn}} = -\frac{1}{k \ln \nu_n t_d} \frac{dQ_R/qA}{dT}\bigg|_{t_d}. \tag{4}$$

Thus measurement of the quantities $\hat{I}$ and Q_R as a function of temperature permits evaluation of σ_n^0 and N_{ss}. The results obtained for σ_n^0 and N_{ss} are shown in figures 4 and 5 respectively.

Also illustrated in figure 5 are measurements of N_{ss} obtained from the thermal relaxation of the capacitance. The principle is the same as the isothermal current relaxa-

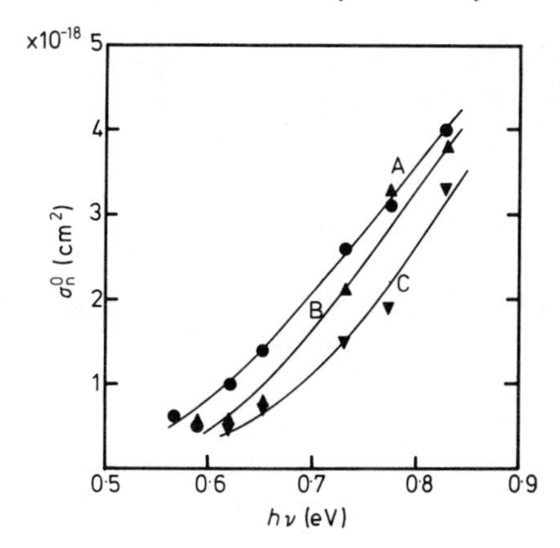

Figure 4. Measured photoionisation cross section σ_n^0 as a function of photon energy $h\nu$ for different trap levels E_t. $E_c - E_t = 0{\cdot}21$, $0{\cdot}28$ and $0{\cdot}36$ eV (curves A, B, C respectively).

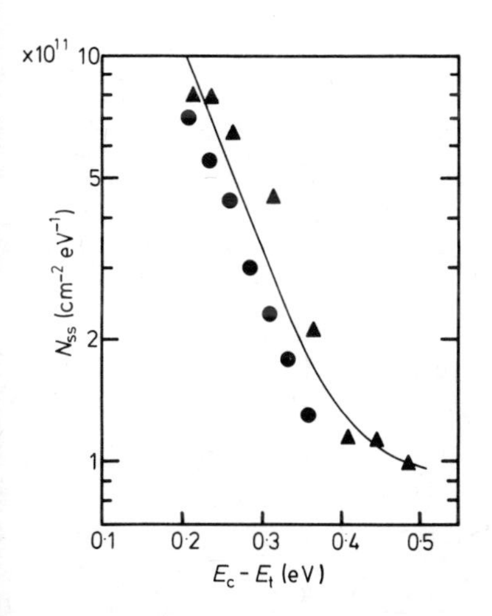

Figure 5. Measured density interface states N_{ss} as a function of the trap level E_t using thermal (▲) and optical (●) relaxation methods.

tion method of Simmons and Wei (1974) except that the change in capacitance is measured instead of the current. The change in interface charge $\Delta Q(t)$ is related to the capacitance through the expression (Zerbst 1966)

$$\Delta Q(t) = Q(t) - Q(t_0) = -\frac{qN_D\epsilon_s C_{ox}}{2}\left(\frac{1}{C(t)^2} - \frac{1}{C(t_0)^2}\right). \tag{5}$$

In the absence of illumination the interface charge changes only by thermal emission so that

$$\Delta Q(t) = -qN_{ss}(E_{tn}(t) - E_{tn}(t_0)) = qkTN_{ss}\ln(t/t_0) \tag{6}$$

where equation (1) has been used for the time dependence of E_{tn}. We then obtain

$$N_{ss}(E_{tn}) = -\frac{\epsilon_s N_D C_{ox}}{2kT}\left(\frac{[1/C(t)^2] - [1/C(t_0)^2]}{\ln(t/t_0)}\right). \tag{7}$$

The density of interface states obtained from the two different methods (optical and thermal relaxation) agree quite well.

The results for both σ_n^0 and N_{ss} disagree with the conclusions of Kamieniecki *et al* (1978, unpublished results). They found that their measurements could be explained by assuming a Lucovsky (1965) cross section and a density of interface states that decreased toward the band edge. Our measurements, however, show a distribution of interface states that increases towards the conduction band, with good agreement between optical and thermal measurements. Furthermore, the monotonically increasing optical cross section observed here is unlike the predictions of the Lucovsky model. Results similar to this were obtained previously (Dahlke and Greve 1979b) in measurements on MOS tunnel devices.

The relation (equation 5) between charge and capacitance can also be used to obtain $\Delta Q(t)$ during pair generation. As shown in figure 6, the rate of pair generation decreases monotonically with time. Generation rates at different photon energies were compared by scaling the time axis so that the slopes matched in the pair generation region. The generation rates fit the relation

$$R \sim (h\nu - E_g/2)^2 \qquad (8)$$

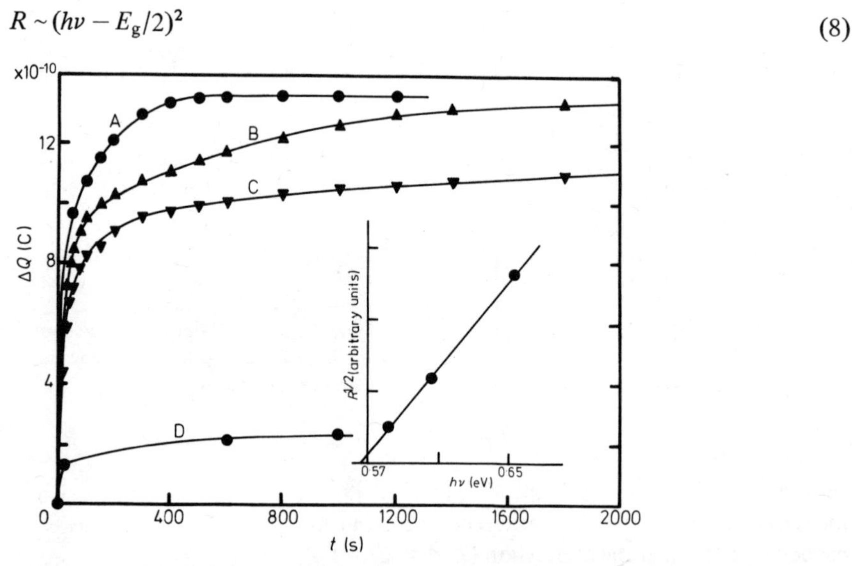

Figure 6. Change in interface charge ΔQ as a function of time for photon energies of 0·65, 0·62 and 0·59 eV (curves A, B, C respectively) and in the dark (curve D). The inset shows the square root of the rate of pair generation R as a function of photon energy $h\nu$.

as shown in the inset to figure 6. A threshold at $h\nu \simeq E_g/2$ is expected, since pair generation can only occur if traps can communicate with both bands.

6. Conclusions

Measurements of the optical relaxation of MOS devices have been analysed to obtain the optical cross section for electron emission and the density of interface states. The results show that the optical cross section increases monotonically with photon energy and the density of states increases toward the conduction band. Both results are in disagreement with previous work. Good agreement is obtained, however, between measurements of the density of interface states by optical and thermal relaxation.

References

Dahlke W E and Greve D W 1978 *Electrochem. Soc. Spring Mtg, Seattle, Washington* Abstract No 127
—— 1979a *Solid St. Electron.* to be published
—— 1979b *Proc. 3rd University/Industry/Government Microelectronics Symp., Texas Technical University, Lubbock, Texas* (New York: IEEE) pp54–6
Kamieniecki E 1973 *Solid St. Electron.* **16** 1487
Kamieniecki E and Nitecki R 1978 *Physics of SiO_2 and its Interfaces* ed S T Pantelides (New York: Pergamon) pp 417–20
Lucovsky G 1965 *Solid St. Commun.* **3** 299
Pierret R F and Roesner B B 1974 *Appl. Phys. Lett.* **24** 366
—— 1976 *Solid St. Electron.* **19** 593
Simmons J G and Wei L S 1974 *Solid St. Electron.* **17** 117
Zerbst M 1966 *Z. Angew. Phys.* **22** 30

Annealing of Si–SiO$_2$ interface states using layers containing hydrogen

L Rischt†, E Pammer‡ and K Friedrich‡
† Siemens AG, Otto-Hahn-Ring 6, 8000 München 83, West Germany
‡ Siemens AG, Balanstrasse 73, 8000 München 80, West Germany

Abstract. Fast surface-state densities as low as 5×10^8 cm^{-2} eV^{-1} ($\approx$ midgap) at the Si–SiO$_2$ interface can be obtained by the deposition of layers containing hydrogen. After layer deposition an annealing step reduces the surface-state density of $(5-10) \times 10^9$ cm^{-2} eV^{-1} by a factor of ten. These low surface-state densities have been measured on poly-Si capacitors by the conductance method and have been calculated from the charge transfer inefficiency, $\epsilon = (2 \pm 1) \times 10^{-5}$ of SCCD's.

1. Introduction

The defect structure of the Si–SiO$_2$ interface is one of the main problems in MOS technology. In simple terms there are three 'classical' defects: the density of fast surface states N_{SS}, the fixed oxide charge Q_{SS} and radiation-induced defects and hole trapping. According to Svensson (1978) and Revesz (1971), these three defects are attributable to three kinds of trivalently bonded Si with slightly different electrical properties (figure 1).

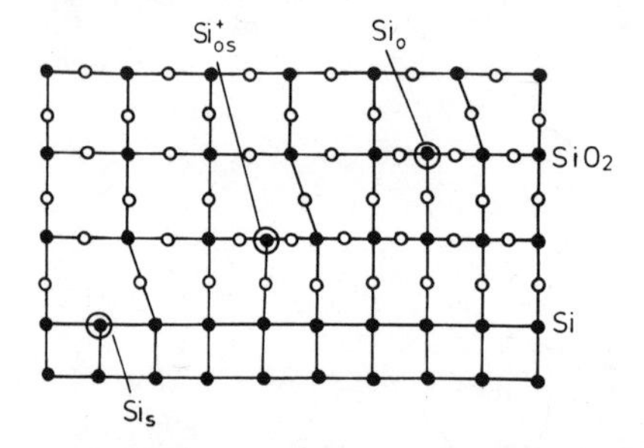

Figure 1. Defect structure of the Si–SiO$_2$ interface according to Svensson (1978). • Silicon atom; o oxygen atom; Si$_S$ fast surface state; Si$^+_{OS}$ fixed oxide charge; Si$_O$ hole trap.

At the Si interface there are Si atoms bonded to three other Si atoms, with one remaining dangling bond. This defect is commonly identified with fast surface states. In the vicinity of the interface there are Si atoms bonded to three oxygen atoms; this state is always positively charged (Q_{SS}). When the same defect occurs deeper in the oxide, it is neutral and is responsible for hole trapping and radiation defects.

After gate oxidation, these three defects are in equilibrium. To reduce the density of fast surface states, the process now usually includes a step that comprises annealing in an atmosphere of hydrogen gas at low temperatures. Hydrogen is known to saturate dangling bonds at the Si–SiO$_2$ interface (Montillo and Balk 1971) according to the two reactions

$$\vdots\,Si^{\cdot} + H_2 \rightleftharpoons SiH + H$$

$$\vdots\,Si + H \rightleftharpoons SiH.$$

0305-2346/80/0050-0114$01.00

The SiH compound is electrically inactive. The residual density of fast surface states obtained by this annealing step is of the order of $(5–10) \times 10^9\ cm^{-2}\ eV^{-1}$.

In SCCD's, in which the electrons are shifted along the interface, it is particularly important that the density of fast surface states is reduced. Transfer inefficiencies ϵ of $(1–2) \times 10^{-4}$ (wide-channel CCD with fat zero) prevent many applications. This type of MOS device needs, more than any other, an advanced technique for reducing the density of fast surface states.

2. Layer preparation

By 'layer annealing' it is possible to reduce N_{ss} by a factor of ten. In addition to a conventional double poly-Si process, dielectric layers deposited by CVD reactions with SiH_4, Si_3N_4, SiO_2, PSG, poly-Si and amorphous Si (see e.g. Pliskin 1977) were investigated:

Si nitride $3SiH_4 + 4NH_3 + (N_2) \rightarrow Si_3N_4 + 12H_2 \uparrow$

Si dioxide $SiH_4 + O_2 \rightarrow SiO_2 + 2H_2 \uparrow$

PSG $2SiH_4 + 4PH_3 + 7O_2 \rightarrow 2P_2O_5 . SiO_2 + 10H_2 \uparrow$

Poly-Si $SiH_4 \rightarrow Si + 2H_2 \uparrow$

Amorphous Si $SiH_4 \rightarrow Si + 2H_2 \uparrow$.

Apart from the deposition reaction itself, where a large concentration of hydrogen is present, a high dose ($\approx 10^{15}\ cm^{-2}$, 35–100 keV) of H^+ was implanted to increase the hydrogen concentration of the layers (Kellner and Goetzberger 1975).

3. Electrical measurements

The low densities of surface states were measured by the conductance method according to Goetzberger *et al* (1976), using a two-phase lock-in amplifier (PAR 5204) on poly-Si capacitors ($d = 80$ nm, p-Si (100), 8 Ω cm, $F = 16 \times 10^{-4}\ cm^{-2}$) and were calculated from the charge losses of SCCD's ($W = 100\ \mu m$, $L = 11\ \mu m$, 400 electrodes, 4ϕ operation) at a clock frequency of 150 kHz according to Carnes and Kosonocky (1972):

$$N_{ss} = (C_{ox}/qF)(G_p/\omega C_{ox})|_{Max}\,(1/f_n(\sigma))$$

$$N_{ss} = \epsilon_T Q / [(\ln 5)\,qkTF].$$

The two techniques are physically different. In the conductance method the charging and discharging of surface states at the Fermi level by holes from the valence band is measured. For a given frequency, the real part of the admittance has a maximum for this surface potential that corresponds to the time constant of N_{ss}.

Figure 2 shows the typical $C(U)$ and $G(U)$ curves of a specimen capacitor (*a*) after conventional hydrogen annealing, (*b*) after layer deposition and (*c*) after subsequent slight annealing in N_2. The variance of the surface potential σ, which influences the width of the conductance peak, was determined by varying the frequency. σ for our oxides is $\approx 0{\cdot}7$; the height of the conductance peak is therefore proportional to N_{ss}. The respective densities of surface states ($\approx 0{\cdot}6$ eV below E_c) are 8×10^9, 6×10^9 and $6 \times 10^8\ cm^{-2}\ eV^{-1}$.

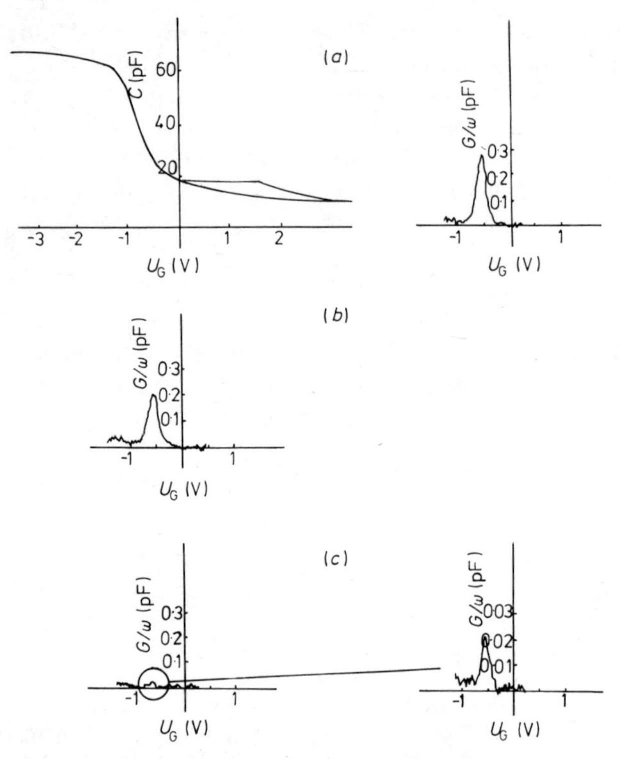

Figure 2. Capacitance and conductance versus gate voltage (*a*) after H_2 annealing; (*b*) after deposition of layer; (*c*) after slight annealing.

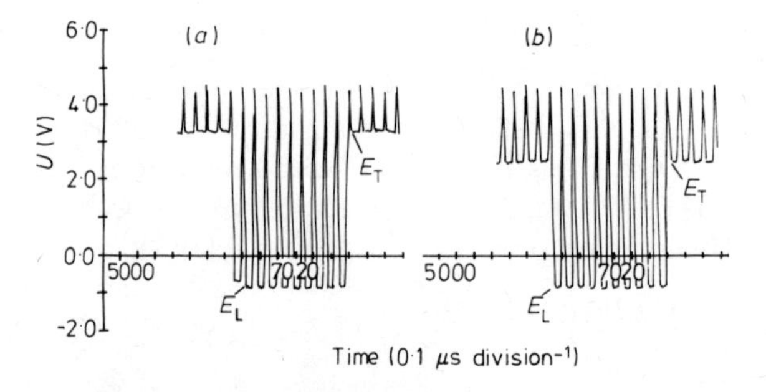

Figure 3. Transfer inefficiency of SCCD's with 0% (*a*) and 20% (*b*) fat zero (with 'layer annealing').

With SCCD's the signal charge (electrons) is shifted along the surface. The depleted fast interface states capture electrons from these charge packets and re-emit them with a time constant that increases exponentially with the energy to the conductance band. If a number of charge packets is shifted through a SCCD, leading and trailing charge losses are observed, which are proportional to N_{SS}.

Figure 3 shows the train of ten binary signals measured automatically after 400 transfers (with 'layer annealing'): the mean transfer inefficiency calculated from these very small losses is $\epsilon_L = 1 \times 10^{-4}$, $\epsilon_T = 3 \times 10^{-5}$ (without fat zero) and $\epsilon = (2 \pm 1) \times 10^{-5}$ (with 20% fat zero). The corresponding values of N_{ss} ($\approx 0{\cdot}3$ eV below E_c) are similarly in the range $(5–10) \times 10^8$ cm^{-2} eV^{-1}.

For the best layers, a value of $N_{ss} \approx 5 \times 10^8$ cm^{-2} eV^{-1} was realised. The residual density of surface states depends on layer preparation and to a greater extent on the annealing temperature and time, which must be optimised for each layer.

Figure 4 shows the typical characteristic of N_{ss}. After conventional hydrogen annealing (*a*), the density is about 8×10^9 cm^{-2} eV^{-1} (approximately at midgap). After layer deposition (*b*), only minor changes are observed, but after subsequent slight annealing in N_2 or H_2 (*c*), the density of surface states drops to $(5–10) \times 10^8$ cm^{-2} eV^{-1}.

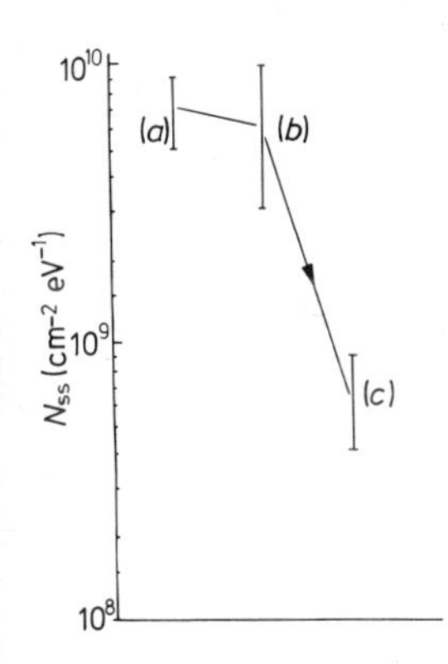

Figure 4. Typical surface-state densities (*a*) after H_2 annealing; (*b*) after deposition of layer; (*c*) after slight annealing.

4. Conclusions

A sharp reduction in N_{ss} occurs during the slight annealing step. As the gas atmosphere (N_2 or H_2) was not observed to have any influence during annealing and hydrogen may also be assumed to reduce the density of surface states in layer annealing, the hydrogen must originate from the layers themselves, in which it will have been incorporated during deposition and subsequent implantation.

During annealing this weakly bonded hydrogen is liberated and diffuses to the Si–SiO_2 interface, where it saturates dangling bonds. We conclude that this layer annealing technique must be more effective because atomic hydrogen is directly available for the reaction

$$\vdots\,\mathrm{Si}^{\bullet} + \mathrm{H} \rightleftharpoons \mathrm{SiH}.$$

No dissociation of the H_2 molecule is necessary, as must occur for normal hydrogen annealing.

These low densities of surface states make it possible to produce SCCD's with transfer inefficiencies of the same order as those of BCCD's at moderate frequencies, with the advantage of higher charge-handling capability and easier fabrication. Better signal : noise ratios, multilevel storage and small-channel SCCD's (which exhibit large edge effects) are now achievable with improved performance. Furthermore, the reduced surface generation is observed to result in a smaller dark current.

References

Carnes J E and Kosonocky W F 1972 *Appl. Phys. Lett.* **20** 261–3
Goetzberger A, Klausmann E and Schulz M 1976 *CRC Crit. Rev. Solid St. Sci.* **6** 15–18
Kellner W and Goetzberger A 1975 *IEEE Trans. Electron. Devices* **ED-22** 531–3
Montillo F and Balk P 1971 *J. Electrochem. Soc.* **118** 1463–8
Pliskin W A 1977 *J. Vac. Sci. Technol.* **14** 1064–81
Revesz A G 1971 *IEEE Trans. Nucl. Sci.* **NS-18** 113–6
Svensson C M 1978 *The Physics of SiO_2 and its Interfaces* ed S T Pantelides (New York: Pergamon) pp328–32

The effect of donors or acceptors on the $Si-SiO_2$ interface

J Snel
Philips Research Laboratories, Eindhoven, The Netherlands

Abstract. The properties of the $Si-SiO_2$ interface have been measured as a function of the type and concentration of dopant in the silicon, by means of a capacitance–voltage measurement at 77 K. It was found that the acceptor and donor impurities widely used in semiconductor devices induce interface states and oxide charges.

Acceptors (B, Ga) cause a higher density of interface states and charges than do donors (P, As). This is found to be independent of the doping process, i.e. whether doping occurs during epitaxial growth, by implantation or by deposition in a furnace. The effect of dopant impurities on oxide charge is probably related to different redistribution behaviour of acceptors and donors across the $Si-SiO_2$ interface.

1. Introduction

The origin of interface states and charges has been the subject of much research on the $Si-SiO_2$ interface. Many investigators mention the increase in the number of interface states and flat-band charges caused by sodium and potassium (Kooi 1966, Schulz *et al* 1975, Wang 1977) and much work has been done to improve MOS technology with regard to this point. As a result the level of contamination with Na has been considerably reduced and this is accompanied by a reduction in the number of interface states and charges.

The effects of other impurities have also been investigated. It has been reported that heavy metals can cause either a decrease or an increase in the number of interface states and charges (Schmidt and Adda 1974, Kaden 1970). Elements with the same valency as silicon do not seem to have any influence on interface properties, as demonstrated by the results with tin (Schulz *et al* 1975) and germanium (Wienhold 1970).

Although Kooi (1966) and Werner (1976) have already noted the influence of the type of silicon on the $Si-SiO_2$ interface, the effect of elements of valency three and five is not really known. Because these impurities are widely used in semiconductor devices, it seemed worthwhile to investigate the dependence of interface states and charges on doping in greater detail.

2. Experimental method

The differential capacitance–voltage curve (1 MHz) of MOS transistors measured at liquid nitrogen temperature was used to determine the properties of the $Si-SiO_2$ interface. This curve (see figure 1) exhibits a hysteresis because of interface states (Goetzberger and Irvin 1968) that are not in thermal equilibrium with the DC voltage. From this voltage

0305-2346/80/0050-0119$01.00

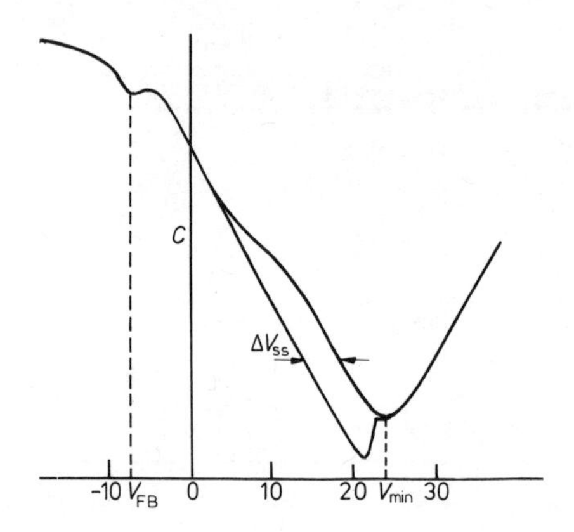

Figure 1. C–V curve of a MOS transistor, measured at 77 K, for an interfacial boron concentration of 2×10^{18} cm^{-3} and an oxide thickness of 1000 Å. Important parameters are the flat-band voltage V_{FB}, the interface state hysteresis ΔV_{ss} and the voltage at which inversion occurs V_{min}.

hysteresis ΔV_{ss}, the number of interface states N_{ss} in the middle part of the gap (approximately 0·2 eV from the edges of both the valence and conduction bands) can be obtained:

$$qN_{ss} = C_{ox}\Delta V_{ss} \tag{1}$$

where q is the elementary charge and C_{ox} is the oxide capacitance measured in accumulation.

Apart from the capacitance minimum at inversion, figure 1 also shows a secondary minimum in the C–V curve. This dip occurs at the flat-band voltage V_{FB} (Gray and Brown 1968) and is only apparent at low temperatures and sufficiently high doping levels. At 77 K the dip is observable for dopant concentrations above 10^{16} cm^{-3}. The bulk Fermi level then lies close to the acceptor (or donor) energy level. As a consequence the number of ionised acceptors at the surface can be easily changed by application of a voltage to the metal gate. This change in the degree of ionisation of the acceptor level takes place around the flat-band condition and gives rise to a secondary minimum in the C–V curve. From V_{FB} the net number N_{FB} of charges in the oxide and in interface states at the flat-band condition can be found:

$$qN_{FB} = C_{ox}V_{FB}. \tag{2}$$

The impurity concentration at the surface may be found by computing the average concentration N in the maximum depletion layer at the inversion voltage. The value of N can be obtained from the voltage $V_{inv} = |V_{min} - V_{FB} - \Delta V_{ss}|$, where V_{min} is the voltage at which the primary capacitance minimum occurs. After calculation of the C–V curves at 77 K using the equations first derived by Brown and Gray (1968), the following numerical relation between the dopant concentration N and V_{min} at a temperature of 77 K may be found:

$$N = (3{\cdot}1 \times 10^{15})\,[(V_{inv} - 1)^2(1000/t_{ox})^2]\ \mathrm{cm}^{-3} \tag{3}$$

where t_{ox} is the oxide thickness in Å.

3. Results and discussion

Figures 2 and 3 show the results obtained for MOS transistors made in 10 μm thick epitaxial layers, doped with boron or phosphorus during epitaxial growth. Substrates of

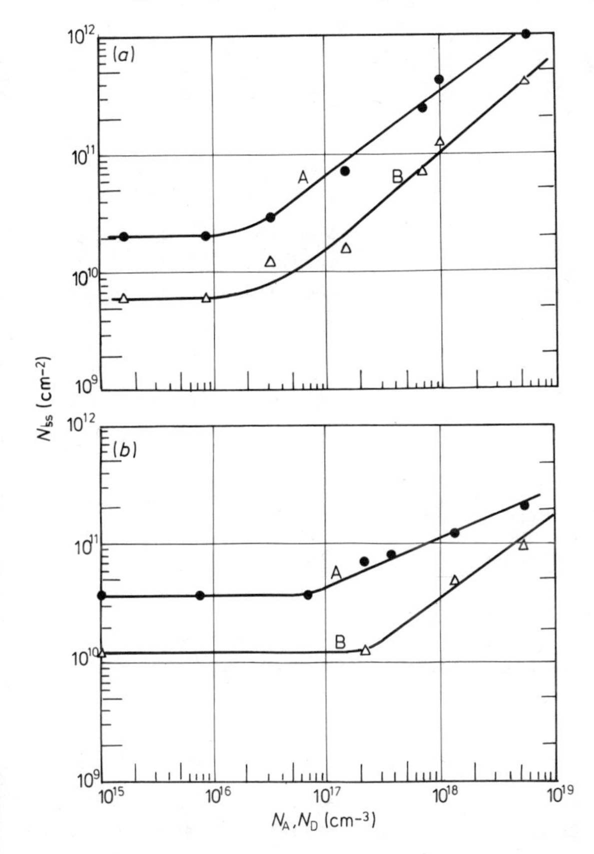

Figure 2. Number of interface states N_{ss} as a function of (*a*) the interfacial boron concentration N_A and (*b*) the interfacial phosphorus concentration N_D before (curve A) and after (curve B) low-temperature annealing at 450 °C in N_2/H_2O. The Si–SiO_2 interface was made by thermal oxidation (1100 °C, dry O_2, followed by N_2 annealing; oxide thickness 1000 Å) of epitaxial layers doped during epitaxial growth.

p^+(B) or n^+(Sb) material with ⟨100⟩ orientation were used. The gate oxide was grown in dry oxygen at 1100 °C to a thickness of 1000 Å, with subsequent N_2 annealing at the same temperature. In each figure two curves are shown giving N_{ss} or V_{FB} before and after annealing at 450 °C for 30 min in N_2/H_2O in the presence of an electron-gun aluminium gate electrode. For boron-doped interfaces, N_{ss} starts to increase above $N_A \cong 10^{16}$ cm^{-3}, reaching a value close to 10^{12} cm^{-2} at $N_A \cong 10^{19}$ cm^{-3}. Low-temperature annealing only partially eliminates these boron-induced states. For phosphorus, an increase in the interface states is also found, but it is much less pronounced than for boron. N_{ss} does not increase until the dopant concentration is higher than 10^{17} cm^{-3} and only reaches 2×10^{11} cm^{-2} for $N_D \cong 10^{19}$ cm^{-3}. From the flat-band voltage data shown in figures 3(*a*) and (*b*), it is also clear that phosphorus gives a better interface.

To assess whether or not doping during epitaxial growth itself creates the interface states and charges, for instance by (dopant-dependent) contamination of the silicon with an unknown impurity, several other doping processes were examined. Almost the same dependence of N_{ss} and V_{FB} on doping as is shown in figure 2(*a*) and (*b*) was found when the silicon was doped by ion implantation of B or P. It made no difference whether the dopant impurities were implanted through the SiO_2 or into the bare silicon, with subsequent formation of the oxide film in the latter case. Doping the silicon by boron deposition in a furnace at 925 °C using a BBr_3 source, followed by oxidation in dry O_2

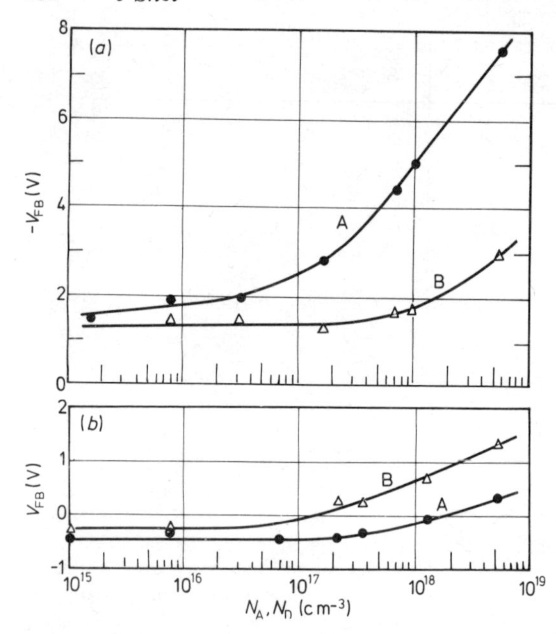

Figure 3. Flat-band voltage V_{FB} as a function of (*a*) the interfacial boron concentration N_A and (*b*) the interfacial phosphorus concentration N_D before (curve A) and after (curve B) low-temperature annealing at 450 °C in N_2/H_2O. The Si–SiO_2 interface was made by thermal oxidation (1100 °C, dry O_2, followed by N_2 annealing; oxide thickness 1000 Å) of epitaxial layers doped during epitaxial growth.

at 1100 °C and annealing in N_2 also gave the same high densities of interface states and flat-band charges.

An increase in $|V_{FB}|$ and ΔV_{ss} with impurity concentration was also found for Si–SiO_2 interfaces where the silicon was doped with gallium or arsenic by ion implantation. Gallium influenced the interface properties in the same way as did boron and arsenic acted in the same way as did phosphorus. This demonstrates that the interface properties do not depend on the choice of the specific donor or acceptor element.

We investigated the dependence of interface states and flat-band charges on the crystal orientation in boron-doped silicon (made by boron deposition in a furnace). After boron deposition, the ⟨100⟩- and ⟨111⟩-oriented silicon wafers were oxidised at 1100 °C in dry O_2 and annealed in N_2 at the same temperature. We found that the number of boron-induced interface states does not depend on the crystal orientation, but only on the boron concentration at the interface. For both orientations the interfacial boron concentration was $2{\cdot}5 \times 10^{18}$ cm^{-2}. Moreover, the flat-band charge is a function of orientation and is largest for ⟨111⟩: $N_{FB}(111) = 1{\cdot}6 N_{FB}(100)$. Colby and Katz (1976) also found the segregation coefficient of boron to be orientation-dependent. They found that at 1100 °C $m(111) = 4$ and $m(100) = 2{\cdot}5$ (m is defined as $N_A(\text{ox})/N_A(\text{Si})$ near the interface). Because the boron concentrations in the silicon were equal for both orientations, this implies that the ⟨111⟩ wafer has 1·6 times more boron in the SiO_2 than has the ⟨100⟩ wafer. It is therefore likely that N_{FB} is made up of oxide charges caused by the boron atoms in the SiO_2.

4. Conclusions

Acceptor and donor atoms induce interface states and oxide charges at the Si–SiO_2 interface. The number of states and charges is determined by the concentration of dopant but is also dependent on the type of dopant. Donors like P and As induce much fewer interface states and charges than acceptors such as B and Ga. For acceptors the effect is apparent for concentrations above 10^{16} cm^{-3} and for donors a concentration of at least 10^{17} cm^{-3} is required. The observed effects are independent of the doping process used. Doping by diffusion, implantation or during epitaxial growth always gives rise to the same doping dependence of the Si–SiO_2 interface properties. As has been verified for boron, the number of interface states is not orientation-dependent, but the flat-band charge qN_{FB} is. Because the segregation of boron between oxide and silicon has the same dependence on orientation as N_{FB}, we are justified in concluding that the flat-band charge resides inside the oxide. Rejection of P and As by the oxide can therefore be the reason for the low oxide charge of n-type Si–SiO_2 interfaces.

References

Brown D M and Gray P V 1968 *J. Electrochem. Soc.* **115** 760–6
Colby J W and Katz L E 1976 *J. Electrochem. Soc.* **123** 409–12
Goetzberger A and Irvin J C 1968 *IEEE Trans. Electron. Devices* **ED-15** 1009–14
Gray P V and Brown D M 1968 *Appl. Phys. Lett.* **13** 247–8
Kaden G 1970 *Phys. Stat. Solidi* **A3** 161–71
Kooi E 1966 *Philips Res. Rep.* **21** 477–95
Schmidt P F and Adda L P 1974 *J. Appl. Phys.* **45** 1826–33
Schulz M, Klausmann E and Hurrle A 1975 *Crit. Rev. Solid St. Sci.* **5** 319–25
Wang K L 1977 *Proc. 3rd Int. Symp. on Silicon Materials Science and Technology* ed H Huff and E Sirtle (Princeton: Electrochemical Society) pp404–13
Werner W M 1976 *J. Electrochem. Soc.* **123** 540–3
Wienhold H 1970 *Phys. Stat. Solidi* **A1** K49–51

Inhomogeneities of surface potential after stress ageing of the thermally grown Si-SiO_2 interface

C Werner, H Bernt and A Eder
Institut für Festkörpertechnologie, Paul-Gerhardt-Allee 42, 8000 München 60, West Germany

Abstract. Measurements of surface potential fluctuations at the Si–SiO_2 interface have been performed before and after application of an electrical bias stress to the gate at an elevated temperature. By comparing the values with theoretical models, it was found that surface charge attributable to mobile ions as well as the so-called 'slow states' appears to be present as small patches rather than distributed homogeneously over the interface area. The fraction of area covered by these patches can be varied between 0·05 and 1·0 by altering the crystal orientation, the oxidation procedure and the stress time. The assumption of negative surface charge, as suggested by other authors, is not necessary for the samples used in the present work.

1. Introduction

A very accurate method for determining densities of surface states at the Si–SiO_2 interface is the conductance technique described by Nicollian and Goetzberger (1967). In this method the equivalent parallel conductance G_p of a MOS capacitor in depletion is measured as a function of the frequency ω. In order to fit the experimental curves to the theoretical data, a non-uniform surface potential must be introduced. Each fit yields a value for the standard deviation σ_s of the potential distribution at the surface.

The origin of these potential fluctuations is assumed to be a statistical distribution of interface charges. With a sophisticated model Brews (1972) obtained a linear relationship between the square of the standard deviation σ_s^2 and the average surface charge density $\bar{Q}$. He assumed that the charges exhibited a Poisson distribution, which is reasonable for charge densities below 10^{12} cm^{-2}, when the probability of a charged centre occurring at a certain place should be independent of the location of other charges.

Measurement of σ_s^2, together with other parameters characteristic of the MOS interface can give additional information about charges and states at the surface. Declerck *et al* (1974) found that their samples contained positive and negative surface charges simultaneously. Ziegler (1978) determined σ_s as a function of temperature and showed that most of the surface states in his sample were of the donor type.

In a recent paper (Werner *et al* 1979), we have shown that deviations of the experimental values of σ_s^2 from Brews' theoretical considerations may be interpreted in terms of inhomogeneities in the distribution of oxide charge. We found that σ_s^2 is smaller than the value predicted by Brews' model if the centroid of oxide charge is situated some distance away from the interface, inside the oxide. On the other hand σ_s^2 will be greater than the predicted value if the charge is not distributed completely statistically but is concentrated on smaller patches of the interface area.

0305-2346/80/0050-0124$02.00

Recently a considerable amount of work has been done to improve the stability of MOS parameters when a high electrical field is applied to the gate (Breed 1975, Hess 1977, Pepper 1977, Jeppson and Svensson 1977, Sinha *et al* 1978). This degradation of the interface is commonly referred to as the slow-state instability and is supposed to be responsible for the shift of the logical window in electrical erasable memory devices (EAROM) during repeated write/erase cycles (El-Dessouky and Balk 1978, Jeppson and Svensson 1977).

From tunnelling experiments (Av-Ron *et al* 1978) and from measurement of the electrical wear-out (Osburn 1974), it has been concluded that the build-up of charge does not take place homogeneously across the area but rather at some localised 'weak points' of the oxide film. Similar inhomogeneities have been found in samples containing mobile ions (Williams and Woods 1972) and in MOS capacitors after exposure to ionising radiation (Zamani and Maserjian 1978).

In the present paper we describe measurements of the surface potential fluctuations before and after application of an electrical bias stress to the gate at an elevated temperature. In this way additional information about the lateral charge distribution at the interface could be obtained. We found that surface charge due to mobile ions as well as the so-called 'slow states' appears to be present as small patches rather than distributed homogeneously over the interface area. Moreover, for the slow-state instability we have shown that the fraction of area covered by these patches is a strong function of crystal orientation, oxidation procedure and stress time. On the other hand, the charge density inside the patches saturates rapidly and shows no correlation with the surface orientation and method of oxide formation.

The assumption of negative surface charge, with a distribution that is statistically independent of the positive charge as considered by Declerck *et al* (1974) and Pepper (1977), is not necessary for our samples.

2. Experimental details

Measurements were performed on silicon wafers after growing a film of dry oxide about 600 Å thick at 1000 °C. Usually the furnace was thoroughly cleaned for some hours by passage of dry O_2/5% HCl before oxidation. After oxidation the wafers were annealed at 1000 °C in dry N_2 for 30 min and received a 0·4 μm layer of Al from an electron gun evaporation source, followed by annealing in the forming gas (60% N_2, 40% H_2) for 45 min at 460 °C. Aluminium gates 0·8 mm in diameter were etched free using conventional photolithography, the reverse side was stripped free from oxide and aluminium and gold layers were evaporated onto this surface in order to obtain a series resistance that was as low as possible. Our values of σ_s^2 were determined from conductance versus frequency curves for 15 frequencies between 3 and 80 kHz. The measurement apparatus consisted of a fully computerised lock-in system and a PDP-11/10 which gave σ_s and N_{ss} values on a line-printer in a time of 20 s. The accuracy obtained was $|\Delta G|/\omega \leqslant 10^{-4}$ C_{ox}, so that reliable measurement of σ_s is possible down to surface state densities of 5×10^9 cm^{-2} eV^{-1}. A detailed description of the system will be given elsewhere (A Eder 1979 unpublished results). The values of σ_s were deduced from the width of the G/ω versus ω curves using data published by Goetzberger *et al* (1976). The values of the oxide charge density Q were derived from the flat-band voltage of the C–V curves at 100 kHz assuming a workfunction difference ϕ_{MS} of 0·3 V.

3. Theoretical considerations

Statistical fluctuations of fixed charges in a MOS device generate lateral inhomogeneities in the surface potential U_s. If we take the charges to be Poisson-distributed, the variance σ_s^2 of the surface potential can be given as a function of the mean interface charge $\bar{Q}$. Brews (1972) has solved the three-dimensional Poisson equation for an MOS capacitor with a statistically varying surface charge. He found that in a linear approximation, the standard deviation σ_s^2 of the surface potential is proportional to the average charge density $\bar{Q}$ at the surface:

$$\begin{aligned} \sigma_s^2 &= \bar{Q}/F \\ F &= (kT/q)^2 [4\pi(\epsilon_{si} + \epsilon_{ox})^2/\ln(1+\alpha)] \qquad (1) \\ \alpha &= [(\epsilon_{si} + \epsilon_{ox})/\lambda(C_{sc} + C_{ox} + C_{ss})]^2 \end{aligned}$$

where C_{sc} = semiconductor capacitance, C_{ox} = oxide capacitance, $C_{ss} = qN_{ss}$ = surface-state capacitance, ϵ_{ox}, ϵ_{si} = dielectric permittivities of the oxide and silicon respectively and λ = minimal fluctuation wavelength (Brews obtained an estimate of 5 Å $\leqslant \lambda \leqslant$ 100 Å).

In a recent paper (Werner *et al* 1979), we extended Brews' model in two aspects regarding the spatial distribution of the oxide charges.

(i) We assumed that the centre of oxide charge is not always located immediately at the interface as it is in Brews' model but may be situated at some distance inside the oxide. In this case equation (1) must be replaced by

$$\sigma_s^2 = f_\sigma(z)\bar{Q}/F. \qquad (2)$$

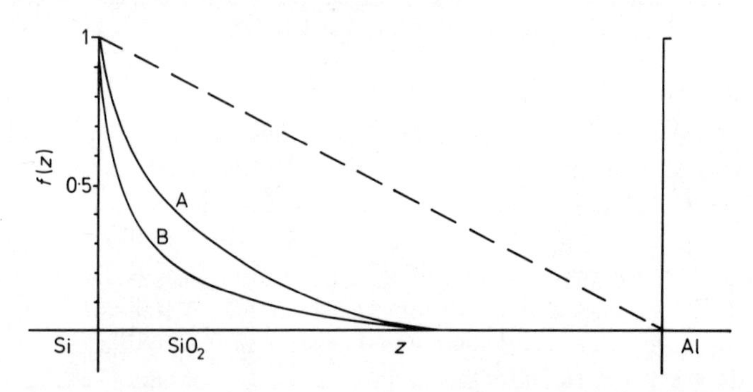

Figure 1. Reduction factor $f_\sigma(z)$ (full curves) for potential fluctuations caused by charge packets inside the oxide. d (the width of the packet) = $0{\cdot}2z_{ox}$ (curve A) and $0{\cdot}01z_{ox}$ (curve B). Broken curve: reduction factor $f_u(z)$ for the voltage shift of the C–V curve.

The function $f_\sigma(z)$ is shown in figure 1 for a rectangular charge packet in the oxide. Details of the calculation are given in our previous paper. It can be seen that σ_s^2 decays rapidly if the charge is located deeper in the oxide. For comparison we note that the voltage shift ΔV of a C–V curve depends linearly on the separation of the charge from the interface. Thus we have

$$\Delta V = f_u(z)\bar{Q}/C_{ox}. \qquad (3)$$

The function f_u is given by the broken line in figure 1. Therefore whenever the experimentally determined value of σ_s is smaller than would be expected from equation (1), we can conclude that the centre of charge is located some distance away from the interface.

(ii) As a second extension of the model we assumed that only part of the area of the MOS capacitor contains charge and interface states, while the remainder will be an ideal MOS interface. This model could take account of the localised peaks of surface charge, which have been found after temperature–bias stress or irradiation of the oxide (Williams and Woods 1972, Zamani and Maserjian 1978, Osburn 1974). We have shown that in this model the measured value of σ_s^2 would result from an average over just those patches that contain interface states, while the flat-band voltage would be an average over the total area. Thus equation (1) must be replaced by

$$\sigma_s^2 = \bar{Q}/(Fv) \tag{4}$$

with

$$v = \text{Area containing charge/Total area} \leqslant 1.$$

This would result in an increase in σ_s^2, compared with equation (1).

Declerck *et al* (1974) suggested another mechanism that could give an increase in σ_s^2 without increasing the displacement ΔV of the C–V curve. If a number of negative charges Q_- are distributed statistically independently from the positive charges Q_+, the standard deviations of both species of charge have to be added:

$$\sigma_s^2 = \sigma_{s,+}^2 + \sigma_{s,-}^2 = (Q_+ + |Q_-|)/F. \tag{5}$$

On the other hand in the C–V curve, the positive and negative charges compensate so that we have

$$\Delta V = (Q_+ - |Q_-|)/C_{ox}. \tag{6}$$

4. Results

4.1. *Measurements prior to temperature–bias stress*

We have determined ΔV and σ_s for 35 wafers (n-material, ⟨111⟩-oriented) for which the net oxide charge density ranged between less than 10^{10} and $1{\cdot}5 \times 10^{11}$ cm^{-2}. Figure 2 shows plots of $\sigma_s^2 F$ against $\Delta V C_{ox}$. If equation (1) were valid, we should find a straight line; this line is shown in figure 2 for the minimum and maximum value of Brews' cut-off parameter λ. It can be seen that the points in figure 2 are distributed completely at random and that there are wafers with σ_s^2 values too high as well as those with σ_s^2 too small, compared with equation (1).

We conclude from §3 that the charge is located some distance away from the interface in the wafers with small values of σ_s^2. The high values of σ_s^2 could arise from a localised concentration of charge as well as from the presence of negative oxide charges Q_-. The high scatter of points in figure 2 might result from different charge distributions from wafer to wafer.

Since it is generally assumed that interface states in the Si–SiO_2 system are correlated with oxide charges near the surface, we have determined the density of interface states

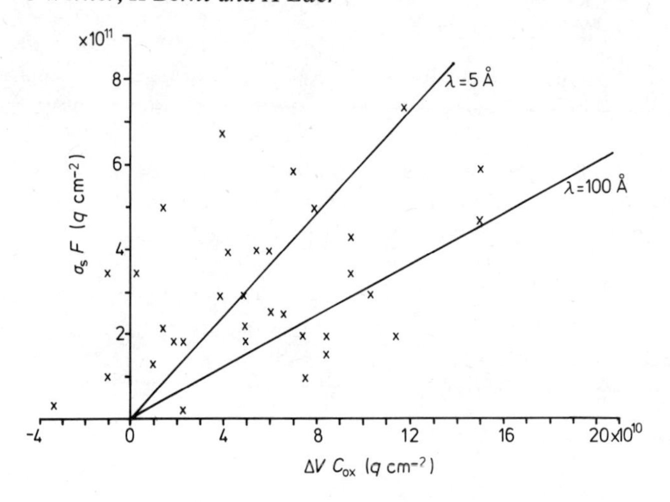

Figure 2. Correlation between oxide charge densities derived from C–V voltage shifts and from σ_s^2. The correlation factor is $r = 0{\cdot}17$. Brews' theory predicts that the points should lie on a line somewhere between the two full lines indicated.

N_{ss} in the middle of the band gap for the wafers used in figure 2. Figure 3 shows the correlation between σ_s^2 and N_{ss}. Compared with figure 2, there are no points with σ_s^2-values too low, whereas there are a number of points with σ_s^2 too high. We argue that the charges located deeper in the oxide did not produce potential fluctuations nor surface states so that there is no possible mechanism to decrease the value of σ_s in figure 3.

4.2. Measurements after drift of positive ions

Further information about the oxide charges was found by performing a temperature–bias stress at 200 °C and 1 MV cm^{-1} on some of the wafers that contained up to 10^{11} mobile ions cm^{-2}. We found an increase in σ_s^2 with stress with the Al gate positive while negative stress caused a decrease in σ_s^2, which is to be expected for a positive species of

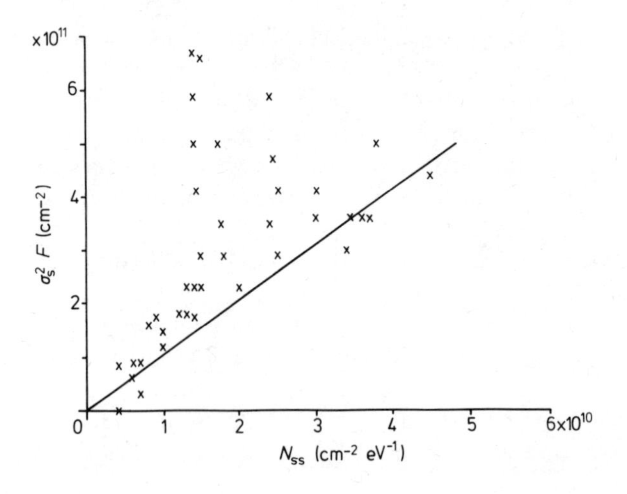

Figure 3. Fluctuations in the surface potential σ_s^2 versus the density of surface states.

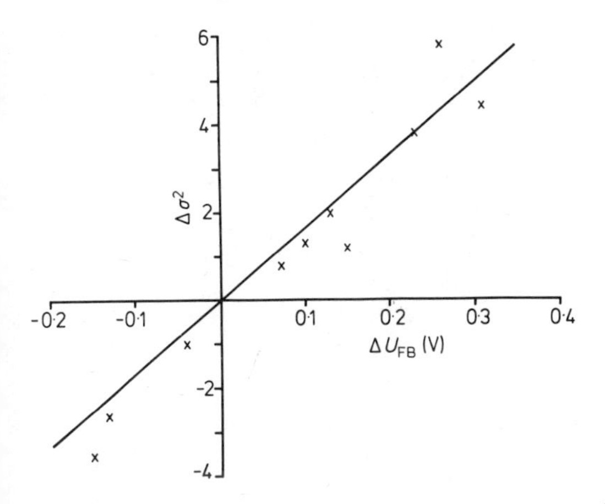

Figure 4. Correlation between changes in σ_s^2 and in the flat-band voltage due to a temperature–bias stress. The correlation factor is $r = 0{\cdot}94$.

mobile ions. Considering only the differences in σ_s^2 and ΔV before and after stress, we can eliminate the influence of immobile charges which are built into the oxide, as distinct from the mobile charges.

In figure 4 the changes $\Delta\sigma_s^2$ that occur after stressing the wafers are plotted against the accompanying shift in the flat-band voltage ΔU_{FB}. To a good approximation we can say that $\Delta\sigma_s^2$ is proportional to ΔU_{FB}.

Quantitative comparison with the theoretical results of §3 shows that the ratio of the values $\Delta\sigma_s^2$ and ΔU_{FB} is three times greater than would be expected from equation (1). Thus we are forced to apply equation (4) and conclude that the mobile charge is confined to some patches that occupy approximately 30% of the total area. The value of $\Delta\sigma_s^2$ cannot be explained by two kinds of charge as it could for unstressed wafers, since negative mobile charges would move in the opposite direction, decreasing $\Delta\sigma_s^2$ and increasing ΔU_{FB}.

4.3. *Slow-state instability*

We will now describe investigations of the instabilities during a bias stress at a raised temperature (200 °C) and an oxide field of 3 MV cm^{-1}. The aluminium gate was biased negatively in this experiment. The charge build-up during such a stress is commonly referred to as the slow-state instability and may be responsible for the degradation of electrically erasable MNOS memories during repeated write/erase cycles (El-Dessouky and Balk 1978, Jeppson and Svensson 1977). The detailed nature of the instability is not known at present (Breed 1975, Jeppson and Svensson 1977, Pepper 1977); most authors agree, however, that there is no drift of ions through the oxide but there is some electrochemical change at the interface.

In our samples we determined σ_s^2 and $\bar{Q}$ from the conductivity and the capacitance versus voltage curves of the MOS capacitors. Again we only consider the changes in the parameters caused by the negative bias stress. We assume that all charges are created at the Si–SiO_2 interface so that we can apply equation (4) and need not consider the corrections given in equations (2) and (3).

Figures 5 and 6 show $\Delta\sigma_s^2$ and the parameter $v = \Delta Q/(\sigma_s^2 F)$ as a function of stress time. From the model described in §3 we know that σ_s^2 represents the build-up of charge in the patches, while v is the fraction of area containing charge and interface states.

From figure 5 it can be seen that in all our samples $\Delta\sigma_s^2$ rises sharply during the first 10 or 20 min of stress and then saturates rapidly. We investigated samples that were oxidised in pure oxygen as well as with the addition of 5% HCl or 10% H_2O. Wafers with both ⟨111⟩ and ⟨100⟩ orientations were used. We conclude from figure 5 that there is no correlation between the values of $\Delta\sigma_s^2$ and the oxidation procedure or the crystal orientation of the surface. That means that the build-up of interface charge in the patches during a negative bias stress is not characteristically influenced by the method of oxide formation or by the wafer orientation. Moreover, the charge saturates after a short time at 3 MV cm^{-1} and 200 °C.

The increase in the overall average of interface-state and surface-charge density, however, exhibits $t^{1/4}$ behaviour over a stress time of 100 min, without saturation. This finding agrees with the observations of Goetzberger *et al* (1973), Sinha and Smith (1978), and Jeppson and Svensson (1977). In our model we can conclude that after the charge

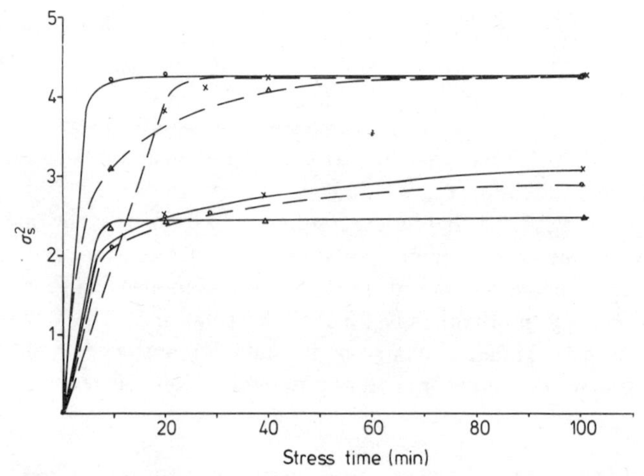

Figure 5. Increase in the potential fluctuations during negative bias stress for samples with different oxidation procedures and wafer orientations. o 10% H_2O; △ 5% HCl; x dry O_2; —— ⟨111⟩ orientation; – – – ⟨100⟩ orientation.

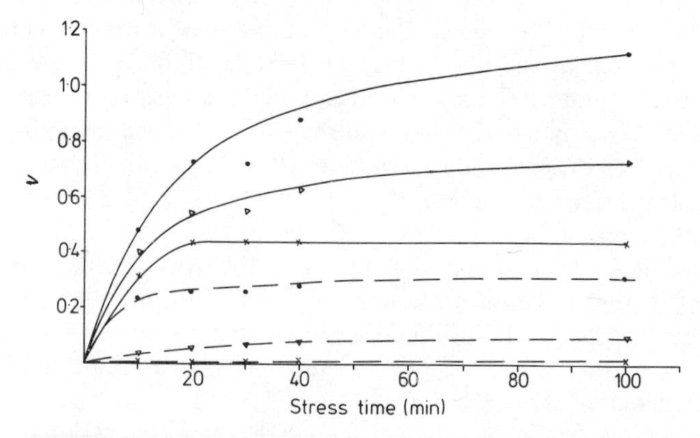

Figure 6. Increase in the area factor $v = \Delta Q/(\Delta\sigma_s F)$ during negative bias stress (3 MV cm^{-1}) at 200 °C for different oxidation procedures and wafer orientations. o 10% H_2O; △ 5% HCl; x dry O_2; —— ⟨111⟩ orientation; – – – ⟨100⟩ orientation.

density in the patches has saturated, the area covered by the patches continues to rise. We cannot decide from our measurements whether this is due to a lateral spread of the patches or whether new patches develop during the stress.

In figure 6 the fraction v of the area covered by the patches is shown as a function of stress time for the samples investigated. Contrary to figure 5, we now have a strong dependence on the oxidation procedure and on the crystal orientation of the surface. It can be seen that the area of the patches is considerably smaller in the ⟨100⟩-oriented samples than in the wafers with ⟨111⟩ orientation. Moreover, the addition of HCl to the oxidising ambient considerably enhances the fraction v. The addition of 10% H_2O gives a further enhancement, so that after a stress time of 100 min almost all the area was covered with charge in the ⟨111⟩ wafer.

5. Discussion

Pepper (1977) has measured the conductance of MOS transistors at liquid helium temperature before and after application of a negative bias stress. At those low temperatures the charge carriers are trapped by the potential fluctuations at the surface. The number of localised electrons N_{loc} found in this work can be compared with our measurement of σ_s^2. Pepper found that the number of localised states decreased during the stress although the net charge in the oxide was drastically increased. Pepper concluded that a considerable amount of negative charge is present in the samples before stress and that this is compensated by a positive charge to give a small net positive charge density. The number of localised states can thus considerably exceed the net charge density. He assumed that during stress, the negative centres are neutralised by the capture of holes. Thus the total charge is decreased while the net charge increases. The negative centres are assumed to be located close to positive centres, giving a dipolar potential.

The number of carriers localised by the dipole depends on the average distance z of the electrons from the interface. This distance was varied by Pepper (1977), who made use of the substrate bias effect. Thus we can write

$$N_{loc} = Q_+ + b(z)\,D \tag{7}$$

where D is the density of dipoles at the surface. In Pepper's model the changes in a bias stress are

$$\Delta D = -\Delta Q_+ \tag{8}$$

and thus

$$\Delta N_{loc} = (1 - b(z))\,\Delta Q_+. \tag{9}$$

Since in Pepper's experiments $\Delta N_{loc} < 0$, we have to conclude $b(z) > 1$.

Similar equations should be valid if N_{loc} is replaced by σ_s^2. We see from equation (7) that the initial value of σ_s^2, before a temperature–bias stress is performed, is increased above the value expected from Brews' model (equation 1) whenever the dipole density D is greater than zero at the interface. This agrees with our experimental findings in §4.1.

However, from equation (9) we conclude that the change in σ_s^2 due to a negative bias stress should be smaller or even negative if dipoles are present. This is in contrast to the experimental results given in §4.3. It must be remembered that $\Delta\sigma_s^2$ was found to be greater than in Brews' model. We want to state that the presence of dipoles cannot be ruled out completely by our measurements. The reduction in $\Delta\sigma_s^2$ due to the dipoles might be overcompensated if the localisation of the slow states in the patches is more pronounced than was derived in §4.3 without the assumption of dipole charge.

References

Av-Ron M, Shatzkes M, DiStefano T H and Cadoff I B 1978 *The Physics of SiO_2 and its Interfaces* ed S T Pantelides (New York: Pergamon) pp46–50
Breed D J 1975 *Appl. Phys. Lett.* **26** 116–8
Brews J R 1972 *J. Appl. Phys.* **43** 2306–13
Declerck G, Van Overstraeken R and Broux G 1974 *J. Appl. Phys.* **45** 2593–5
El-Dessouky A and Balk P 1978 *Rev. Physique Appl.* **13** 833–5
Goetzberger A, Klausmann E and Schulz M J 1976 *CRC Crit. Rev. Solid St. Sci.* **6** 1–27
Goetzberger A, Lopez A D and Strain R J 1973 *J. Electrochem. Soc.* **120** 90–6
Hess D W 1977 *J. Electrochem. Soc.* **124** 740–3
Jeppson K O and Svensson C M 1977 *J. Appl. Phys.* **48** 2004–14
Nicollian E H and Goetzberger A 1967 *Bell Syst. Tech. J.* **46** 1055–69
Osburn C M 1974 *J. Electrochem. Soc.* **121** 809–15
Pepper M 1977 *Proc. R. Soc.* **A353** 225–46
Sinha A K, Levinskin H J, Adda L P, Fuls E N and Povilonis E 1978 *Solid St. Electron.* **21** 531–5
Sinha A K and Smith T E 1978 *J. Electrochem. Soc.* **125** 743–6
Werner C, Bernt H and Eder A 1979 *J. Appl. Phys.* accepted for publication
Williams R and Woods M H 1972 *J. Appl. Phys.* **43** 4142–7
Zamani N and Maserjian J 1978 *The Physics of SiO_2 and its Interfaces* ed S T Pantelides (New York: Pergamon) pp443–8
Ziegler K 1978 *Appl. Phys. Lett.* **32** 249–51

Influence of the MOS surface channel on a channel-to-contact diode characteristic

W Hönlein†, K von Klitzing† and G Landwehr‡
† Physikalisches Institut der Universität Würzburg, Rontgenring 8, 8700 Würzburg, Germany
‡ Max-Planck-Institut für Festkörperforschung, Grenoble, France

Abstract. Current–voltage characteristics of Si–SiO_2–Al devices with the source and drain contacts doped with material of opposite polarity were studied. At room temperature the system can be described as a double injecting structure bypassed by a surface channel, forming a pn junction with one contact. At high gate voltages the channel determines the characteristic and tunnelling occurs through the pn junction. The reverse breakdown of the pn junction can be modulated by the gate voltage over a wide range of values. At low temperatures the tunnelling characteristic yields pronounced structures; this is attributed to indirect tunnelling involving intermediate states located at the Si–SiO_2 interface.

1. Introduction

We present here the characteristics of devices that differ from MOS field-effect transistors in that the doping of the source contact is of opposite polarity to that of the drain contact. By application of an appropriate voltage to the gate electrode, a channel of higher conductance is formed at the surface of the semiconductor. Depending on the doping of the substrate, either p^+n n^+- or p^+p n^+-structures are obtained. Analogous to the MOS transistor accumulation or inversion surface channels can be generated by changing the polarity of the gate voltage. In contrast to the ordinary MOS transistor, however, there is always a pn junction, formed by either the source or the drain contact and the surface channel. The location of this junction depends on whether an accumulation or inversion surface channel has been established (figure 1). In addition, there is a highly unsymmetrical pn junction, located at the interface between the substrate and the contact doped with material of opposite polarity, which is not affected by the gate voltage.

Figure 2(*a*) shows a two-dimensional plot of the energy-band diagram of a device with a p-substrate without the application of either a gate voltage or a source–drain voltage (initial band bending caused by surface states, oxide charges and by the difference between the workfunctions of the gate electrode and the bulk semiconductor is neglected). In this case, the pn junction is located at the interface between the substrate and the n^+-contact. By application of a large positive gate voltage, an inversion layer is formed at the Si–SiO_2 interface, which can be described by bending the energy bands downwards (figure 2*b*). (Band bending in the contact regions because of the stray field of the gate electrode can be neglected because of the degeneracy of the doping.) Therefore an additional pn junction between the inversion layer and the p^+-contact is built up. If the gate voltage is high enough to cause large carrier concentrations in the

0305-2346/80/0050-0133$01.00

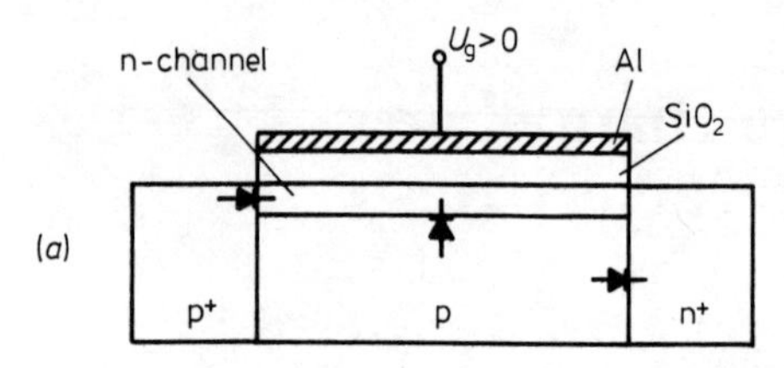

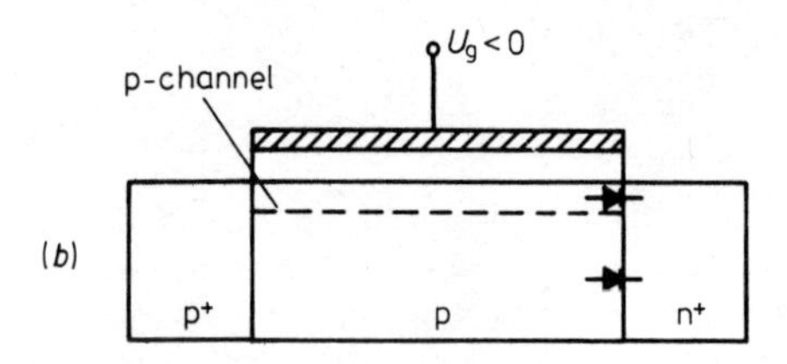

Figure 1. Schematic diagram of a MOS pn device with a p-type substrate. Diode symbols indicate the location of the pn junctions for inversion (*a*) and accumulation (*b*).

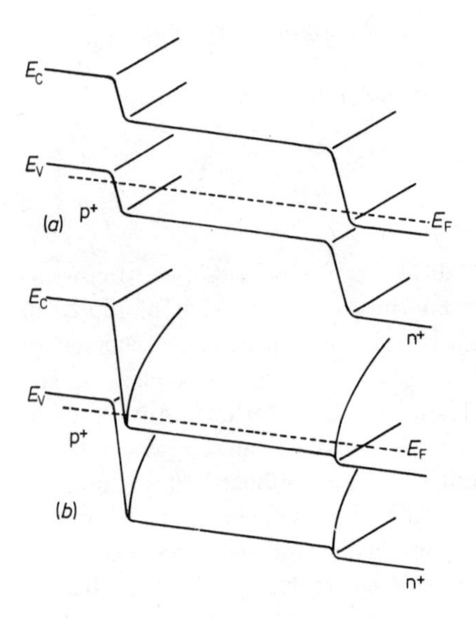

Figure 2. Energy-band diagram at the surface of a MOS pn device with a p-type substrate. (*a*) Flat bands at the substrate surface; (*b*) inversion channel due to a large positive gate voltage.

surface channel, the contact between channel and drain becomes Ohmic (n^+n^+) and a p^+n^+ junction is formed between the channel and source, through which the tunnelling of electrons might occur under suitable experimental conditions.

Quinn *et al* (1978) pointed out that tunnelling spectroscopy using such junctions could yield valuable information about the energy and other data of electric sub-bands in inversion and accumulation layers. Whenever electrons can tunnel into an electric sub-band (which arises from boundary quantisation of charge carriers in a narrow surface channel), structure is expected in the current–voltage characteristic and its derivatives. The direct determination of sub-band splitting is of interest, since optical data require sophisticated corrections because of plasma and exciton effects. Moreover, the application of a strong magnetic field should produce Landau quantisation and allow direct determination of the carrier effective masses and their dependence on energy.

2. Experiments and discussion

The experiments were carried out on two different kinds of samples:

(i) Linear devices with p^+- and n^+-contacts separated by a channel region of ~8 μm with either a p- or n-type substrate epitaxially grown in the (100) direction on spinel. The substrate thickness was 0·7 μm and the doping concentration $\sim 2 \times 10^{15}\ cm^{-3}$. The average surface-channel cross section was 0·25 μm^2, assuming an average channel depth of 5 nm. The oxide thickness was 120 nm and the gate electrode material was Al. The concentration of dopant in the contact exceeded $10^{20}\ cm^{-3}$. The samples were provided by the Siemens Forschungslaboratorien, München.

(ii) Circular devices with p^+- and n^+-contacts at a distance of 10–13 μm with an n-type substrate (thickness 0·65 μm, doping concentration $2 \times 10^{15}\ cm^{-3}$) epitaxially grown on sapphire in a (100) direction. The concentration of dopant in the contacts exceeded $10^{20}\ cm^{-3}$; the average surface-channel cross section was 0·5 μm^2, assuming an average channel depth of 5 nm. The gate electrode material was Al and the oxide thickness was 100 nm. The samples were provided by LETI, Grenoble.

2.1. *Room-temperature characteristics*

Figure 3 shows a typical plot of the current through a MOS pn device on spinel at different constant source–drain voltages U_{sd} as a function of the gate voltage U_g at 300 K. The current minimum close to zero gate voltage coincides with the threshold voltage of an ordinary MOS transistor fabricated on the same substrate. This indicates that the reduction in the current in this part of the characteristic is caused by the absence of a surface channel between the accumulation and inversion thresholds. Assuming a flat-band condition at the surface, double injection of carriers from the contact regions causes a current through the substrate, characterised by the minimum in figure 3. As soon as the formation of the surface channel is accomplished, additional forward current is observed. The increase in the current as a function of the gate voltage cannot be attributed to a variation in the substrate current caused by a change in the depletion

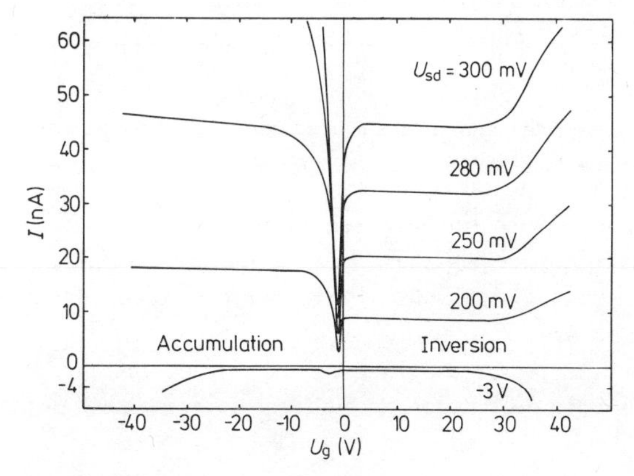

Figure 3. Forward current (upper curves) and backward current (lower curve) of a MOS pn device on spinel with a p-type substrate as a function of the gate voltage U_g at different constant source–drain voltages U_{sd} (T = 300 K).

width, since such an effect could not explain the increase in the forward current for both accumulation and inversion. Over this range of gate voltages above the threshold, the forward characteristic is defined mainly by the surface channel. As a function of gate voltage, the following devices can be realised: $p^+n^+n^+$, $p^+n\,n^+$, $p^+i\,n^+$, $p^+p\,n^+$, $p^+p^+n^+$.

The experimental results in figure 3 demonstrate that for surface carrier concentrations that are not too high, a forward current that is almost independent of the gate voltage may be observed. This behaviour originates from increases in both the height of the potential barrier U_{bi} and the forward voltage bias U_f of the pn junction. For strong injection and neglecting the Ohmic voltage drop across the channel, the changes in U_{bi} and U_f are the same (Herlet and Spenke 1955a, b, c), which results in a constant current.

At higher gate voltages an additional increase in the forward current is observed (figure 3). For this range of voltages the carrier concentration in the channel can become very high and the device can be described as a $p^+n^+n^+$ or $p^+p^+n^+$ junction, depending on the polarity of the gate voltage. The voltage drop across the Ohmic p^+p^+ or n^+n^+ part of the junction becomes negligibly small ($U_{sd} = U_f$) and the n^+p^+ or p^+n^+ junction space-charge layer might be narrow enough for the onset of an additional tunnelling current.

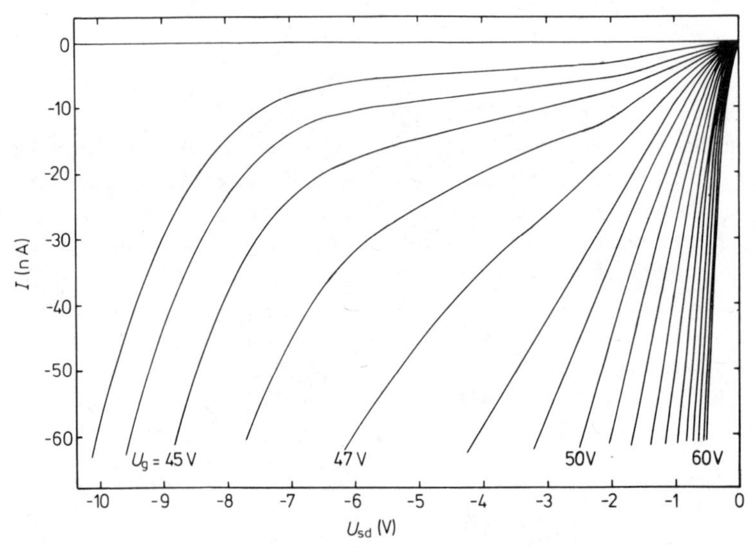

Figure 4. Reverse breakdown behaviour of a $p^+n^+n^+$ device at room temperature (300 K). Each curve corresponds to a 1 V change in the gate voltage U_g.

This interpretation is in agreement with the observed reverse breakdown characteristic of the device (figure 4). Again considering the region of the gate voltage, where the conduction through the surface channel dominates, the breakdown voltage decreases with increasing gate voltage. This behaviour is analogous to the breakdown of differently doped conventional diodes. With increasing carrier concentration on the channel side of the pn junction, the rate of ionisation of avalanche multiplication processes in the space-charge layer increases, which leads to lower breakdown voltages. Some of the samples investigated showed instabilities at reverse biases above $U_{sd} \approx -3$ V, which may be attributed to the generation of microplasmas and transit time effects. These phenomena, which are often observed in the avalanche breakdown region, disappear at breakdown voltages lower than $U_{sd} \approx -3$ V. For voltages in this range avalanche breakdown becomes less favourable, whereas internal field emission or Zener tunnelling increases (Lukaszek *et al* 1976). Figure 4 also shows that a characteristic change is observed in the breakdown curves for voltages in this region.

2.2. *Low-temperature characteristics*

Tunnelling and avalanche breakdown are majority-carrier effects and should be observed even at low temperatures. In contrast to these processes, generation and recombination currents are minority-carrier effects, which should not be present at low temperatures because the minority carriers are 'frozen out'. Figure 5 shows that at $T = 4{\cdot}2$ K, there is no longer a minimum current, but there is a well-defined region where no conduction occurs between the two thresholds. These can be determined precisely if the forward source–drain voltage U_{sd} exceeds the built-in voltage U_{bi} of the pn junction. The difference between the two thresholds is determined by the density of surface states Q_{ss} and the charge density Q_s in the depletion layer.

Figure 6 shows equivalent measurements on another sample for smaller source–drain voltages U_{sd}. Beyond the threshold region the constant-current plateau is no longer visible and no reverse saturation current flows because of the absence of minority carriers. The tunnelling current, however, is still present as is the avalanche breakdown of the junction. The shift in the onset of the tunnelling current as a function of the applied constant forward voltage is caused by the variation in the width of the junction space-charge layer, which decreases with increasing forward voltage.

As shown by figure 6, the tunnelling current yields pronounced structures at $T = 4{\cdot}2$ K. These structures are not reproducible in measurements on different samples and are not dependent on transverse magnetic fields up to 9 T. Tidey *et al* (1974), Pals and van Heck (1973) and Cole *et al* (1976) observed similar structure in the field-effect mobility of

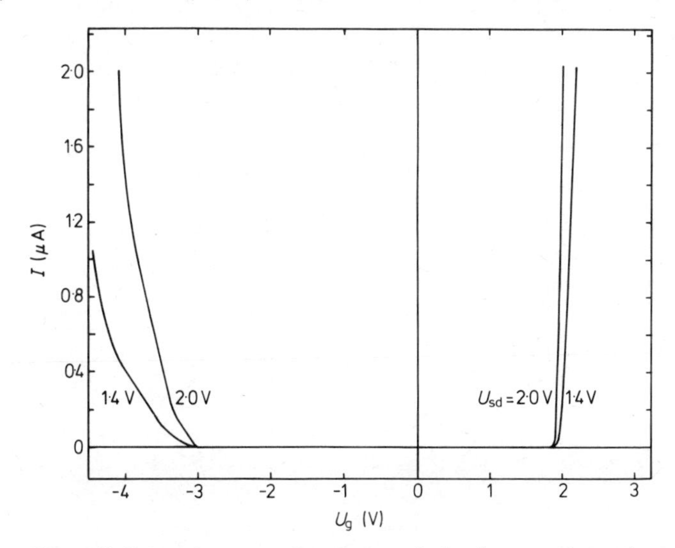

Figure 5. Forward current of a MOS pn device (n-type substrate) as a function of the gate voltage U_g for two different constant forward voltages $U_{sd} > U_{bi}$ ($T = 4{\cdot}2$ K).

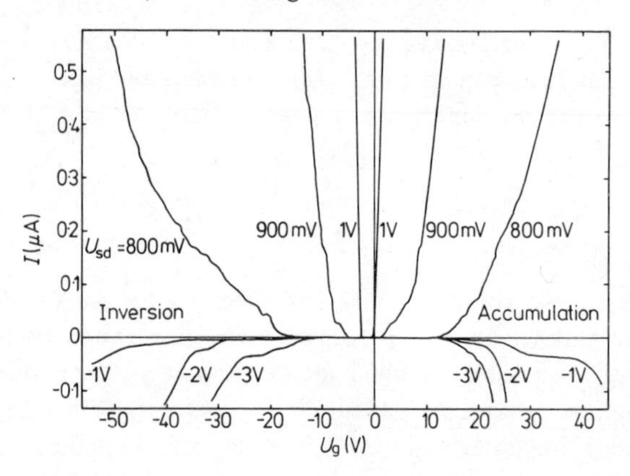

Figure 6. Low-temperature (4·2 K) forward and backward current of a MOS pn device (n-type substrate) as a function of the gate voltage U_g at different constant source–drain voltages $U_{sd} < U_{bi}$.

MOS transistors at low temperatures. These authors attributed the structure to contact inhomogeneities, trapping and scattering by surface states, localised states, surface strains and surface roughness. Since at low temperatures the direct band-to-band tunnelling at forward biases shown in figure 6 is not possible, the observed current is interpreted in terms of an excess tunnelling current that flows over some intermediate states (Chynoweth *et al* 1961). In the tunnelling junctions considered here, these states reflect the properties of the surface, because the junction is restricted to areas close to the semiconductor–insulator interface. The structure in the tunnelling characteristic can therefore arise from indirect tunnelling processes that involve intermediate surface states located in the energy gap. In addition, levels within the band gap, originating from bulk impurities, may influence the tunnelling probability. Thus the structure can be interpreted as fluctuations in the energy distribution of the density of surface states and bulk impurity levels throughout the band gap.

Figure 7 shows a typical tunnelling characteristic for low bias voltages and a high and constant gate voltage. This curve demonstrates that the current is very small at bias

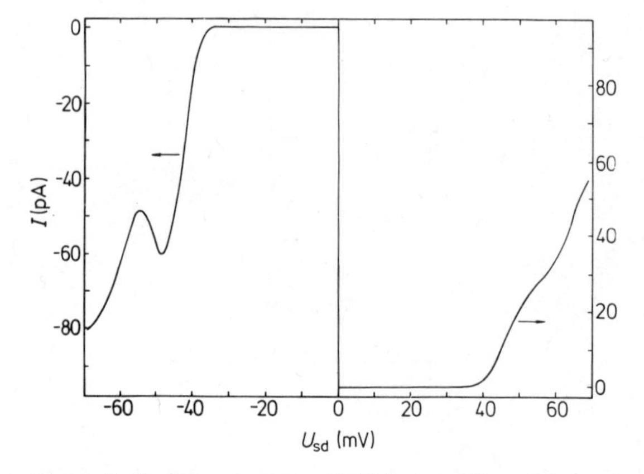

Figure 7. Low-temperature (4·2 K) tunnelling characteristic of a $p^+n^+n^+$ device (accumulation) at low bias voltages U_{sd} and a constant gate voltage $U_g = 45$ V. $T = 4{\cdot}2$ K.

voltages below ~40 mV and increases strongly at higher values in both the forward and reverse directions. This result is similar to measurements on Si and Ge tunnelling diodes, where the threshold for tunnelling processes appears at bias voltages that correspond to the energy of phonons necessary for momentum conservation. Moreover, for small bias voltages the tunnelling probability could be strongly reduced because one side of the tunnelling junction consists of a quasi-two-dimensional charge-carrier system. In this case, the increase in the current under forward bias might arise from the tunnelling of electrons from localised states in the band gap of the n-channel side into free states of the p^+ contact. The threshold under reverse bias is interpreted as tunnelling from the contact into three-dimensional states above the highest energies of the electric sub-band system.

Our experiments have shown that the initial expectation that tunnelling into electric surface sub-bands would be observed has not yet been met. Obviously the quality of the tunnelling junctions leaves something to be desired. It seems that the gradient in the impurity concentration is not high enough to meet the stringent conditions for a tunnelling experiment. It is hoped, however, that devices with ion-implanted source and drain regions may be more suitable for these experiments.

Acknowledgments

The authors gratefully acknowledge financial support for this work from the Deutsche Forschungsgemeinschaft.

References

Chynoweth A G, Feldmann W L and Logan R A 1961 *Phys. Rev.* **121** 684
Cole T, Lakhani A A and Stiles P J 1976 *Surface Sci.* **58** 56–9
Herlet A and Spenke E 1955a *Z. Angew. Phys.* **7** 99–107
—— 1955b *Z. Angew. Phys.* **7** 149–63
—— 1955c *Z. Angew. Phys.* **7** 195–212
Lukaszek W A, van der Ziel A and Chenette E R 1976 *Solid St. Electron.* **19** 57–71
Pals J A and van Heck W J J A 1973 *Appl. Phys. Lett.* **23** 550–2
Quinn J J, Kawamoto G and McCombe B D 1978 *Surf. Sci.* **73** 190–6
Tidey R J, Stradling R A and Pepper M 1974 *J. Phys. C: Solid St. Phys.* **7** L353

Tunnelling MIS structures

H C Card
Columbia Radiation Laboratory, Columbia University, New York, New York 10027, USA

Abstract. When the insulating layer of an MIS structure is made very thin (less than $\simeq 50$ Å), appreciable currents can be made to flow between the metal and semiconductor by quantum-mechanical tunnelling, even for modest electric fields. Elastic tunnelling is formulated within the WKB approximation; a Franz dispersion relation between the (imaginary) wavevector and energy within the energy gap of the insulator is used and different effective masses in the conduction and valence bands of the insulator are assumed. Experimental techniques are described which provide for the separation of majority- and minority-carrier tunnel currents and for the independent measurement on the same MOS sample of the tunnelling barriers to electrons and holes from the semiconductor. The present data indicate that the holes see a greater barrier in the metal–SiO_2–Si system, that majority carriers dominate the tunnel current in Au–SiO_2–nSi devices and that minority carriers dominate in Al–SiO_2–pSi devices. The Schottky barrier to MOS transition may be observed by variation of oxide thickness, or, for a constant thickness, by variation of the level of illumination with visible light, for example. The application of tunnelling MIS structures in photovoltaic energy conversion, in optoelectronics and in negative-resistance devices is explained. It is shown, for instance, how the oxide thickness can control the degree of minority-carrier injection in an electroluminescence application, or the open-circuit voltage of a photovoltaic solar cell.

1. Introduction

Tunnelling in metal–insulator–semiconductor (MIS) structures with ultrathin insulating layers was first studied by Gray (1965) in an attempt to probe the interface states at a semiconductor surface. More detailed studies by Waxman *et al* (1967), Shewchun *et al* (1967) and Dahlke and Sze (1967) with well-defined thermally grown silicon dioxide (SiO_2) layers, and the theoretical studies of Freeman and Dahlke (1970) have elucidated the various tunnelling currents present in these structures. Subsequent studies by Card and Rhoderick (1971a, b), Clarke and Shewchun (1971) and Kar and Dahlke (1972a, b), which take into account the dependence of the statistics of minority carriers and electrons in interface states on the thickness d of the insulating layer, have resolved most of the remaining questions about the electrical characteristics.

The fundamental MIS tunnelling structure is shown in figure 1(*a*) and the electron energy-band diagram in (*b*), for the particular example of an n-type substrate. The tunnel currents that flow between the metal and the semiconductor under an applied bias voltage are shown in figure 2: J_n is associated with tunnelling transitions between the metal and the conduction band of the semiconductor, J_p is due to transitions involving the valence band and J_{sm} is due to transitions to and from the interface states. As in MOS structures with thick oxide layers, interface states are localised electronic states associated with the semiconductor–insulator interface and are caused by the bond defect structure or by impurities.

0305-2346/80/0050-0140$03.00

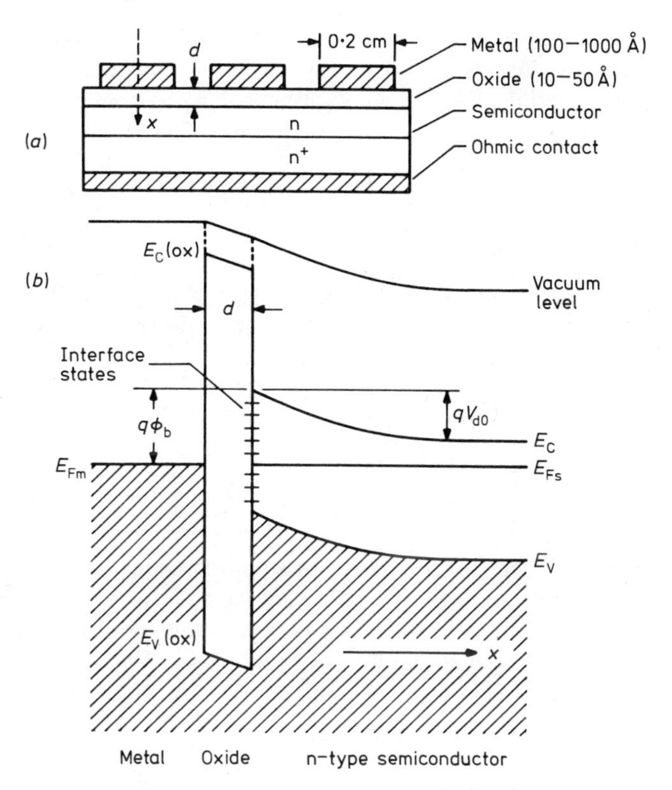

Figure 1. The tunnelling MIS structure for an n-type semiconductor.

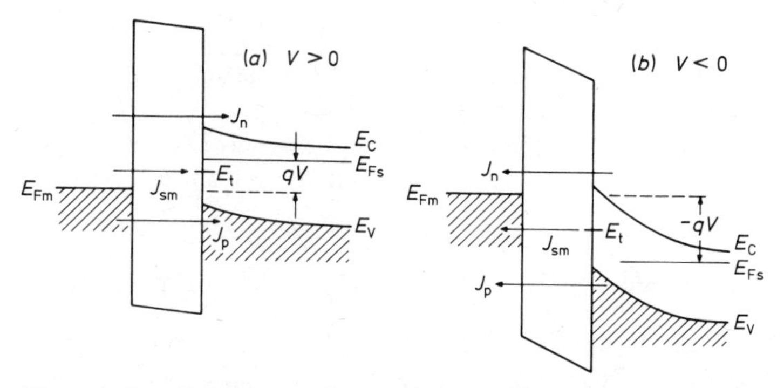

Figure 2. Tunnel currents to the majority-carrier band (J_n), the minority-carrier band (J_p) and to interface states (J_{sm}). Electron particle flows are in opposite directions to the conventional currents shown. E_t, energy of interface state.

We begin in this paper with a theoretical description of MIS tunnelling that involves the WKB approximation and a Franz dispersion relation between the electron energy and wavevector in the energy gap of the insulating thin film. Experimental results are then discussed separately for systems where transitions to the majority-carrier band dominate the tunnelling current (as in Au–SiO_2–nSi structures) and where transitions to the minority-carrier band dominate (as in Al–SiO_2–pSi structures). Experimental determinations of tunnelling probabilities for both types of transitions are described and the factors affecting the statistical energy distributions of electrons in the conduction and valence bands at the surface of the semiconductor are explained. Finally, applications of MIS tunnelling in optoelectronics, in negative-resistance devices and in photovoltaic energy conversion are described, and the advantages obtained by the unique characteristics of these devices are considered.

2. Band-to-band tunnelling currents

The tunnel currents are formulated following the independent-particle treatment of Harrision (1961). Elastic tunnelling is assumed in all cases; we do not deal in this paper with tunnelling mechanisms involving the interaction of the tunnelling electron with phonons or other quasi-particles. We can write for the magnitude of the tunnel current density (Card and Rhoderick 1971a, b)

$$J = (2mq/\hbar^3) \int_{E_x} \int_{E_t} |M_{sm}|^2 \rho_s \rho_m (f_s - f_m) \, dE_t \, dE_x \tag{1}$$

where $|M_{sm}|^2$ is the matrix element for the transition from the semiconductor (s) to the metal (m), ρ_s and ρ_m are the density-of-states factors, f_s and f_m are the Fermi functions, E_x and E_t are the components of energy due to, respectively, crystal momenta perpendicular and transverse to the tunnelling barrier of the insulating film and the integrals are over the allowed energy states.

Using the WKB approximation for the tunnelling transmission coefficient of the insulating barrier we obtain

$$|M_{sm}|^2 = \frac{1}{\rho_s \rho_m (2\pi)^2} \exp\left(-2 \int_{X_s}^{X_m} K \, dX\right) \tag{2}$$

where X_s and X_m are the classical turning points and K is the magnitude of the imaginary wavevector in the forbidden gap of the insulator.

If we now consider specifically tunnelling from the semiconductor conduction band into the metal when the population of the conduction band is non-degenerate, we have

$$f_s \simeq \exp -[(E - E_{Fn})/KT] = \exp -[(E_x + E_t - E_{Fn})/KT] \tag{3}$$

where the conduction band edge $E_c = 0$ so that E_{Fn} is negative in expression (3). Also

$$f_m \simeq \exp[-(E_x + E_t - E_{Fm})/KT]. \tag{4}$$

Substitution of equations (2), (3) and (4) into equation (1) gives

$$J_n = \frac{mq}{2\pi^2\hbar^3} \int_0^\infty \int_0^\infty \exp\left(-2 \int_{X_s}^{X_m} K \, dX\right) \left[\exp\left(\frac{E_{Fn}}{KT}\right) - \exp\left(\frac{E_{Fm}}{KT}\right)\right] \exp\left(-\frac{E_x}{KT}\right)$$

$$\times \exp\left(-\frac{E_t}{KT}\right) dE_x \, dE_t$$

$$= \frac{4\pi mq}{h^3} (KT)^2 \exp\left(-2 \int_{X_s}^{X_m} K \, dX\right) \left[\exp\left(\frac{E_{Fn}}{KT}\right) - \exp\left(\frac{E_{Fm}}{KT}\right)\right] \tag{5}$$

where E_{Fn}, E_{Fm} are the quasi-Fermi levels at the surfaces for electrons in the semiconductor conduction band and in the metal respectively. The energy dependence of the tunnelling exponent that enters through $K(E)$ has been assumed to be sufficiently weak to remove this factor from the integral. Equation (5) may be rewritten as

$$J_n = \frac{4\pi mq}{h^3}(KT)^2 \exp\left(-2\int_{X_s}^{X_m} K\,dX\right) \exp\left(\frac{E_{Fn}}{KT}\right)\left[1-\exp\left(\frac{E_{Fm}-E_{Fn}}{KT}\right)\right]$$

$$= \frac{4\pi mq(KT)^2}{h^3 N_C}\, n(0) \exp\left(-2\int_{X_s}^{X_m} K\,dX\right)\left[1-\exp\left(\frac{E_{Fm}-E_{Fn}}{KT}\right)\right] \quad (6)$$

where N_C is the effective density of states in the conduction band and $n(0) = N_C \exp(E_{Fn}/KT)$ is the concentration of conduction electrons at the semiconductor surface.

A similar expression is obtained for tunnelling transitions between the metal and the valence band of the semiconductor. We refer to this process as hole tunnelling and the corresponding expression is

$$J_p = \frac{4\pi mq(KT)^2}{h^3 N_V}\, p(0) \exp\left(-2\int_{X_s}^{X_m} K\,dX\right)\left[\exp\left(\frac{E_{Fp}-E_{Fm}}{KT}\right)-1\right] \quad (7)$$

where N_V is the effective density of states in the valence band of the semiconductor, $p(0) = N_V \exp(-E_{Fp}|KT)$ is the surface hole concentration and the quasi-Fermi levels for holes, and for electrons in the metal, E_{Fp} and E_{Fm}, are in this case measured with respect to the edge of the valence band E_V at the surface of the semiconductor.

3. Currents to interface states

We consider an energy distribution of localised interface states that extends over the energy gap of the semiconductor. There may be interface states with energies outside this range, but their charge state will not be affected by changes in the bias voltage. The potential well that creates the interface state (figure 3) is considered to be strongly confined to the insulator–semiconductor interface. Apart from the spread in their wavefunctions (which depends on the energy in the gap), these states can therefore be said to exist at the insulator–semiconductor interface.

Interface states exchange charge with the conduction and valence bands in the semiconductor via the Shockley–Read–Hall generation–recombination mechanisms. The rate of capture of electrons by interface states is given by (Shockley and Read 1952, Hall 1952)

$$U_n = \sigma_n V n(0)(1-f) \quad (8)$$

where σ_n is the capture cross section for electrons by the interface state, V is the thermal velocity of electrons, n(0) is the concentration of electrons at the surface of the semiconductor ($x = 0$) and f is the fraction of interface states in question that are occupied. The rate of electron emission from interface states (to the conduction band) is

$$G_n = e_n f = \sigma_n V n_i \exp[(E_t - E_i)/KT] f \quad (9)$$

where e_n is the emission probability of electrons and the right-hand side of equation (9)

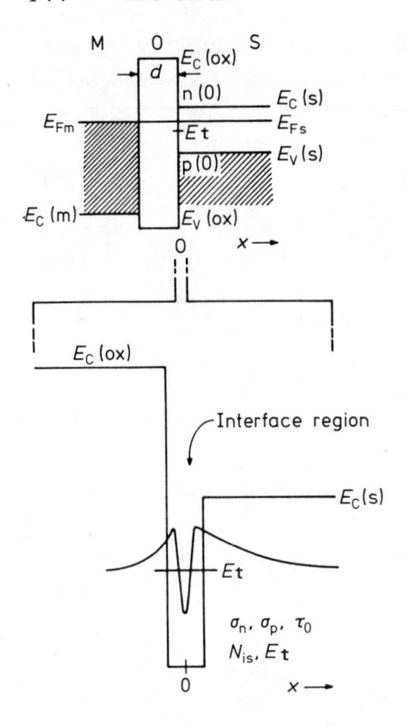

Figure 3. Potential well at oxide–semiconductor interface creates a localised interface state of energy E_t, capture cross sections σ_n, σ_p for electrons, holes and characteristic time constant τ_0 for tunnelling.

follows from detailed balance by equating expressions (8) and (9) under thermal equilibrium conditions. E_t is the energy of the interface state and E_i is the intrinsic (midgap) energy level in the semiconductor.

The net current into interface states from the conduction band is given by

$$J_{sc} = -q(U_n - G_n) = q\sigma_n V\left[n_i \exp\left(\frac{E_t - E_i}{KT}\right) f - n(0)(1-f)\right]. \tag{10}$$

Similarly the rate of capture of holes is

$$U_p = \sigma_p V p(0) f \tag{11}$$

and the rate of emission of holes is

$$G_p = e_p(1-f) = \sigma_p V n_i \exp\left[(E_i - E_t)/KT\right](1-f) \tag{12}$$

so that the net current into interface states from the valence band is given by

$$J_{sv} = q(U_p - G_p)$$

$$= q\sigma_p V\left[p(0) f - n_i \exp\left(\frac{E_i - E_t}{KT}\right)(1-f)\right]. \tag{13}$$

In the above expressions σ_p is the capture cross section of the interface states for holes, V is the thermal velocity of holes (assumed equal to that for electrons), p(0) is the hole concentration at the semiconductor surface and f is again the fraction of interface states of energy E_t that are occupied by electrons.

Provided that the insulating layer between the metal and semiconductor is sufficiently thin, there will also be a tunnel current into the interface states from the metal. This is described by (Freeman and Dahlke 1970)

$$J_{sm} = \frac{q}{\tau_0} \exp\left(-2\int_{X_s}^{X_m} K\,dX\right)(f - f_m) \tag{14}$$

where τ_0 is a time constant that is characteristic of the interface state potential and depends weakly on energy. For the creation of interface states by a three-dimensional delta-function potential, $\tau_0 \simeq 10^{-13}$ s (Lundström and Svensson, 1972).

4. Evaluation of the tunnelling exponent

In the previous two sections, we have formulated the tunnelling current between the metal and the conduction band, the valence band and the interface states all at the surface of the semiconductor. In each case a tunnelling exponent of the form

$$\exp\left(-2\int_{X_s}^{X_m} K\,dX\right) \tag{15}$$

was obtained within the WKB approximation. The magnitude of the imaginary wavevector K in equation (15) is a function of energy within the energy gap of the insulating film between the metal and semiconductor. This K will be different for transitions between the metal and the conduction band, the valence band and interface states at the semiconductor surface, since these states are opposite different energies in the insulator gap.

In the presence of an electric field in the insulator, K will also be a function of the position X between the classical turning points X_s and X_m. This is because tunnelling occurs at a constant total energy, whereas the conduction- and valence-band edges of the insulator are tilted by the electric field.

Detailed knowledge of the dispersion relation between the energy and the (imaginary) wavevector (E against K) within the energy gap of the insulating film is not generally available. In some cases the E–K relation has been established over a small portion of the energy gap by studies on relatively thick insulating layers, for example by Fowler–Nordheim tunnelling. One such series of studies has been performed for ultrathin SiO_2 layers by Maserjian (1974), Maserjian and Petersson (1974), Petersson *et al* (1975) and Lewicki and Maserjian (1975). These studies were for Cr–SiO_2–pSi structures with SiO_2 thicknesses in the range 20–40 Å, in which the currents measured were attributed to tunnelling between the metal and the conduction band of the semiconductor. In other words this interpretation corresponds to a minority-carrier tunnelling structure.

We can write for the dependence of the magnitude of K on E in the energy gap of an insulator (Franz 1956)

$$K^{-2} = K_V^{-2} + K_C^{-2} = (\hbar^2/2m_V^* E) + [\hbar^2/2m_C^*(E_g - E)] \tag{16}$$

where m_C^*, m_V^* are the effective masses in the conduction and valence bands of the insulator, E is the energy measured relative to the valence-band edge and E_g is the energy gap of the insulator. This is shown in figure 4. Equation (16) reduces to the normal parabolic expressions near the conduction- and valence-band edges. If m_C^* and m_V^* are assumed to be equal (in the absence of further knowledge) and we call this value m^*, expression (16) reduces to

$$K(E) = (2m^*/\hbar^2)^{1/2}E^{1/2}[1-(E/E_g)]^{1/2}. \tag{17}$$

The tunnelling exponent given by expression (15) is commonly written as

$$\exp(-2Kd) = \exp\{-2(2m/\hbar^2)^{1/2}(m^*/m)^{1/2}E^{1/2}[1-(E/E_g)]^{1/2}d\}$$
$$= \exp(-\chi^{1/2}d) \tag{18}$$

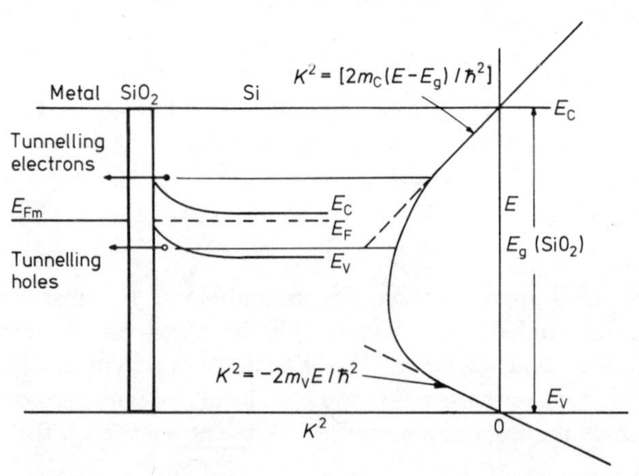

Figure 4. Franz dispersion relation between electron energy E and (imaginary) wavevector K in energy gap of an insulator such as SiO_2. A two-band model has been assumed.

where d is the thickness of the insulating layer (the distance between the classical turning points X_s and X_m) and χ is an effective tunnelling barrier. In the final form of expression (18), χ is expressed in eV and d in Å and there is a constant $[2(2m/\hbar^2)]^{1/2}$ of magnitude very close to unity (1·01) and of dimensions $(\text{eV})^{-1/2}$ $(\text{Å})^{-1}$ that has been omitted from the exponent for convenience, but which makes this exponent dimensionless. From equation (18) we have

$$\chi = (m^*/m)E[1-(E/E_g)]. \tag{19}$$

Determination of the tunnelling probability $\exp(-\chi^{1/2}d)$ therefore becomes a matter of obtaining the value of χ appropriate to the energy of the tunnelling carrier, within the energy gap of the insulator. A calculation of this dependence of χ on energy E in the energy gap of SiO_2 for $m^* = 0{\cdot}4\, m_0$ (a value obtained by Maserjian (1974) and Maserjian and Petersson (1974) for SiO_2) and $E_g = 8$ eV is shown in figure 5. The dependence of χ on E is symmetrical for the lower half of the energy gap (not shown), provided a single m^* is assumed as it is here. Recent experimental work has suggested that perhaps the energy gap of SiO_2 should be increased to $E_g = 9$ eV in these calculations (Williams 1977), but some debate exists regarding this value.

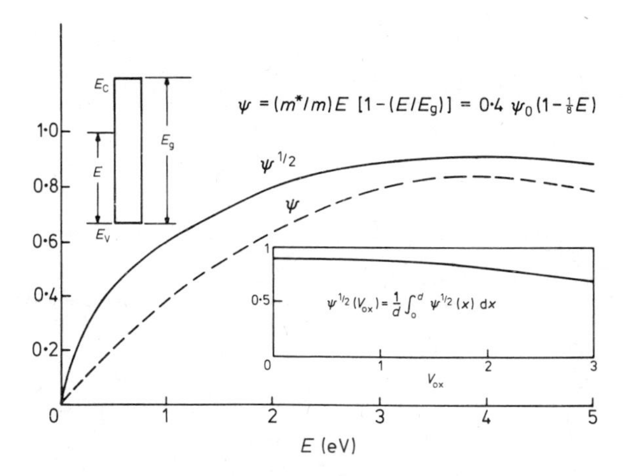

Figure 5. Effective tunnelling barrier ψ and dependence on energy E in the energy gap of SiO_2 assuming $m_C^* = m_V^* = 0{\cdot}4m_0$.

5. Au–SiO_2–nSi (majority-carrier) structures

In this section it will be shown experimentally that for Au–SiO_2–n-type silicon structures with ultrathin SiO_2 layers (d in the range 10–50 Å), the dominant currents (in the absence of optical illumination) are due to tunnelling transitions between the metal and the conduction band of the semiconductor (J_n of equation 6). Since we are considering an n-type semiconductor, this implies a majority-carrier tunnelling junction. In the experiments described in this section, oxidation was performed at relatively low temperatures; typical oxidation data are shown in figure 6. In all cases, the thickness of the SiO_2 layers was measured by ellipsometry (Archer 1962).

The expression for J_n can be rewritten by making the substitutions (Card and Rhoderick 1971a, b)

$$\begin{aligned}\exp(E_{Fn}/KT) &= -q(V_d + \phi_n)\\ &= -q(\phi_b - V/n)\end{aligned} \tag{20}$$

where V_d is the bias-dependent diffusion potential (band bending) in the semiconductor, V_{d0} is the value of V_d at zero bias, $\phi_b = V_{d0} + \phi_n$ is the so-called Schottky barrier height, ϕ_n is the Fermi potential $= (E_C - E_F)/q$ in the bulk, and from the electrostatics of the junction

$$n = (-\mathrm{d}V_d/\mathrm{d}V)^{-1} = 1 + \left(\frac{(d/\epsilon_i)[(\epsilon_s/W) + qD_{sb}]}{1 + (d/\epsilon_i)qD_{sa}}\right) \tag{21}$$

where d is the thickness and ϵ_i the permittivity of the insulating layer, W is the width of the depletion region in the semiconductor, ϵ_s is the permittivity of the semiconductor and D_{sa} and D_{sb} are, respectively, the densities of interface states (in $cm^{-2}\,eV^{-1}$) in equilibrium with the metal and with the majority carriers in the semiconductor. With these substitutions equation (6) becomes (Card and Rhoderick 1971a, b)

$$J_n = AT^2 \exp(-\chi^{1/2}d)\exp(-q\phi_b/KT)[\exp(qV/nKT) - 1] \tag{22}$$

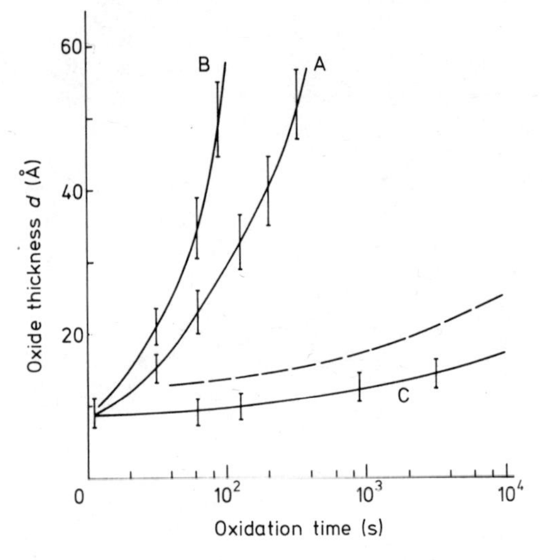

Figure 6. Representative data for oxidation in 100% dry O_2 at 700 °C (curve A, present study) and 800 °C (curve B) and in air at 25 °C (curve C). The broken curve data obtained by Goodman and Breece (1970) at 600 °C. Oxide thicknesses were measured by ellipsometry.

where $A = 4\pi mqK^2/h^3$ is the Richardson constant and $\chi^{1/2}d$ has been substituted for $2\int_{X_s}^{X_m} K\, \mathrm{d}X$ as discussed in a previous section.

Representative experimental data for Au–SiO_2–nSi junctions are given in figure 7. This figure shows that as the thickness of the SiO_2 layer is increased from $\simeq 10$ to 19 Å, the value of n increases from 1·03 to 1·2 because of the term $qD_{sb}d/\epsilon_i$ in equation (21) and the saturation current decreases from 10^{-7} to 10^{-9} A cm^{-2} because of the increase in the tunnelling exponent $\chi^{1/2}d$ in equation (22).

It is observed that as the thickness of SiO_2 increases in these Au–SiO_2–nSi structures, the zero-bias surface potential and hence the barrier height ϕ_b decreases with d. The dependence of ϕ_b on d is shown in figure 8. These results were obtained by an analysis

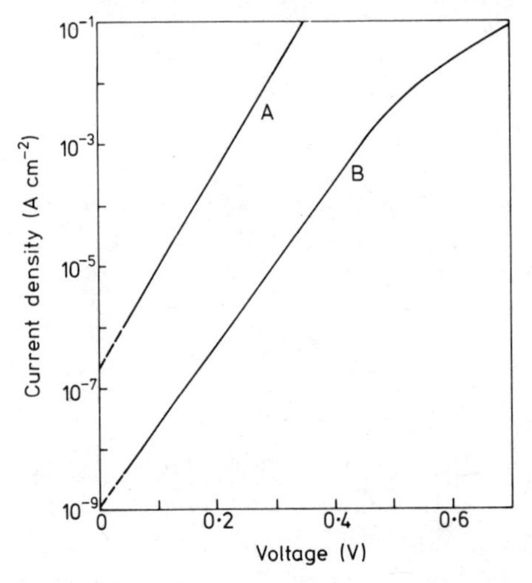

Figure 7. Experimental dependence of current density $J \simeq J_n$ on applied bias voltage (metal positive) for Au–SiO_2–nSi structures. Oxidation in dry O_2 at 700 °C; $J = AT^2 \exp(-q\phi_b/KT) \exp(-\psi^{1/2}d) \exp(qV/nKT)$; $n = 1 + \{(d/\epsilon_i)[(\epsilon_s/W) + qD_{sb}]/[1 + (d/\epsilon_i)qD_{sa}]\}$. Curve A, $d < 10$ Å, $n = 1·03$; curve B, $d = 19$ Å, $n = 1·2$.

of the dependence of the capacitance–voltage (C–V) characteristics on SiO_2 thickness, so that the determination of ϕ_b is made independently of data such as those given in figure 7. The origin of the decrease in ϕ_b with d is the dipole associated with the presence of positive charge in the SiO_2 layer, in the vicinity of the Si–SiO_2 interface. This has been allowed for in figure 7, where the results for both $d = 10$ Å and $d = 19$ Å are referred to the same $\phi_b = 0{\cdot}8$ V. Thus the difference in saturation currents is due solely to the tunnelling term $\exp(-\chi^{1/2}d)$ in equation (22).

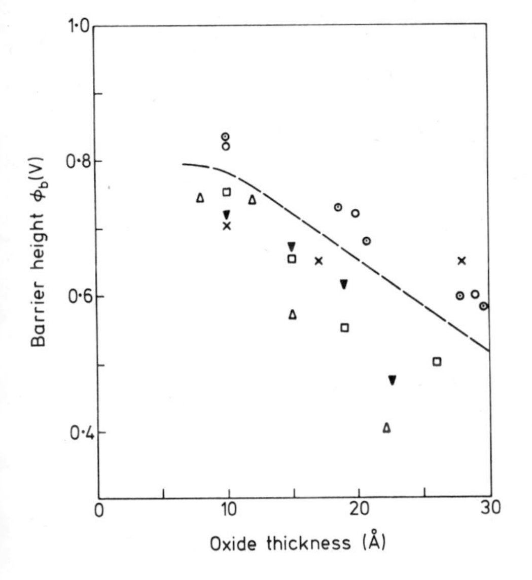

Figure 8. Experimental dependence of Schottky barrier height ϕ_b on SiO_2 thickness d for Au–SiO_2–nSi structures.

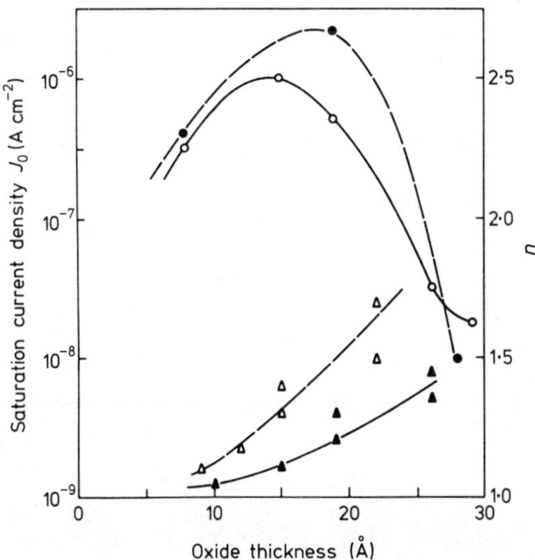

Figure 9. Experimental dependence of saturation current density $J_0 = AT^2\exp[-(q\phi_b/KT)]$ $\exp(-\psi^{1/2}d)$ (●, ○, upper curves) and the value of n (▲, △, lower curves) on SiO_2 thickness d Au–SiO_2–nSi structures. – – – oxidation with wet O_2, 900 °C; —— oxidation with dry O_2, 700 °C.

The raw data for the dependence of the saturation current $J_0 = AT^2 \exp(-\chi^{1/2}d) \exp(-q\phi_b/KT)$ and the value of n on SiO_2 thickness d is shown in figure 9 for oxides grown in both dry and wet O_2. The growth rate for wet oxides at 900 °C is much faster than that for dry oxides at 700 °C. It can be seen from figure 9 that there is a monotonic increase in the value of n with oxide thickness in the range 10–30 Å. However, the saturation current J_0 initially increases because of a rapid increase in $\exp(-q\phi_b/KT)$ caused by decreasing values of ϕ_b, and then decreases above ~20 Å as the decrease in $\exp(-\chi^{1/2}d)$ overcompensates for the further increase in $\exp(-q\phi_b/KT)$ in equation (22).

The experimental dependence of $\chi^{1/2}d$ on SiO_2 thickness for Au–SiO_2–nSi structures is shown in figure 10. The full circles are experimental points obtained at Columbia University, whereas the broken curve represents the results of an earlier study performed at the University of Manchester, UK. The reproducibility is quite good. These data represent the effective tunnelling exponents for electrons, that is for transitions between the conduction band of the silicon and the metal. The open circles in figure 10 represent the tunnelling exponents for holes, i.e. for transitions between the valence band of the semiconductor and the metal, for the same Au–SiO_2–nSi samples. These results (for holes) were obtained recently in our laboratory (Ny and Card 1979) by the following method.

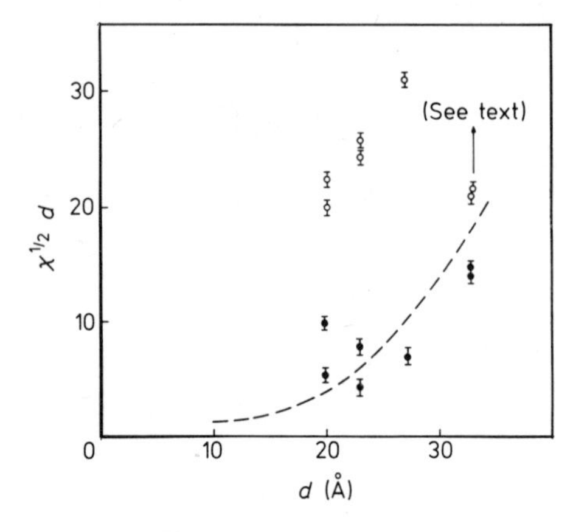

Figure 10. Experimental dependence of tunnelling exponents for electrons $\psi_e^{1/2}d$ (•) and for holes $\psi_h^{1/2}d$ (○) tunnelling into the metal on SiO_2 thickness d for Au–SiO_2–nSi structures.

The characteristics of the Au–SiO_2–nSi samples (current–voltage and capacitance–voltage measurements) were first obtained in the dark to deduce the dependence of $\chi^{1/2}d$ on SiO_2 thickness for electron tunnelling, by the method described above. The identical samples (all of which were prepared with semitransparent Au electrodes) were then illuminated by optical radiation with $h\nu > E_g$, the energy gap of the silicon. This produced hole–electron pairs and the holes (minority carriers) were collected as a photocurrent. It was found that for $d \gtrsim 20$ Å, the photocurrent for zero-bias J_{sc} was reduced, compared with that of an ideal Schottky barrier ($d < 10$ Å) prepared under identical conditions (metal electrode fabricated in the same evaporation step). We refer to this as photocurrent suppression and observe in passing that this implies that MIS solar cells require $d < 20$ Å for unsuppressed photocurrent and hence for reasonable conversion efficiency.

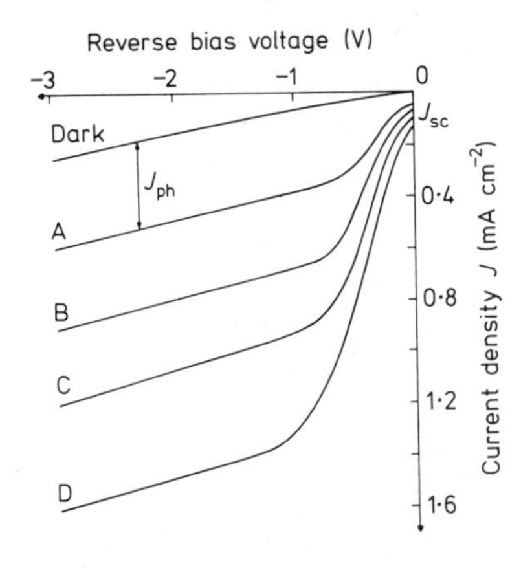

Figure 11. Experimental dependence of current on voltage (metal negative) for Au–SiO_2–nSi structures with SiO_2 thickness d = 25 Å and with semitransparent Au electrodes. Curves A to D represent increasing levels of optical illumination.

Even for $d > 20$ Å, it is found that an appreciable reverse bias will remove the photocurrent suppression so that the photocurrent density J_{ph} will be exactly equal to that of the ideal Schottky barrier under the same illumination. This is illustrated by the experimental data of figure 11, which were obtained on a Au–SiO_2–nSi sample with $d = 25$ Å. The photocurrent suppression and its removal by reverse bias has been explained (Ng and Card 1977) in terms of the hole tunnelling expression of equation (7). For zero bias, the photogeneration causes an accumulation of holes at the semiconductor surface and an increase in p(0). This continues until in the steady state the difference between photogeneration and the reverse diffusion current of holes from the surface into the bulk of the semiconductor exactly balances the tunnel current of holes (equation 7) for the steady-state value of p(0). For $d > 20$ Å, this balance results in a (short-circuit) photocurrent which is less than the total photogeneration in the silicon. Application of a reverse bias, however, reduces the tunnelling barrier of the SiO_2 layer to the point where this layer no longer suppresses the photocurrent and the total photogeneration of holes (within a diffusion length of the silicon surface) is collected by the metal.

A comparison of the short-circuit photocurrent density J_{sc} with the total photogeneration J_{ph} provides an experimental method for determining the tunnelling exponent for holes $\exp(-\chi^{1/2}d) = \exp(-2\int_{X_s}^{X_m} K\, dX)$ in equation (7). Moreover, this determination is made on the identical MIS samples, for which the corresponding information on electron tunnelling was obtained previously, as described above. J_p from equation (7) is equated to the net supply of holes to the semiconductor surface, given by

$$J_p = J_{ph} - (qD_p p(0)/L_p) \exp(-qV_{d0}/KT) = J_{sc} \tag{23}$$

where the second term is the reverse diffusion current into the bulk. The hole diffusion length L_p is determined independently by a method due to Kar (1974) and the zero-bias diffusion potential V_{d0} is obtained, as for electrons, from capacitance–voltage measurements.

We observe from figure 10 that the tunnelling exponents for holes (transitions between the metal and the silicon valence band) are, for the same thickness of SiO_2, appreciably larger than the values for electrons (transitions between the metal and the silicon conduction band). This is not to be expected from the theoretical calculations given in figure 5 which involved a Franz dispersion relation for E against K in the SiO_2 energy gap, since tunnelling at an energy $E \simeq 4$ eV in figure 5 (the energy for hole tunnelling) corresponds to approximately the same χ as for electron tunnelling at $E \simeq 5$ eV. (These values of 4 and 5 eV from the SiO_2 valence band have been established for thicker SiO_2 layers, $\simeq$1000 Å, but the present state of knowledge is that the gross aspects of band structure are preserved for ultrathin films such as those studied here.)

Our interpretation of the asymmetry in the tunnelling barriers for holes and electrons shown in figure 10 is that it arises from the presence of positive charge in the SiO_2 layer. Further evidence for its existence is the rapid decrease in barrier height ϕ_b with SiO_2 thickness d shown earlier in figure 8. Positive charge arising from excess silicon in the SiO_2 is known to exist over a region extending 20–40 Å into the oxide from the Si–SiO_2 interface (see review by Williams 1977). This charge will cause the energy bands in the SiO_2 to become distorted; the distortion increases the effective energy in the SiO_2 energy gap for both electrons and holes tunnelling into the metal. If we assume that the increase in the effective tunnelling energy is of the order of ~2 eV, then the tunnelling energies for holes and electrons change from ~4 and 5 eV to ~6 eV for holes and 7 eV for electrons. This can account for the asymmetry in $\chi^{1/2}d$ discussed above. Since $\chi^{1/2}$ is relatively flat from ~2 to 6 eV, the probability of hole tunnelling is not appreciably affected by the distortion of the energy bands caused by the positive charge. The electron tunnelling probability is greatly enhanced, however, as the energy approaches

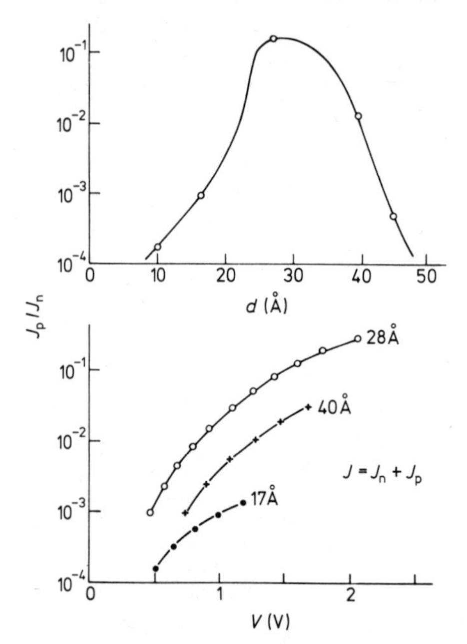

Figure 12. Experimental results for the ratio of minority- to majority-carrier current (minority-carrier injection ratio) for Au–SiO_2–nSi structures for positive voltage on metal.

to within $\simeq$1 eV of the SiO_2 conduction band. From figure 5, the appropriate values of $\chi^{1/2}$ are ~0·8 and 0·55, respectively, for holes and electrons. For an energy gap of 9 eV (rather than 8 eV as in figure 5), the asymmetry in $\chi^{1/2}$ between electrons and holes is still greater. The results of figure 10 are consistent with a band distortion Δ due to positive charge of the order of 2 eV or slightly greater; this quantity is an average over the path of travel of the tunnelling carrier. This magnitude of Δ arises from a positive surface charge density of the order of 10^{13} cm^{-2} for a 20–40 Å thick SiO_2 layer. An alternative explanation of the increased tunnelling probability for electrons (as compared to holes) is that trap-assisted electron tunnelling occurs.

It has been assumed in this section that the dark characteristics of Au–SiO_2–nSi structures are dominated by the electron tunnelling current J_n. We have verified this experimentally by fabricating transistor structures (using n-type epilayers on p-type substrates) in which a Au–SiO_2–nSi contact to the n-type epilayer was the emitter junction and the p–n junction was the collector for the minority carriers injected at the emitter (Card and Rhoderick 1973). The majority-carrier current J_n flowed out of the base through an n^+ Ohmic contact to the n-type epilayer. Experiments were performed with SiO_2 layers of thickness up to 50 Å in the emitter contact and the results are shown in figure 12. The ratio of minority-carrier to majority-carrier current never exceeds ~10% for these Au–SiO_2–nSi devices, so that we are justified in interpreting them as majority-carrier MIS junctions.

6. Al–SiO_2–pSi (minority-carrier) structures

It has been established experimentally that, unlike the Au–SiO_2–nSi structures discussed in the previous section, the Al–SiO_2–pSi structure has a dark current dominated by minority carriers, i.e. by tunnelling transitions between the metal and the conduction band of the silicon. This arises because the Al–SiO_2–pSi structure naturally has a large barrier height ϕ_b as a result of the difference in work function between Al and p-type

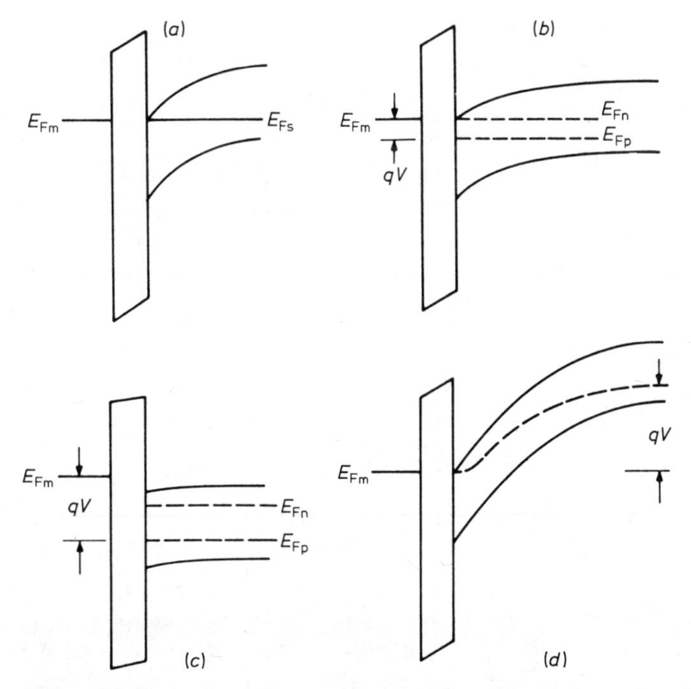

Figure 13. Energy-band diagrams for Al–SiO_2–pSi minority-carrier structures. (*a*) $V = 0$; (*b*) moderate forward bias voltage, current is diffusion-limited; (*c*) large forward voltage, current becomes tunnel-limited; (*d*) reverse bias.

silicon, which is made still larger as a consequence of the positive charges in the SiO_2 referred to in the previous section. The silicon surface is strongly inverted for zero bias as shown in figure 13 and this n^+ inversion region acts as a source of electrons, so that the device resembles an n^+p junction. The resemblance to an n^+p junction only occurs, of course, if the metal can readily supply electrons to the inversion layer by tunnelling and this tunnelling does not itself limit the current. The tunnelling current is described by equation (6) where n(0) is large as a consequence of the strongly inverted semiconductor surface.

The process of tunnelling between the metal and the inversion layer is in series with the transport of electrons within the conduction band of the semiconductor. For a negative bias on the metal, electrons tunnel into the inversion layer from the metal and in the steady state, this tunnel current exactly matches the diffusion current from the inversion layer into the bulk of the semiconductor. If the SiO_2 layer is thin enough for the tunnelling not to limit the current (the term $\exp(-2\int K\,dX) = \exp(-\chi^{1/2}d)$ in equation 6 is small), then the current–voltage characteristics obey the diffusion equation of an n^+p junction, namely

$$J_n = (qD_n n_i^2/L_n N_a)[\exp(qV/KT) - 1] \tag{24}$$

where D_n and L_n are the diffusion coefficient and diffusion length of electrons in the p-type semiconductor, n_i is the intrinsic carrier concentration, N_a is the doping concentration and V is the applied voltage. J_n in equation (24) must be equal to J_n of equation (6) and this equality determines the drop in the electron quasi-Fermi level $E_{Fm} - E_{Fn}$ across the oxide layer. For a thin oxide in which the current is diffusion-limited and not tunnel-limited, $E_{Fm} - E_{Fn}$ will be $\ll KT$, which implies that a negligible fraction of the applied voltage appears across the oxide layer and the current density is therefore given by equation (24) with V representing the total applied voltage. This is shown in figure 13(*b*).

As the bias voltage increases, the diffusion current (equation 24) increases exponentially so that eventually the supply of electrons by tunnelling will limit the current. The voltage at which the current becomes tunnel-limited will be larger for junctions with (i) thin oxides, since then the term $\exp(-2\int K\,dX) = \exp(-\chi^{1/2}d)$ is large and (ii)

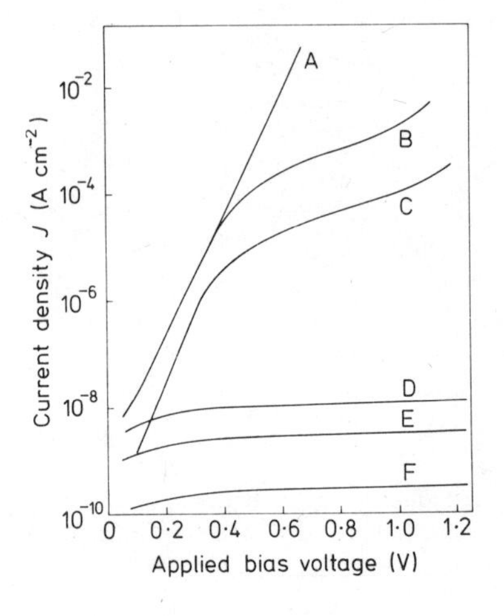

Figure 14. Experimental dependence of current density $J \simeq J_n$ on applied bias voltage for Al–SiO_2–pSi minority-carrier structures (after Shewchun *et al* 1974). $N_a = 7 \times 10^{15}\ cm^{-3}$. Curves A, B, C, negative voltage on metal; curves D, E, F, positive voltage on metal; curves A, E, d = 17 Å; curves B, D, d = 22 Å; curves C, F, d = 24 Å.

strongly inverted surfaces, since then n(0) is large. It is found experimentally that for the Al–SiO_2–pSi system, an SiO_2 layer of thickness d less than 15 or 20 Å will be diffusion-limited over the normal range of bias voltages ($V \lesssim 0{\cdot}6$ V) and current densities ($J \lesssim 10^{-1}$ A cm^{-2}) and will therefore behave very similarly to an n$^+$p junction (Green and Shewchun 1974, Green *et al* 1974).

One method of distinguishing between majority-carrier and minority-carrier behaviour (without using transistor structures as in the previous section) is to look for a dependence of the current density on the doping concentration of the semiconductor. This will be present for minority-carrier devices (equation 24) but will not be found for majority-carrier devices, where instead a dependence on barrier height ϕ_b is expected (equation 22).

Experimental current–voltage characteristics for Al–SiO_2–pSi structures with various SiO_2 thicknesses are shown in figure 14 (Shewchun *et al* 1974). The current at low voltages is described approximately by equation (24), although some excess current is present, associated with recombination in the space-charge region (this excess current is also expected for diffused n$^+$p junctions). Note that for $d = 22$ Å, the current above $V \simeq 0{\cdot}5$ V becomes tunnelling-limited and the characteristic departs from that of an n$^+$p junction. For $V \gtrsim 0{\cdot}5$ V and $d = 22$ Å, the tunnel current J_n of equation (6) cannot supply the number of electrons required by the diffusion current (equation 24), without a substantial drop in the quasi-Fermi level $E_{Fm} - E_{Fn}$ across the oxide layer. A portion of the applied voltage $(E_{Fm} - E_{Fn})/q$ is lost to the oxide layer and only the remaining voltage $V - (E_{Fm} - E_{Fn})/q$ is available to drive the diffusion current. This corresponds to the condition shown in figure 13(*c*). The value of $E_{Fm} - E_{Fn}$ for a given value of V is determined by the condition that equations (6) and (24) are equal in the steady state. The characteristics for a positive voltage on the metal (reverse-biased n$^+$p junction) are also shown in figure 14 and the corresponding energy diagram in figure 13(*d*).

7. Statistics of tunnelling electrons

Under equilibrium conditions (in the absence of applied voltage or optical illumination), the population of the electron Bloch states in the metal and in the conduction and valence bands of the semiconductor, and of the localised states at the Si–SiO_2 interface is governed by Fermi–Dirac statistics with a common Fermi energy E_F. When a bias voltage is applied, or when the structure is exposed to optical radiation, the equilibrium is upset and we must describe the statistics of the electrons in the metal, in the conduction band of the semiconductor, in the valence band of the semiconductor and in the interface states separately. Following the example of Shockley (1950), we associate a separate quasi-Fermi level with each of these groups of states and assume that under the non-equilibrium conditions, the electrons in each of the four groups are nevertheless in equilibrium with themselves (although not with those in the other groups). Actually this is a considerable simplification of a picture which is in reality much more complicated. Consider the electrons in the metal, for example, when a large positive voltage is applied to the metal (figure 15*a*). Electrons tunnel into the metal from the conduction and valence bands of the semiconductor and from filled interface states. Near the surface of the metal, the electrons supplied by the steady-state tunnel currents will enrich the population of the higher-energy states. Deeper into the metal ($\gtrsim 100$ Å), the electron distribution returns to the normal Fermi–Dirac distribution since the hot electrons have

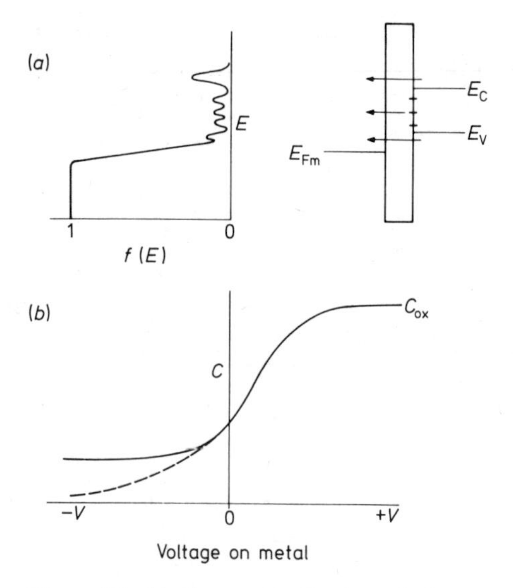

Figure 15. (*a*) Energy distribution of electrons in metal for conditions of pronounced tunnelling from the semiconductor bands and from interface states. (*b*) Schematic dependence of capacitance on voltage for tunnelling MIS structures on n-type semiconductors: minority carriers in equilibrium with majority carriers (——) or with electrons in the metal (– – –).

lost their excess energy by collisions and have come into equilibrium with the majority. Provided the number of hot electrons is small ($f(E) \ll 1$ for these high-energy states), we are justified in retaining E_{Fm} as the metal Fermi level even under non-equilibrium conditions such as those shown in figure 15(*a*).

It is found that under most conditions the majority carriers at the semiconductor surface may be assumed to be in approximate equilibrium with those in the neutral region (the bulk) of the semiconductor. This is because the large concentration of majority carriers in the neutral region can readily supply (or remove) electrons to the surface by diffusion. They are not exactly in equilibrium of course, or there would be no majority-carrier current, but for the purposes of determining the number of occupied states in the majority-carrier band at the surface this departure from equilibrium may be ignored. This means that for an n-type semiconductor, the population of the conduction-band states can be determined by extrapolating the bulk Fermi level from the neutral region up to the surface (horizontally on an energy diagram), even in the presence of an applied voltage.

The population of the minority-carrier band at the semiconductor surface is determined by equating all of the currents into these states in the steady state for the appropriate conditions of bias voltage and/or illumination. Let us consider again an n-type semiconductor and initially assume dark conditions. In this case, for a positive voltage on the metal, the tunnelling current to the minority-carrier (valence) band of the semiconductor is given by equation (7). The diffusion current of holes into the semiconductor may be written as (Card and Rhoderick 1973).

$$J_p = (qD_p/L_p)\, p(0) \exp(-qV_d/KT) \tag{25}$$

where V_d is the surface potential which is determined by the applied bias voltage. By equating expressions (7) and (25) we can find E_{Fp}, the quasi-Fermi level for holes at the semiconductor surface. This in turn defines $p(0) = N_V \exp(-E_{Fp}/KT)$ where E_{Fp} is measured from E_V, the valence-band edge at the surface. Knowing p(0) we can determine the current using equations (7) or (25).

For a very thin oxide layer, tunnelling is relatively easy; the term $\exp(-2\int K\,dX)$ in equation (7) is large and when this is equated with expression (25), we find $E_{Fp} - E_{Fm} \ll KT$. This means that p(0) is determined by the Fermi energy E_{Fm} in the metal; the holes at the semiconductor surface (i.e. the electrons of the valence band) are in equilibrium with the electrons in the metal. Thus the MIS tunnelling structure behaves as a Schottky barrier with the minority carriers diffusion-limited and equation (25) reduces to equation (24). When the oxide layer becomes sufficiently thick for tunnelling to limit the hole current rather than diffusion, then $E_{Fp} - E_{Fm}$ becomes large and the MIS structure resembles the normal thick oxide device with no tunnelling, insofar as the statistics of the minority carriers are concerned.

It has been found experimentally that for the Au–SiO_2–nSi structure, the minority carriers (holes) are in equilibrium with the metal over the bias range of interest (positive voltage on the metal) for $d \lesssim 15$–20 Å. For the opposite bias polarity (negative voltage on the metal), the tunnelling current equation (7) must balance only the small generation current density:

$$J_p = qD_p n_i^2 / L_p N_d. \tag{26}$$

In practice, the holes are in approximate equilibrium with the metal for this polarity for SiO_2 layers of thickness $d \lesssim 30$–35 Å in the Au–SiO_2–nSi system (Card and Rhoderick 1971b).

Under optical illumination, the generation current equation (26) is substantially increased by the photogeneration of electron–hole pairs (assuming that $h\nu > E_g$) in the semiconductor. If we assume this photogeneration J_{ph} occurs within a minority-carrier diffusion length of the semiconductor surface, the net current of holes to the surface is given by (Ng and Card 1979)

$$J_p = J_{ph} - (qD_p/L_p)\, p(0) \exp(-qV_d/KT) \tag{27}$$

which must be equal to the tunnel current equation (7) in the steady state. p(0) is again determined by this balance. The holes at the semiconductor surface will be in equilibrium with the metal for very thin oxides, but then the critical thickness is determined by the intensity of the illumination, which determines J_{ph}, as well as by the bias voltage.

It is possible to recognise the dependence of the surface concentration of minority carriers p(0) on the applied bias voltage by examining the capacitance–voltage characteristics. This provides insight into the rate-limiting mechanism, diffusion or tunnelling, for the minority-carrier (hole) current. In figure 15(*b*) we show schematically two possibilities for the dependence of C on V with a negative voltage on the metal (n-type semiconductor). Actual experimental data of this nature may be found in the papers by Card and Rhoderick (1971*b*), Clarke and Shewchun (1971) and Kar and Dahlke (1972*a*). For a very thin oxide, the holes are in equilibrium with the metal so that on the application of a (reverse) negative bias on the metal p(0) remains constant, determined by E_{Fm}. Since the charge at the semiconductor surface associated with these holes does not increase with the negative bias, very little voltage is developed across the oxide layer. Instead, the entire voltage appears across the space-charge region of the semiconductor, which expands with the reverse bias in the manner of a Schottky barrier so that the capacitance exhibits the deep-depletion characteristic shown by the broken curve in figure 15(*b*).

For a sufficiently thick oxide layer, on the other hand, the hole current is limited by tunnelling and the thermal generation in the semiconductor (equation 26) is able to supply the small number of holes required. In this case, the hole concentration at the semiconductor surface is in equilibrium with the electron (majority-carrier) concentration as in a thick-oxide MOS structure. The tunnel current is small enough not to disturb the statistics of the carriers in the semiconductor appreciably and the usual MOS characteristic (full curve in figure 15*b*) is obtained. As we have stated, the Schottky barrier to MOS transition can be observed experimentally in the Au–SiO_2–nSi system by increasing the SiO_2 thickness. The transition occurs over a fairly small range of d, from ~30 to 35 Å, as a consequence of the exponential dependence of the tunnelling probability on d.

The Schottky barrier to MOS transition can also be observed for a fixed oxide thickness by varying the illumination level or the temperature. For example, with an SiO_2 thickness of ~25 Å in a Au–SiO_2–nSi structure, the broken curve in figure 15(*b*) is obtained in the dark. As the intensity of optical illumination is increased, the photogeneration term J_{ph} in equation (27) increases, the hole current then eventually becomes tunnel-limited and the C–V curve shifts up to the full curve of figure 15(*b*). This transition may also be observed by increasing the temperature in the dark, in which case thermal generation is enhanced because of the term n_i^2 in equation (26), where $n_i^2 \propto \exp(-E_g/KT)$, so that the minority-carrier current becomes tunnel-limited at sufficiently high temperatures (Green and Shewchun 1975, Shewchun and Green 1975).

As for the statistics of electrons in interface states, this has been discussed in detail in a recent review (Card 1979). In the interest of brevity, we merely point out here that the quasi-Fermi level of midgap interface states undergoes a transition from the metal Fermi level to that of the majority carriers in the semiconductor in the range of oxide thicknesses 15 Å $\lesssim d \lesssim$ 25 Å as determined experimentally for the Au–SiO_2–nSi system (Card and Rhoderick 1972). The transition occurs at lower values of d for larger forward bias (Card 1975, Walker 1974). It is of interest to note that the density of interface states in tunnelling MIS junctions is found to increase for a given metal with decreasing whickness of SiO_2 in the range 10 Å $\lesssim d \lesssim$ 50 Å (data of Kar and Dahlke, 1971, 1972a,b, Kumar and Dahlke 1977a,b, Ma and Barker 1974a,b, Hunter *et al* 1970, Kar 1978, Barret and Vapaille 1975, Deneuville 1974).

8. Applications

The MIS tunnelling structure is of academic importance for the understanding of transport mechanisms and interface states and has also been able to account for the non-ideal behaviour of Schottky barriers (Card and Rhoderick 1971a, Card 1976). In the course of these studies, however, several new device applications have arisen based upon the ability to control the tunnelling currents by means of the oxide thickness. These applications are summarised in table 1.

We have already discussed the dependence of the minority-carrier injection ratio J_p/J_n for MIS tunnelling junctions. The dependence of J_p/J_n on d for Au–SiO_2–nSi

Table 1. Applications of MIS tunnelling.

Application	References
Enhanced minority-carrier injection and injection electroluminescence	Jaklevic *et al* (1963), Fischer and Moss (1963), O'Sullivan and Malarkey (1965), Card and Smith (1971), Card and Rhoderick (1973), Haeri and Rhoderick (1974), Walker and Pratt (1976), Nicollian *et al* (1976), Clark *et al* (1976), Lagerstedt *et al* (1978), Livingstone *et al* (1973), Pankove (1975)
Controlled-inversion layer devices	
(*a*) Negative resistance and switching in p–n–I–m structure	Yamamoto and Morimoto (1972), Yamamoto *et al* (1974, 1976). Kroger and Wegener (1978a, b) Habib and Simmons (1979) Buxo *et al* (1978) Chik and Simmons (1979)
(*b*) Surface oxide transistor using majority-carrier MIS structures as emitter and collector	Shewchun and Clarke (1973)
(*c*) Low-voltage photo-multiplication	Green and Shewchun (1974), Green *et al* (1975)
Cold-cathode emitters	Martin (1975)
Radiation detectors	Shewchun *et al* (1974), Ma and Barker (1974a, b), Share *et al* (1974)
Hydrogen sensors	Shivaraman *et al* (1976), Keramati and Zemel (1978)
MIS solar cells (*a*) Majority-carrier cells	Reviews: Hovel (1975), Pulfrey (1978), Wilson *et al* (1978a, b); see also Stirn and Yeh (1975, 1977), Charlson and Lien (1975), Anderson *et al* (1974, 1977), Lillington and Townsend (1976), Fonash (1975, 1977), Ponpon and Siffert (1978), Card and Yang (1976), Fabre (1976), Olsen (1977), Kipperman and Omar (1976), Kar (1977, 1978), Card (1977), Ghosh *et al* (1978), Van Halen *et al* (1978), Wilson *et al* (1978a, b), Mizrah and Adler (1977), DuBow *et al* (1976), Viktorovitch and Kamarinos (1977), ng and Card (1980), Srivastava *et al* (1979), Landsberg and Klimpke (1977)
(*b*) Minority-carrier cells	Shewchun *et al* (1974, 1977), Green *et al* (1974), Singh and Shewchun (1976), Green and Godfrey (1976), Tarr and Pulfrey (1979), Van Halen *et al* (1978), Shewchun *et al* (1978), Pulfrey (1977), Ng and Card (1980)

structures was shown in figure 13 (Card and Rhoderick 1973), where it was found that the oxide layer increases the degree of minority-carrier injection compared with a Schottky barrier and there is an optimum value of d (~30 Å for this system). This enhancement of the minority-carrier injection has been used to obtain injection electroluminescence in MIS structures on CdS (Jaklevic *et al* 1963, Fisher and Moss 1963, O'Sullivan and Malarkey 1965, Clark *et al* 1976), on ZnS and ZnSe (Livingstone *et al* 1973, Walker and Pratt 1976), on GaP and GaAs (Card and Smith 1971, Haeri and Rhoderick 1974, Nicollian *et al* 1976) and on GaN (Pankove 1975, Lagerstedt *et al* 1978).

Another class of devices is based upon the control of the minority-carrier concentration at the semiconductor surface in reverse bias (negative voltage on the metal for an MIS structure on an n-type semiconductor). These are referred to as controlled-inversion-layer devices and the basic principle is illustrated in figure 16. A voltage V applied to the structure divides between the oxide layer V_{ox} and the space-charge region of the semiconductor V_d. For $d \lesssim 30$ Å, the minority carriers (holes) tunnel into the metal as rapidly as they are supplied by thermal generation in the semiconductor with the result that no inversion layer forms in reverse bias. Virtually all of the applied voltage in this case appears across the semiconductor: $V_d \simeq V$ and $V_{ox} \simeq 0$. If, on the other hand, holes are supplied in abundance to the surface by photogeneration under optical illumination (assuming an optically transparent metal electrode), an inversion layer does form, and a redistribution of voltages occurs; V_{ox} becomes large at the expense of V_d. The tunnel current, which we assume is predominantly the majority-carrier current J_n, increases dramatically as a result of the increase in V_{ox} (which raises E_{Fm} up to the conduction band edge E_C of the semiconductor). The result is that the increase in J_n is much larger than the increase in J_p because of photogeneration and a photomultiplication takes place. It has been found experimentally that a gain of 10^2 to 10^3 may be obtained for Al–SiO_2–nSi majority-carrier photomultipliers with $d \simeq 20$–25 Å (Green *et al* 1975). A disadvantage of this device is that the response time becomes very long for the detection of weak optical signals.

The same principle is used to obtain transistor action by injecting the required holes from another MIS junction located within a diffusion length of the MIS collector. Current gains of the order of 10 or 20 could be obtained with Al–SiO_2–nSi majority-carrier emitters and collectors, as a result of the multiplication at the collector, which overcompensates for the low minority-carrier injection ratio at the emitter junction. Another example of a controlled-inversion layer device is the p–n–I–m geometry shown in figure 17. For low reverse voltages (metal negative), the absence of an inversion layer causes deep depletion and a small and saturating reverse current (*a*). When the depletion region reaches through to the p-layer, however, the p–n junction injects holes into the n-region, an inversion layer is formed and the voltage across the oxide increases dramatically. The current J_n becomes large as in (*b*) and we have regenerative behaviour. Such a structure can be operated as either a two- or three-terminal device, depending on whether the n-region is contacted. This was first observed by Yamamoto and Morimoto (1972); later detailed studies by other authors are given in table 1.

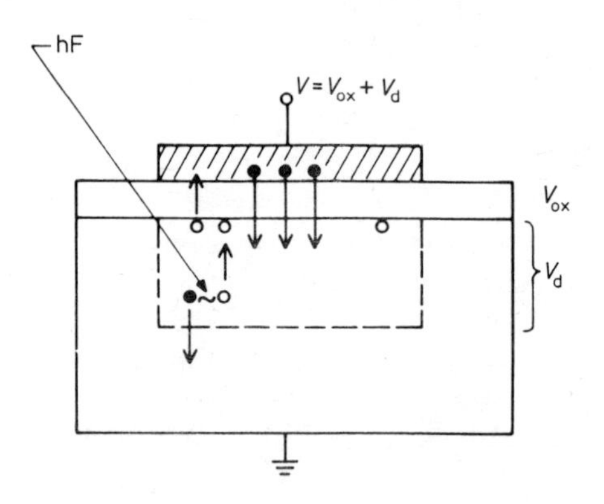

Figure 16. Control of inversion-layer charge and hence voltage distribution between the oxide and the space-charge region by photogeneration of minority carriers in tunnelling MIS structures. • electrons; ○ holes.

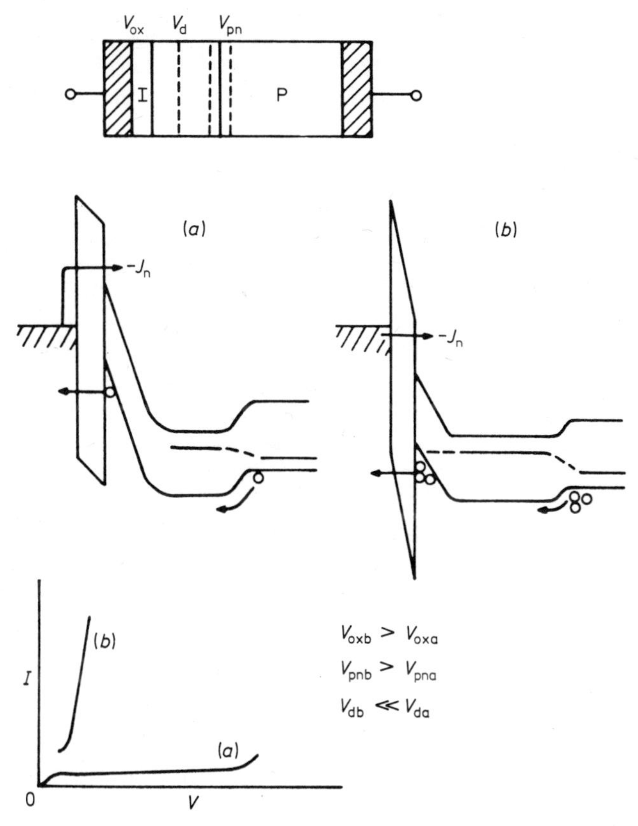

Figure 17. Operating principles underlying negative resistance and switching phenomena in p–n–I–m tunnelling structures. Holes injected from a p–n junction cause an inversion layer, producing a large oxide voltage which controls the majority-carrier current $J_{\mathbf{n}}$.

Potentially the most important applications of MIS tunnelling are in photovoltaic energy conversion. It has been shown that the presence of the insulating layer can improve the conversion efficiency of a Schottky barrier solar cell. This was first demonstrated for GaAs by Stirn and Yeh (1975), for silicon by Shewchun *et al* (1974) for minority-carrier MIS structures, and by Charlson and Lien (1975) and Lillington and Townsend (1976) for majority-carrier MIS structures. An abundance of experimental and theoretical work has confirmed these findings (table 1). It is consistently observed that the open-circuit voltage of the MIS cell is improved over that of the ideal Schottky barrier and in the best cases approaches that of a p–n junction. Typical data from the earlier work referred to above are shown in figure 18. This may be understood in terms of the suppression of the dark current by the insulating layer. The problem is to understand the experimental observation that the photocurrent is the same as that of a Schottky barrier for $d \lesssim 20$ Å; unlike the dark current this is not suppressed by the insulating layer.

There are two explanations for these observations; one is correct if the dark current of the MIS tunnelling structure is dominated by transitions to the minority-carrier band (minority-carrier cells); the other is correct for majority-carrier cells.

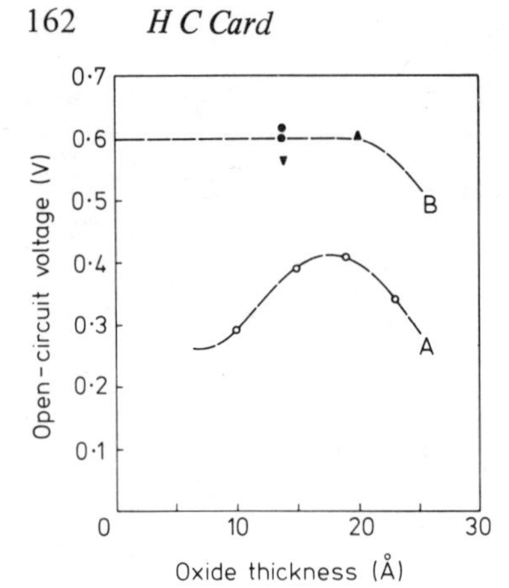

Figure 18. Dependence of open-circuit voltage on SiO_2 thickness d for majority carrier (Au–SiO_2–nSi, $\phi_b = 0{\cdot}75$ eV, curve A) and minority-carrier (Al–SiO_2–pSi, curve B) MIS solar cells. Short-circuit current (not shown) is independent of d for $d \lesssim 20$ Å (after Singh and Shewchun 1976, Lillington and Townsend 1976).

For minority-carrier MIS solar cells such as the Al–SiO_2–pSi structure (Singh and Shewchun 1976, Green and Godfrey 1976, Green *et al* 1976) the dark current is described by equation (1) and behaviour identical to that of a p–n junction is observed. The open-circuit voltage is obtained by equating formula (24) to the short-circuit current density J_{sc} for $V = V_{oc}$ and this gives

$$V_{oc} \simeq (KT/q) \ln (J_{sc} L_n N_a / q D_n n_i^2) \tag{28}$$

for a p-type substrate. For majority-carrier cells, the dark current is dominated by tunnelling transitions between the metal and the majority-carrier band. For an n-type substrate as in the Au–SiO_2–nSi majority-carrier solar cell (Card 1977), the open-circuit voltage is obtained by equating J_n from expression (22) with the short-circuit current J_{sc} for $V = V_{oc}$, and we then have

$$V_{oc} \simeq (nKT/q) [\ln (J_{sc}/AT^2) + (q\phi_b/KT) + \chi^{1/2} d]. \tag{29}$$

The open-circuit voltage has been increased compared with that of a Schottky barrier because of the increases in n and $\chi^{1/2}d$ in equation (29). For $d \lesssim 20$ Å, suppression of the photocurrent, i.e. of the photogenerated holes tunnelling into the metal (equations 7 and 23), does not occur. J_{sc} is independent of d over this range of SiO_2 thicknesses for the Au–SiO_2–nSi structure and the reason is that the surface concentration of holes p(0) is sufficiently large for the current not to be limited by tunnelling for the values of J_{sc} encountered under solar illumination (~30 mA cm^{-2}). For stronger illumination levels, or for MIS majority-carrier cells with a smaller ϕ_b and hence a smaller p(0), the photocurrent will be suppressed by SiO_2 layers thinner than 20 Å (Card 1977).

9. Summary

First-order agreement may be obtained between the experimental characteristics of tunnelling MIS structures and a theoretical description of the carrier transport based upon the WKB approximation and a Franz dispersion relation in the insulator. Qualitatively different behaviour is observed for structures in which the current is dominated by transitions to either the majority- or the minority-carrier band of the semiconductor. The experimental tunnelling probability for electrons from the conduction band is appreciably higher than for electrons from the valence band for tunnelling through ultrathin SiO_2 layers. The population of the minority-carrier band at the semiconductor surface for a given bias voltage may be controlled by the thickness of the insulating layer, or for a given thickness by optical illumination, by temperature, or by injection from a nearby MIS structure or p–n junction. These characteristics of tunnelling MIS structures provide for applications in optoelectronics, for example as quantum detectors with internal multiplication or as light-emitting diodes, as photovoltaic solar cells and as switching devices similar in behaviour to pnpn structures.

Acknowledgments

The author is grateful for the comments and suggestions of his students George Ng, Paul Panayotatos, Eric Chan, Tommy Poon and of his colleagues Edward Yang, Wei Hwang, Malvin Teich and Yannis Tsividis, who contributed either directly or indirectly to this paper. The financial support of the Joint Services Electronics Program under Contracts DAAG29-77-C-0019 and DAAG29-79-C-0079 and of the US Department of Energy under Contract DOE ET 78R 031876 is also very much appreciated.

References

Anderson W A, Delahoy A T and Milano R A 1974 *J. Appl. Phys.* **45** 3914–15
Anderson W A, Kim J K and Delahoy A E 1977 *IEEE Trans. Electron. Devices* **ED-24** 453–7
Archer R J 1962 *J. Opt. Soc. Am.* **52** 970–7
Barret C and Vapaille A 1975 *Solid St. Electron.* **18** 25
Buxo J, Owen A E, Sarrabayrouse G and Sebaa J-P 1978 *Rev. Phys. Appl.* **13** 767–70
Card H C 1975 *Solid St. Electron.* **18** 881–3
—— 1976 *IEEE Trans. Electron. Devices* **ED-23** 538–44
—— 1977 *Solid St. Electron.* **20** 971–6
—— 1979 *Solid St. Electron.* in press
Card H C and Rhoderick E H 1971a *J. Phys. D: Appl. Phys.* **4** 1589–601
—— 1971b *J. Phys. D: Appl. Phys.* **4** 1602–11
—— 1972 *Solid St. Electron.* **15** 993–8
—— 1973 *Solid St. Electron.* **16** 365–74
Card H C and Smith B L 1971 *J. Appl. Phys.* **42** 5863–5
Card H C and Yang E S 1976 *Appl. Phys. Lett.* **29** 51–3
Charlson E J and Lien J C 1975 *J. Appl. Phys.* **46** 3982–7
Chik K C and Simmons J G 1979 *Solid St. Electron.* **22** 589–94
Clark M D, Baidyaroy S, Ryan F and Ballantyne J M 1976 *Appl. Phys. Lett.* **28** 36–8
Clarke R A and Shewchun J 1971 *Solid St. Electron.* **14** 957–73
Dahlke W E and Sze S M 1967 *Solid St. Electron.* **10** 865–73
Deneuville A 1974 *J. Appl. Phys.* **45** 3079
DuBow J B, Burk D E and Sites J R 1976 *Appl. Phys. Lett.* **29** 494–6
Fabre E 1976 *Appl. Phys. Lett.* **29** 607–10
Fischer A G and Moss H J 1963 *J. Appl. Phys.* **34** 2112
Fonash S J 1975 *J. Appl. Phys.* **46** 1286
—— 1977 *J. Appl. Phys.* **48** 3953–8
Franz W 1956 *Handbuch der Physik* (Berlin: Springer)
Freeman L B and Dahlke W E 1970 *Solid St. Electron.* **13** 1483–503
Ghosh A K, Fishman C and Feng T 1978 *J. Appl. Phys.* **49** 3490–8
Gray P V 1965 *Phys. Rev.* **140** A179–86
Green M A and Godfrey R B 1976 *Appl. Phys. Lett.* **29** 610–12
Green M A, Godfrey R B and Davies L W 1976 *12th IEEE Photovoltaic Specialists Conf.* p896

Green M A, King F D and Shewchun J 1974 *Solid St. Electron.* **17** 551–61
Green M A and Shewchun J 1974 *Solid St. Electron.* **17** 349–65
Green M A and Shewchun J 1975 *J. Appl. Phys.* **46** 5185–90
Green M A, Temple V A K and Shewchun J 1975 *Solid St. Electron.* **18** 745–52
Goodman A M and Breece J 1970 *J. Electrochem. Soc.* **117** 982–4
Harrison W A 1961 *Phys. Rev.* **123** 85–9
Habib S E D and Simmons J G 1979 *Solid St. Electron.* **22** 181–92
Haeri S Y and Rhoderick E H 1974 *Metal-Semiconductor Contacts* (Inst. Phys. Conf. Ser. 22) pp84–90
Hall R N 1952 *Phys. Rev.* **87** 837
Hovel H J 1975 *Solar Cells* (New York: Academic)
Hunter W R, Eaton D H and Sah C T 1970 *Appl. Phys. Lett.* **17** 211–13
Jaklevic R C, Donald D K, Lambe J and Vassell W C 1963 *Appl. Phys. Lett.* **2** 7
Kar S 1974 *Appl. Phys. Lett.* **25** 587–9
—— 1977 *IEEE–IEDM Digest Tech. Papers* 56A–56H
—— 1978 *J. Appl. Phys.* **49** 5278–83
Kar S and Dahlke W E 1971 *Appl. Phys. Lett.* **18** 401–3
—— 1972a *Solid St. Electron.* **15** 221–37
—— 1972b *Solid St. Electron.* **15** 869–75
Keramati B and Zemel J N 1978 *Physics of SiO_2 and its Interfaces* ed S T Pantelides (New York: Pergamon) pp459–63
Kipperman A H M and Omar M H 1976 *Appl. Phys. Lett.* **28** 620–1
Kroger H and Wegener H A R 1978a *Solid St. Electron.* **21** 643–54
—— 1978b *Solid St. Electron.* **21** 655–61
Kumar V and Dahlke W E 1977a *Solid St. Electron.* **20** 143–52
—— 1977b *IEEE Trans. Electron Devices* **ED-24** 146–53
Lagerstedt O, Monemar B and Gislason H 1978 *J. Appl. Phys.* **49** 2953–7
Landsberg P T and Klimpke C 1977 *Proc. R. Soc.* **A354** 101–18
Lewicki G and Maserjian J 1975 *J. Appl. Phys.* **46** 3032–8
Lillington D R and Townsend W G 1976 *Appl. Phys. Lett.* **28** 97–9
Livingstone A W, Turvey K and Allen J W 1973 *Solid St. Electron.* **16** 351
Lundström I and Svensson C 1972 *J. Appl. Phys.* **43** 5045–7
Ma T P and Barker R C 1974a *Solid St. Electron.* **17** 913–29
—— 1974b *J. Appl. Phys.* **45** 317–21
Martin P 1975 *MSc Thesis* University of Manchester
Maserjian J 1974 *J. Vac. Sci. Technol.* **11** 996–1003
Maserjian J and Petersson G P 1974 *Appl. Phys. Lett.* **25** 50–2
Mizrah T and Adler D 1977 *IEEE Trans. Electron Devices* **ED-24** 458–62
Nicollian E H, Schwartz B, Coleman D J Jr, Ryder R M and Brews J R 1976 *J. Vac. Sci. Technol.* **13** 1047–55
Ng K K and Card H C 1977 *IEEE–IEDM Digest Tech. Papers* 57–61
—— 1979 *Solid St. Commun.* in press
—— 1980 *IEEE Trans. Electron. Devices* in press
Olsen L C 1977 *Solid St. Electron.* **20** 741–51
O'Sullivan D D and Malarkey E C 1965 *Appl. Phys. Lett.* **6** 5
Pankove J I 1975 *IEEE Trans. Electron Devices* **ED-22** 721
Petersson G P, Svensson C M and Maserjian J 1975 *Solid St. Electron.* **18** 449–51
Ponpon J P and Siffert P 1978 *13th IEEE Photovoltaic Specialists Conf.* 639–44
Pulfrey D L 1978 *IEEE Trans. Electron Devices* **ED-25** 1308–16
—— 1977 *Solid St. Electron.* **20** 455–7
Share S, Epstein A S, Kumar V, Dahlke W E and Haller W 1974 *J. Appl. Phys.* **45** 4894–8
Shewchun J and Clarke R A 1973 *Solid St. Electron.* **16** 213–19
Shewchun J and Green M A 1975 *J. Appl. Phys.* **46** 5179–84
Shewchun J, Green M A and King F D 1974 *Solid St. Electron.* **17** 563–72
Shewchun J, Singh R, Burk D, Spitzer M, Loferski J and DuBow J 1978 *13th IEEE Photovoltaic Specialists Conf.* pp528–35

Shewchun J, Singh R and Green M A 1977 *J. Appl. Phys.* **48** 765–70
Shewchun J, Waxman A and Warfield G 1967 *Solid St. Electron.* **10** 1165–86
Shivaraman M S, Lundström I, Svensson C M and Hammarsten H 1976 *Electron. Lett.* **12** 483–4
Shockley W 1950 *Electrons and Holes in Semiconductors* (New York: Van Nostrand)
Shockley W and Read W T 1952 *Phys. Rev.* **87** 835
Singh R and Shewchun J 1976 *Appl. Phys. Lett.* **28** 512–13
Srivastava G P, Bhatnager P K and Dhariwal S R 1979 *Solid St. Electron.* **22** 581–7
Stirn R J and Yeh Y C M 1975 *Appl. Phys. Lett.* **27** 95–8
—— 1977 *IEEE Trans. Electron Devices* **ED-24** 476–83
Tarr N G and Pulfrey D L 1979 *Appl. Phys. Lett.* **34** 295–7
Van Halen P, Mertens R P, Van Overstraeten R J, Thomas R E and Van Meerbergen J 1978 *IEEE Trans. Electron Devices* **ED-25** 507–11
Viktorovitch P and Kamarinos G 1977 *J. Appl. Phys.* **48** 3060–4
Walker L G 1974 *Solid St. Electron.* **17** 763–5
Walker L G and Pratt G W Jr 1976 *J. Appl. Phys.* **47** 2129–33
Waxman A, Shewchun J and Warfield G 1967 *Solid St. Electron.* **10** 1187–98
Williams R 1977 *J. Vac. Sci. Technol.* **14** 1106–11
Wilson J I B, McGill J and Robinson P 1978a *13th IEEE Photovoltaic Specialists Conf.* pp755–60
Wilson J I B, McGill J and Weaire D 1978b *Adv. Phys.* **27** 365–85
Yamamoto T, Kawamura K and Shimizu H 1976 *Solid St. Electron.* **19** 701–6
Yamamoto T and Morimoto M 1972 *Appl. Phys. Lett.* **20** 269–70
Yamamoto T and Shimizu H 1974 *Proc. 5th Coll. on Microwave Communications, Budapest* pp363–72

Self-consistent theoretical study of injection and extraction effects in a MIS tunnel structure

G Kamarinos, G Pananakakis and P Viktorovitch
Laboratoire 'Physique des Composants à Semiconducteurs', ERA CNRS No 659, ENSER, 23 Avenue des Martyrs, 38031 Grenoble Cedex, France

Abstract. A self-consistent theoretical analysis of the working of a MIS tunnel device is presented. Attention is focused on the influence of the intermediate insulating layer and the action of interface states. Electrostatic and kinetic action (recombination and tunnelling) of interface states are analysed and typical examples involving metal–silicon oxide–n-type silicon devices are given.

1. Introduction

It is well known that injection phenomena are very important for the working of MIS tunnel devices (Sze 1969, Pananakakis *et al* 1977a; for a discussion of the recovery time of HF Schottky diodes and the efficiency of solar cells, see Pananakakis *et al* 1977b). Intimate metal–semiconductor contacts have already been analysed (Scharfetter 1965, Yu and Snow 1969) and the complex behaviour that arises when an interfacial layer is present was soon noticed (Card and Rhoderick 1973). Indeed, in the presence of an insulating layer, injection and extraction effects are complicated and only after complete numerical analysis can the phenomena that are involved be understood (Green *et al* 1974, Green and Shewchun 1973).

We have investigated a self-consistent model that describes the working of a MIS diode and takes into account the interfacial parameters of the structure: the insulator thickness δ, the barrier heights of the insulating layer for semiconductor electrons and holes, the semiconductor doping, the difference between the workfunctions of the metal and the semiconductor ϕ_{ms} and finally, the density and the nature of interface states that are supposed to interact with the three carrier reservoirs: the valence and the conduction bands of the semiconductor and the metal. In this way we can examine the kinetic and electrostatic action of the interface states.

Assuming the validity of thermionic theory (Rhoderick 1978) we can establish and solve (numerically) the exact equations for the transport of carriers across the MIS structure.

The voltage drop across the insulator is given by

$$U_s = -(\delta/\epsilon_i)(Q_{sc} + Q_{ss})$$

where Q_{sc} and Q_{ss} are the complete expressions for the charges of the semiconductor and interface states:

0305-2346/80/0050-0166$01.00

$$Q_{sc} = -(\text{sgn}\, V_s)(2\epsilon_s kT)^{1/2}\left\{n\left[\exp\left(\frac{qV_s}{kT}\right) - 1\right] + p\left[\exp\left(-\frac{qV_s}{kT}\right) - 1\right] - (N_D - N_A)\frac{q}{kT}V_s\right\}^{1/2}.$$

The form of Q_{ss} depends on the nature of the surface states.

(i) For uniformly distributed traps of density D_{ss} ($m^{-2}\,eV^{-1}$)

$$Q_{ss} = -qD_{ss}(E_{Fj} - \phi_0)$$

where E_{Fj} is the quasi-Fermi level that controls the traps and ϕ_0 is the neutral level.

(ii) For localised traps of density N_t (m^{-2})

$$Q_{ss} = -qN_t f \qquad \text{(acceptors)}$$

$$Q_{ss} = qN_t(1-f) \qquad \text{(donors)}$$

where f is the occupancy of the states.

Furthermore, the continuity of electron and hole fluxes between the metal and semiconductor neutral bulk boundary is written as

$$J_{nc} + J_{nt} - J_R = J_n$$

$$J_{pc} + J_{pt} - J_R = J_p$$

where J_{nc} and J_{pc} are the direct collection fluxes at the semiconductor–insulator interface, J_{nt} and J_{pt} are electron and hole fluxes from the metal to the semiconductor through the interface states, J_R is the recombination current and J_n and J_p are the electron and hole flows through the neutral bulk of the semiconductor.

Finally the external bias is given by:

$$U = \delta V_s + \delta U_s + V_{sc} \qquad \delta V_s = V_s - V_{s0},\ \delta U_s = U_s - U_{s0}$$

where V_{sc} is the voltage drop across the neutral region. Details of the equations given above and the symbols used in this paper have been published recently (Pananakakis *et al* 1979).

Our computations allow us to follow simultaneously variations in the following parameters with the external applied bias:

(i) The current across the diode.
(ii) The injection ratio $\gamma = J_p(J_p - J_n)$.
(iii) The excess carrier densities Δn and Δp (at the boundary of the neutral bulk semiconductor; $\Delta n = \Delta p$).
(iv) The position of the three quasi-Fermi levels with respect to the band-gap edges of the semiconductor.
(v) The distribution of the external applied voltage across the structure.
(vi) The occupation ratio of the interface states.

Our results are presented in two sections: in §2 we analyse the working of the ideal MIS tunnel diode and in §3 we examine the influence of the interface states. The results described are for metal–silicon oxide–n-type silicon structures.

2. Ideal structure (without interface states)

The injection ratio γ is a very important parameter for a MIS tunnel device and determines whether the diode is a majority- or a minority-carrier device. The injection ratio γ is given by

$$\gamma = J_{pc}/(J_{pc} - J_{nc}) = \frac{J_{pc}/J_{nc}}{(J_{pc}/J_{nc}) - 1} = \Gamma/(\Gamma + 1). \tag{1}$$

The variation of Γ will be studied.

We note that γ can be calculated either at the semiconductor surface or at the neutral bulk boundary if electron and hole fluxes are conserved: this is true only if interface states have no kinetic effect. When the semiconductor surface holes are in equilibrium with the electrons of the metal ($E_{Fn} \cong E_{Fp}$), Δn is given by

$$\Delta n = p_0 [\exp (qV/kT) - 1]. \tag{2}$$

Therefore Γ is given by

$$\Gamma \cong \frac{D_p/L_p}{V_{cn}} \frac{p_0}{n_0} \exp \left[q \left(\frac{\delta \overline{U_s}}{kT} \right) \right] \exp (\chi^{1/2} \delta) \tag{3}$$

where V_{cn} is the collection velocity of electrons at the insulator–semiconductor interface. From equation (3), we deduce that Γ, and consequently γ, are exponentially increasing functions of δ: very thin insulating layer diodes (i.e. $\delta \leqslant 20$ Å for the figures presented here) are majority-carrier devices whereas for the thicker insulating layers, the diodes are minority-carrier devices.

When semiconductor holes and metal electrons are not in equilibrium, Γ is given by an expression that is generally valid:

$$\Gamma = \frac{p_0}{n_0} \exp \left[-\left(\frac{2q}{kT} \right) V_{s_0} \right] \exp \left[\left(\frac{2q}{kT} \right) \delta U_s \right] \left(\frac{1 - [1 + (\Delta n/p_0)] \exp (-qV/kT)}{\exp (qV/kT) - 1} \right). \tag{4a}$$

If $[(\Delta n/p_0) + 1] \exp [-(qV/kT)] \ll 1$

$$\Gamma \simeq \frac{p_0}{n_0} \exp \left[-\left(\frac{2q}{kT} \right) V_{s_0} \right] \exp \left(\frac{q \delta U_s}{kT} \right) \exp \left(-\frac{q \delta V_s}{kT} \right). \tag{4b}$$

Equations (4*a*) and (4*b*) are easily deduced from the detailed expressions for J_{nc} and J_{pc} given by Pananakakis *et al* (1979). Equation (4*b*) shows that Γ is determined by the distribution of the external potential U between the surface barrier and the insulating layer. Therefore Γ can increase or decrease according to the relative values of δU_s and δV_s.

Figures 1 and 2 illustrate the above analysis for two characteristic thicknesses $\delta = 5$ and 35 Å; figure 3 shows the variation of γ for various values of the parameter δ for different values of ϕ_{MS} and doping N_D. From figure 3 it can be seen that for an intimate metal–semiconductor contact, γ increases because the drift component of the hole current increases and ceases to be negligible in comparison with its diffusion component. For MIS tunnel diodes and for insulating layer thicknesses of the order of $\delta \simeq 20$ Å, γ is not a monotonic function of U. For low voltages, when $E_{Fm} \simeq E_{Fp}$, $\Gamma \sim \exp (q\delta U_s/kT)$ and γ increases (equation 3). For larger values of U the metal electrons and the holes of

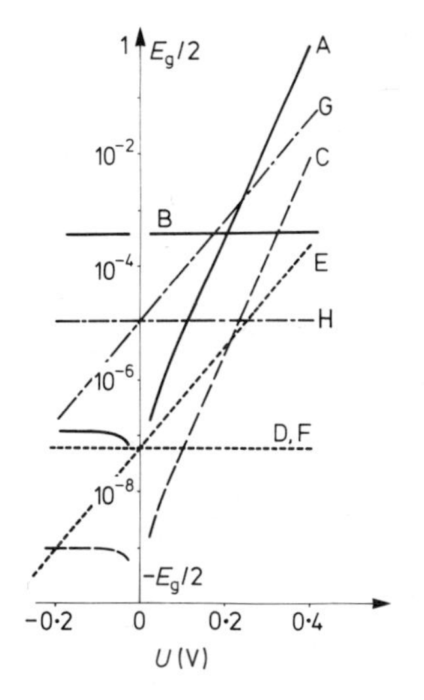

Figure 1. Variation with external bias U of the current density J ($10^2\,\mathrm{mA\,cm^{-2}}$) across the structure (absolute values, curve A), the injection ratio γ (curve B), the excess carrier densities Δn ($10^{19}\,\mathrm{m^{-3}}$) at the boundary of the neutral bulk semiconductor (absolute values, curve C), the E_{Fm} Fermi level with respect to the band-gap edges (eV, $E_g/2$ = 0·55 eV, curve D) and the E_{Fn} (curve E) and E_{Fp} (curve F) quasi-Fermi levels with respect to the band-gap edges (eV, $E_g/2 = 0{\cdot}55$), the differences $\delta V_s = V_s - V_{s_0}$ (V, curve G), $\delta U_s = U_s - U_{s_0}$ (V, curve H) for an ideal structure (n-type Si) with δ = 5 Å (intimate contact). For further details see Pananakakis *et al* (1979). $N_D = 10^{22}\,\mathrm{m^{-3}}$, $\phi_{MS} = 0{\cdot}6$ eV, $D_{ss} = 0$.

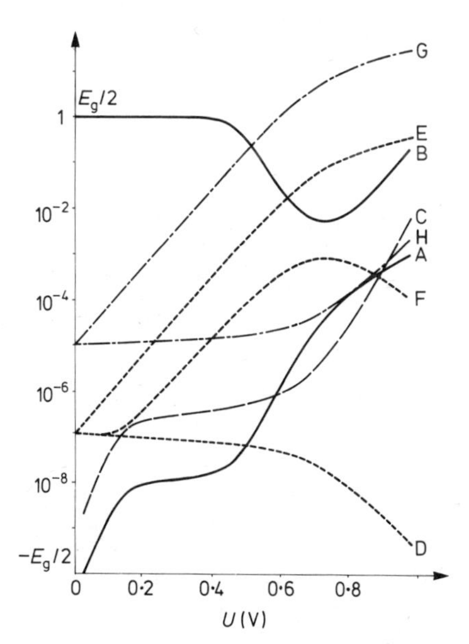

Figure 2. As figure 1 but for an ideal structure with δ = 35 Å.

the semiconductor surface are not in equilibrium; the term $\exp[-(q\delta V_s/kT)]$ that appears in equation (4*b*) dominates and consequently γ decreases towards a minimum value. After this γ increases again, and the term $\exp(q\delta U_s/kT)$ then governs the expression for Γ (see equation 4*b* and figure 3). For thicker insulating layers ($\delta > 20$ Å), γ reaches a plateau ($\gamma = 1$) before decreasing.

3. Influence of interface states

The action of interface states is both electrostatic, because of charge storage and kinetic, because a path is provided between the various carrier reservoirs.

3.1. *Interface states connected with only one reservoir*

The action of these states is only electrostatic. In other words, the capture cross sections for the carriers of the two other reservoirs are supposed to be negligible. Depending on the connection with one of the three free-carrier reservoirs, the interface states will be referred to as metal electron traps (MET), semiconductor free-electron traps (SET) or semiconductor hole traps (SHT).

According to Cowley and Sze (1965), the charge stored at the interface in each case is expressed by, respectively

$$Q_{ss} = -qD_{ss}(E_{Fm} - \phi_0) \quad (5a)$$

$$Q_{ss} = -qD_{ss}(E_{Fn} - \phi_0) \quad (5b)$$

$$Q_{ss} = -qD_{ss}(E_{Fp} - \phi_0) \quad (5c)$$

A MIS diode usually exhibits a carrier-injection effect ($\Delta n > 0$) when in forward bias and an extraction effect ($\Delta n < 0$) when reverse-biased. Nevertheless, in the presence of interface states, this behaviour can be dramatically modified; indeed for large densities ($D_{ss} \simeq 10^{18}\,m^{-2}\,eV^{-1}$) an extraction effect when in forward bias, or an injection effect when in reverse bias can be obtained.

These unusual effects, initially observed by Harrick (1959), can occur for MET or SET interface states as explained previously (Viktorovitch and Kamarinos 1974, 1975, Viktorovitch 1975, Pananakakis *et al* 1977a,b). It was pointed out in these papers that MET states provoke the collapse of the surface barrier (V_s) in forward bias, favouring an extraction effect. For SET states the 'pinning' of E_{Fn} leads to the collapse of the barrier ϕ_{Bn} in reverse bias and so to an injection effect. For SHT states (which were not examined previously), analogous phenomena (surface-barrier collapse and injection effects in reverse bias) can be observed (figure 4).

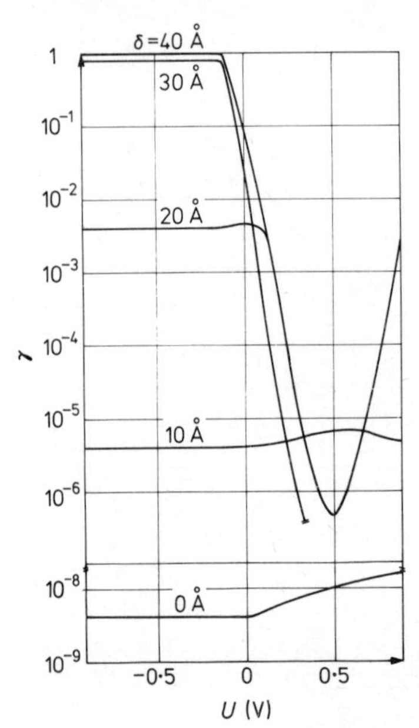

Figure 3. Variation of the injection ratio γ with the external applied voltage U for different thicknesses δ of the intermediate insulating layer. n-type Si, $N_D = 10^{21}\,m^{-3}$, $\phi_{MS} = 0{\cdot}3$ eV, $D_{ss} = 0$.

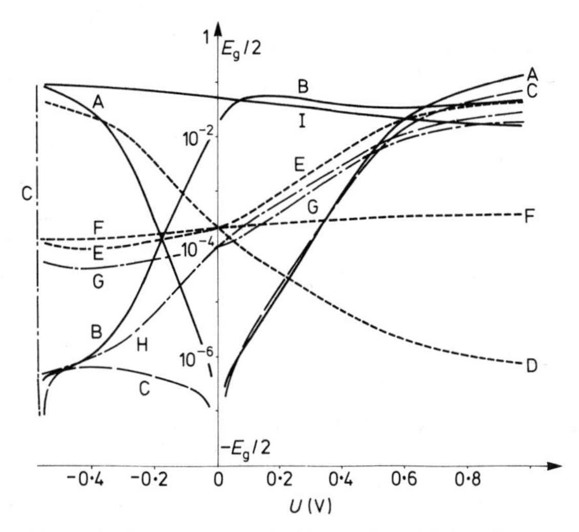

Figure 4. Curves A–H as in figures 1 and 2 but for a MIS structure (n-type Si) with semiconductor hole traps. $\delta = 20$ Å, $\phi_0 = 0{\cdot}1$ eV, $N_D = 10^{19}\,m^{-3}$, $\phi_{MS} = 0{\cdot}6$ eV, $D_{ss} = 10^{18}\,m^{-2}\,eV^{-1}$. Also shown is the variation with U of the occupation ratio of the surface states (curve I).

In the examples analysed, the neutral level ϕ_0 is located 0·1 eV above the midgap; when ϕ_0 moves towards the semiconductor conduction band anomalous effects caused by traps will be enhanced. The collapse of the surface barrier mentioned above generally induces a decrease in the injection ratio γ provided that the carrier-injection effect ($\Delta n > 0$) does not appear (see also equation 4*a*). When injection becomes important ($\Delta n/p_0 \gg 1$), γ increases (equation 4*a*).

It is also worth noting that the onset of the injection effect (in reverse bias) occurs when the drift component of the minority current density J_p (normally negligible for the doping used here) greatly exceeds the diffusion component of the same current.

3.2. Interface states connected with two or three reservoirs

We now consider single-energy surface states that have now both an electrostatic and a kinetic effect. Figures 5 and 6 present typical examples of the influence of these states for two different thicknesses of the insulating layer. These states are connected only with the semiconductor bands, they behave as acceptors and their energy level is located in the midgap. In order to emphasise their recombination properties, we have chosen a low density and large cross section for semiconductor electrons and holes.

γ is then given by

$$\gamma = J_p/(J_p - J_n) = \gamma^* - (J_R/J)$$

where γ^* can be expressed by one of the equations given in §2, depending on the example under consideration. Generally for low values of forward bias, the recombination current J_R dominates the carrier transport (see figure 5 where the total current appears as a plateau for this reason). Consequently γ decreases (in comparison with the ideal structure) and for large values of δ ($\delta > 15$ Å) the diode becomes a majority-carrier device. For

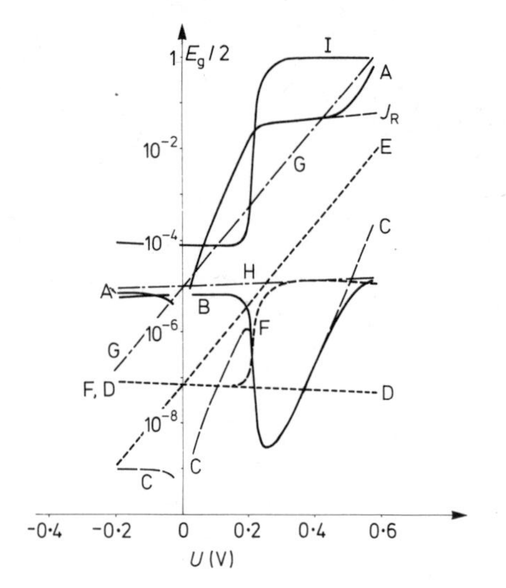

Figure 5. Curves A–I as for figures 1, 2 and 4 but for an MIS structure with interface states acting as recombination centres. $\delta = 15$ Å, $N_D = 10^{22}\,\mathrm{m}^{-3}$, $\phi_{MS} = 0{\cdot}6$ eV, $N_t = 10^{15}\,\mathrm{m}^{-2}$. Electron and hole capture coefficients $C_n = C_p = 5 \times 10^{-11}\,\mathrm{m}^3\,\mathrm{s}^{-1}$; $E_c - E_t = 0{\cdot}55$ eV.

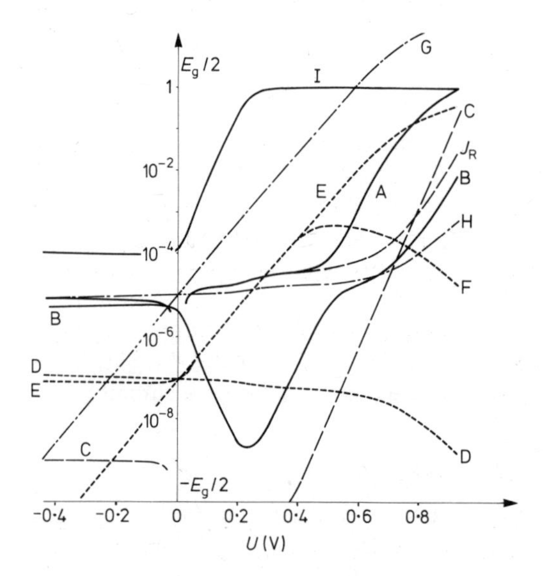

Figure 6. As figure 5 but for an MIS structure with interface states acting as recombination centres and with $\delta = 25$ Å.

larger values of bias, J_R does not dominate the whole current and the behaviour of γ the same as for the ideal structure (§2). For example, a MIS structure with $\delta = 25$ Å (and the recombination states given above) has a value of $\gamma < 10^{-5}$ for a large range of forward and reverse biases (for comparison the value of γ for an ideal structure with $\delta = 25$ Å equals unity, $\gamma = 1$).

It is also observed that the rate of occupation of the interface states is equal to 0·5 when the energy level of the states is equidistant from E_{Fn} and E_{Fp}; we deduce that when the MIS structure conducts, the occupancy of interface states is not governed by just one of the three quasi-Fermi levels (see figure 5).

Finally it can be seen that the excess carrier density Δn is not a monotonic function of the external applied voltage. We remark that the sharp decrease in Δn is due to the enhancement of recombination processes which, for the example, are accompanied by an increase in the rate of occupation of interface states. For larger biases Δn increases again because it is no longer controlled by recombination.

4. Conclusions

Our computations show that the injection and extraction phenomena in a MIS tunnel structure are very complex. It is obvious that without a model such as that described here that allows a detailed analysis of each example examined, it is very difficult to predict the behaviour of a given structure under an external bias.

References

Card H C and Rhoderick E H 1973 *Solid St. Electron.* **16** 365–74
Cowley A M and Sze S M 1965 *J. Appl. Phys.* **36** 3213–20
Green M A, King F D and Shewchun J 1974 *Solid St. Electron.* **17** 551–61
Green M A and Shewchun J 1973 *Solid St. Electron.* **16** 1141–50
Harrick N J 1959 *Phys. Rev.* **115** 876–82
Pananakakis G, Kamarinos G and Viktorovitch P 1979 *Rev. Phys. Appl.* **14** 639–47
Pananakakis G, Viktorovitch P and Kamarinos G 1977a Paper presented at *7th European Solid State Device Research Conference (ESSDERC), Brighton, September 1977* unpublished
—— 1977b *C.R. Acad. Sci.* **284B** 471–4
Rhoderick E H 1978 *Metal Semiconductor Contacts* (Oxford: Clarendon)
Scharfetter D L 1965 *Solid St. Electron.* **8** 299–311
Sze S M 1969 *Physics of Semiconductor Devices* (London: Wiley)
Viktorovitch P 1975 *Thèse* Grenoble
Viktorovitch P and Kamarinos G 1974 in *Metal Semiconductor Contacts* (Inst. Phys. Conf. Ser. 22) pp36–48
—— 1975 *J. Phys. D.: Appl. Phys.* **8** 246–53
Yu A Y C and Snow E H 1969 *Solid St. Electron.* **12** 155–60

Operational characteristics and structure of the Si-SiO_2 interface of MOS solar cells

F A Abou-Elfotouh and M A Almassari
Physics Department, Faculty of Science, Riyadh University, Riyadh, Saudi Arabia

Abstract. The operational characteristics of p-type silicon MOS solar cells were investigated after various heat treatments. Post-oxidation annealing was found to have a substantial effect on the properties of the interfacial transitional layer and the relative concentrations of the various oxide states observed. The values of the open circuit voltage V_{oc} and fill factor F_f can be maximised by low-temperature annealing (<550 °C) for a long time (>10 h); the interface states are found to behave as donor-like hole traps. Under these conditions maximum values of V_{oc} and F_f can be retained over a very wide range of oxide thicknesses t_{ox} (20 Å $< t_{ox} <$ 75 Å). Substrates with high doping levels ($>10^{16}$ cm^{-3}) showed better performance.

1. Introduction

The properties of surfaces and interfaces between insulating films and semiconductors have a strong influence on the operational characteristics of a wide range of devices that involve semiconducting materials. A very interesting example of this type of device is the MOS solar cell. It is well known that the photovoltaic conversion efficiency of a MOS solar cell can be considerably enhanced by optimisation of the interfacial layer (see for example Kar *et al* 1978, Lillington and Townsend 1976). Much work has been done on the properties of MOS solar cells (see e.g. Stirn and Yeh 1975, Green and Godfrey 1976) and others. Fonash (1975) and Kar (1978) have discussed some mechanisms to explain current transport. However, the exact role played by interface states is not well established. The primary objectives of this work are to investigate the effect of preparative procedures on the density and properties of interface states, and to relate the results to other theoretical and experimental work.

In this way the dependence of the performance of MOS structures on the action of interface states can be analysed in more detail, and on the basis of this analysis, the properties of such devices can be controlled more effectively.

2. Experimental details

MOS solar cells were prepared from p-type boron-doped (5×10^{15}–5×10^{17} cm^{-3}) Si wafers. Substrates with (111)-oriented surfaces were carefully polished, chemically cleaned and then immersed in HF to remove natural oxides present on the surface. Oxide films were then grown immediately by heating in an alumina tube in a diffusion furnace. Annealing treatments were carried out in the same furnace in an atmosphere of nitrogen to reduce the number of interface states. The thicknesses of the oxide films were determined with an AME 500 ellipsometer. MOS tunnel devices were prepared with aluminium (evaporated at a pressure about 10^{-5} Torr) as both the barrier metal and the back Ohmic contact (after removal of the oxide layer). The distribution of the density of interface states was estimated from capacitance–voltage measurements; data were analysed as described by Whelan (1965). MOS and Schottky barrier solar cells with a semitransparent aluminium film on the front surface were prepared in an identical manner. The Schottky barrier cell was used as a control sample for comparison between the operational characteristics (as determined from I–V measurements) of the two types of cells to determine the effect of oxide layer.

Surface analyses were carried out on a LHS 10 system using XPS, UPS and SIMS to identify the transition layer at the interface between Si and the oxide film and to investigate various oxide states and chemical species in this layer. In this study samples were fabricated in the preparation chamber of the machine and could then be transferred directly to the analysis chamber.

3. Results and discussion

Figure 1 shows the total number of interface states N_s as a function of oxidation temperature T. It is clear that N_s increases steadily with T up to 1200 °C and then rises more sharply at higher temperatures. When oxidation was performed in wet oxygen, values of N_s higher than those determined after oxidation in a dry atmosphere were obtained for $T \geqslant 1100$ °C.

Before annealing, the distribution of the density of interface states with energy was a non-uniform U-shape, with a peak about 0·25 eV above the valence band. In regions near the conduction band and midgap, the magnitude of N_s was smaller and its variation with energy was slow. Figure 2 indicates that the energy in the Si gap at which the peak value of N_s is obtained shifts after annealing from the valence-band edge towards the conduction-band edge. The maximum value of this shift was observed after annealing at 450 °C for periods longer than 10 h; annealing at 800 °C did not give rise to such a considerable shift.

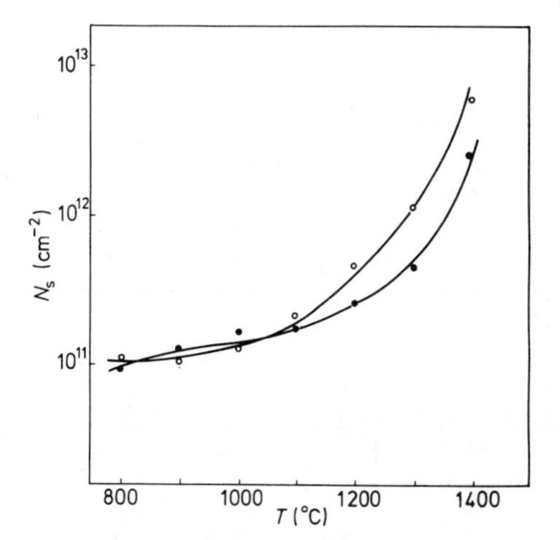

Figure 1. Number of interface states as a function of oxidation temperature for oxide layers formed in wet (○) and dry (●) oxygen. Oxide thickness t_{ox} = 130–170 Å.

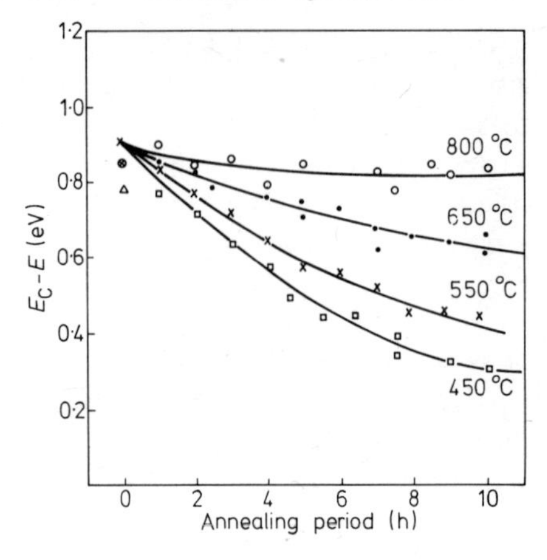

Figure 2. Shifts in energy at which the peak value of N_s is obtained as a function of the post-oxidation annealing period and temperature. Oxidation temperatures were 1200 °C (×), 1100 °C (⊕) and 1000 °C (△).

Under the same preparation conditions, the peak density of interface states appears to be dependent on the thickness of the oxide film, but no definite behaviour has been established.

Figure 3 shows the total number of interface states as a function of the annealing conditions. It can be seen that N_s decreases significantly as the temperature and duration of annealing increase.

It has been shown by Kar (1978) that both N_s and the product $N_s t_{ox}$ (t_{ox} is the thickness of the oxide layer) decrease with t_{ox} and Shewchun *et al* (1977) found that the effect of t_{ox} on the conversion efficiency η was dependent on the range of t_{ox} under investigation. It was shown that better values of η were obtained with N_s as low as possible for film thickness less than 20 Å. For thicker films, η decreases with N_s. Our results showed that higher values of η, V_{oc} and F_f are obtained as the value of the product $t_{ox}N_s$ decreases and that the relation between t_{ox} and the product $t_{ox}N_s$ is dictated by the preparation conditions which determine the effect and behaviour of the interface states.

Viktorovitch *et al* (1978) computed the *I–V* characteristics of MOS solar cells for two sets of conditions, when the interface states behave as donors and when they act as

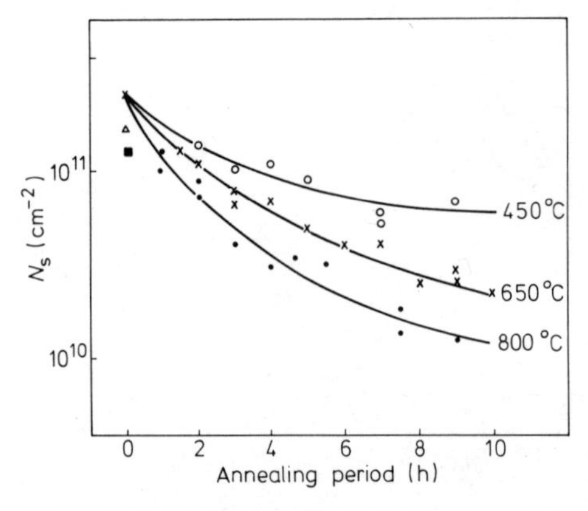

Figure 3. Total density of interface states as a function of annealing time for three annealing temperatures. Oxidation temperatures were 1200 °C (×), 1000 °C (△) and 900 °C (■).

acceptors. A comparison of their results with our data led to the conclusion that the interface states in samples oxidised at 1200 °C in dry oxygen and annealed at 450 °C behave as donors. The interface states in unannealed samples or in samples annealed at high temperatures (800 °C) behave as acceptors. Figure 4 shows that maximum values of V_{oc} and F_f can be maintained over a wide range of oxide thickness (20–70 Å) when donor-like interface states are present and the product $t_{ox}N_s$ is kept almost at its minimum for this range of t_{ox}. For acceptor-like interface states, not only are values of V_{oc} and F_f smaller but the effect of t_{ox} is also very critical. An optimum value around 20 Å was obtained for t_{ox}. It must be mentioned that the values of V_{oc} and F_f are always higher for MIS structures than the equivalent values for MS structures.

Measurements obtained from SIMS studies in conjunction with other surface analysis techniques showed that the oxide layer formed on the Si surface is largely of the SiO_2 structure. However, a transitional layer with various chemical species and oxidation states is formed between the oxide film and Si. The existence of this defective transition layer was first indicated by Deal (1974) in his model for the origin of interface states. Figure 5 shows the different structures observed in the transition layer formed with a 0 Å thick oxide film. These structures include (C, OH, H) bonded to Si and (SiO, Si_2O_3, Si_2O_5 and SiO_4) oxidation states.

The chemical shifts arising from some of these bonding states, observed by XPS measurements in this work and also detected by Garner *et al* (1978), are in good agreement with the value of 2·6 eV obtained by Goddard *et al* (1978) and attributed by them to oxidation states with decreased O–O bonding. These confirm the existence of oxide states with loosely bound oxygen atoms as proposed by Cheng (1970) in his oxygen model.

The transitional layer was found to be thicker when the rate of oxidation was lower. Annealing at 800 °C reduces the thickness of this layer but treatment at 450 °C has no effect. The results also demonstrated some reduction in the concentration of various oxidation states after annealing. The major reduction was seen in the concentration of the SiO_4 (about three orders of magnitude). Schwarz *et al* (1978) in their work on phosphorus-doped n-type Si attributed the increase in the width of the transition layer to the

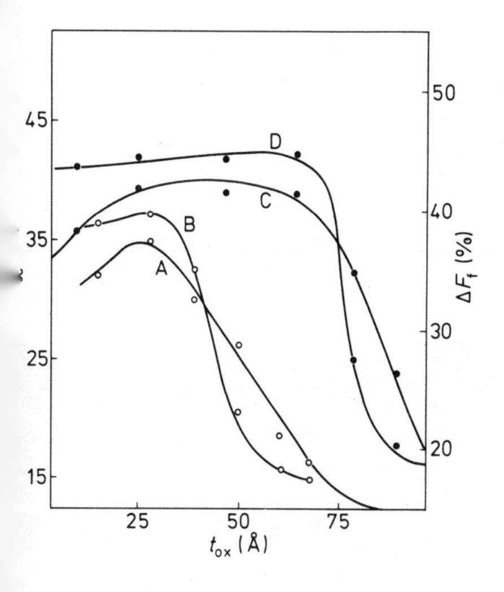

Figure 4. Percentage increases in the values of V_{oc} (curves A, C) and F_f (curves B, D) for an MOS solar cell with respect to a Schottky barrier cell. •, Donor-like interface states; ○, acceptor-like interface states.

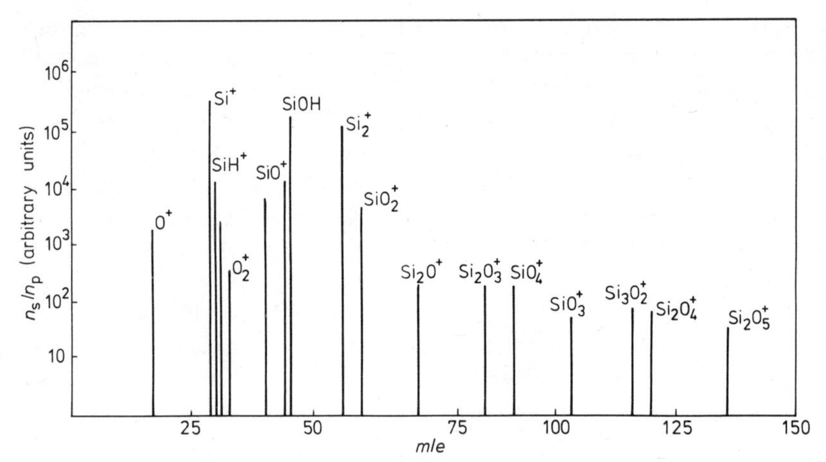

Figure 5. Chemical species and oxidation states observed in the transition layer of an oxid film 50 Å thick.

formation of a SiO_xP_y phase. Formation of a SiO_4 A-phase in p-type Si (A-accepto impurity) was also suggested by Fuller *et al* (1960) among others, to explain the forma tion of shallow thermal donors by annealing. In the present experiments, the reductio in the concentration of SiO_4 after annealing at 450 °C and increasing the doping leve could be attributed to the formation of a SiO_4 B-phase. The presence of such a structur at the interface layer could account for the increase in the number of localised donor in this region. These centres could be positively charged and act as hole traps.

4. Conclusions

In order to improve the operational characteristics of MOS solar cells, the preparatio conditions and material parameters must be adjusted to obtain a minimum value of th product $N_s t_{ox}$ and donor-like behaviour of the interface states. These conditions can b achieved by using low-resistivity materials with an oxide film grown at a low rate c oxidation and annealed at 450 °C.

References

Cheng Y C 1970 *Surface Sci.* **20** 434–6

Deal B E 1974 *J. Electrochem. Soc.* **121** 198c

Fonash S J 1975 *J. Appl. Phys.* **46** 1286–9

Fuller C S, Doleiden F H and Wolfstirn K 1960 *J. Phys. Chem. Solids* **13** 187–203

Goddard W A, Barton J J, Redondo A and McGill T C 1978 *J. Vac. Sci. Technol.* **15** 1274–85

Garner C M, Lindau I, Su C Y, Pianetta P and Spicer W E 1978 *J. Vac. Sci. Technol.* **15** 1290–1

Green M A and Godfrey R B 1976 *Appl. Phys. Lett.* **29** 610–2

Kar S 1978 *J. Appl. Phys.* **49** 5278–83

Kar S, Shanker D and Joshi S P 1978 *Proc. Int. Solar Energy Congr., New Delhi* (Oxford: Pergamon) pp625–8

Lillington D R and Townsend W G 1976 *Appl. Phys. Lett.* **28** 97–8

Schwarz S A, Helms C R and Spicer W E 1978 *J. Vac. Sci. Technol.* **15** 1519

Shewchun J, Singh R and Green M A 1977 *J. Appl. Phys.* **48** 765–70

Stirn R J and Yeh Y C M 1975 *Appl. Phys. Lett.* **27** 95–8

Viktorovitch P, Kamarinos G, Even P and Fabre E 1978 *Phys. Stat. Solidi* **A48** 137–45

Whelan M V 1965 *Philips Res. Rep.* **20** 562–6

Dynamic properties of switching in MISS structures and applications to charge transfer devices

G Sarrabayrouse, J Buxó, A Muñoz-Yagüe, A E Owen† and J-P Sebaa
Laboratoire d'Automatique et d'Analyse des Systems du CNRS, 7 Avenue du Colonel Roche, 31400 Toulouse, France

Abstract. This paper is concerned with the dynamics of metal–thin insulator (tunnelling)–n–p^+ silicon devices in which electrical switching occurs when an inversion layer is formed in the silicon under the thin insulator. Two structures are considered. A single MISS device, or cell, in which the inversion charge is provided by minority carrier injection from the p^+–n junction and a two-cell system, in which diffusion of charge from a previously switched cell causes the adjacent cell to switch. Analytical solutions are presented for the different dynamic regimes and criteria for switching and the delay time at turn-on are formulated.

1. Introduction

Electrical switching at a critical voltage V_S has been observed by several authors in metal–thin insulator (tunnelling)–n–p^+ silicon devices when the MIS Schottky barrier is reverse-biased (Yamamoto and Morimoto 1972, Kroger and Wegener 1973, Yamamoto *et al* 1976, Simmons and El Badry 1977, Buxo *et al* 1978). The possibility of using an array of such devices as a shift register has also been demonstrated (Yamamoto *et al* 1976). We have recently developed a mechanistic model, based on a regenerative process, in which switching occurs as soon as an inversion layer is created in the silicon immediately below the thin oxide layer (Sarrabayrouse *et al* 1979). The aim of this paper is to investigate the dynamic 'turn-on' properties of (i) a single isolated switch or cell and (ii) two adjacent cells. The switching delay time for an isolated device is related to the rate at which the inversion layer is established and analysis shows that when an adjacent device is also present, a succession of different regimes govern the charge-transfer mechanism from one cell to the other. The mathematical analysis provides useful guidelines to evaluate the turn-on performance of switching devices in terms of their geometrical and physical parameters.

2. Delay time before switching of a single cell

2.1. Device fabrication

MISS devices with the structure shown schematically in figure 1(*a*) (a single cell) were fabricated from n-type epitaxial layers of silicon (7 μm thick and $\rho \approx 2\ \Omega$ cm) on p^+-substrates. The thin oxide on the n-layer was grown by thermal oxidation at approxi-

† Present address: Department of Electrical Engineering, University of Edinburgh, King's Buildings, Edinburgh EH9 3JL, UK.

0305-2346/80/0050-0179$01.00

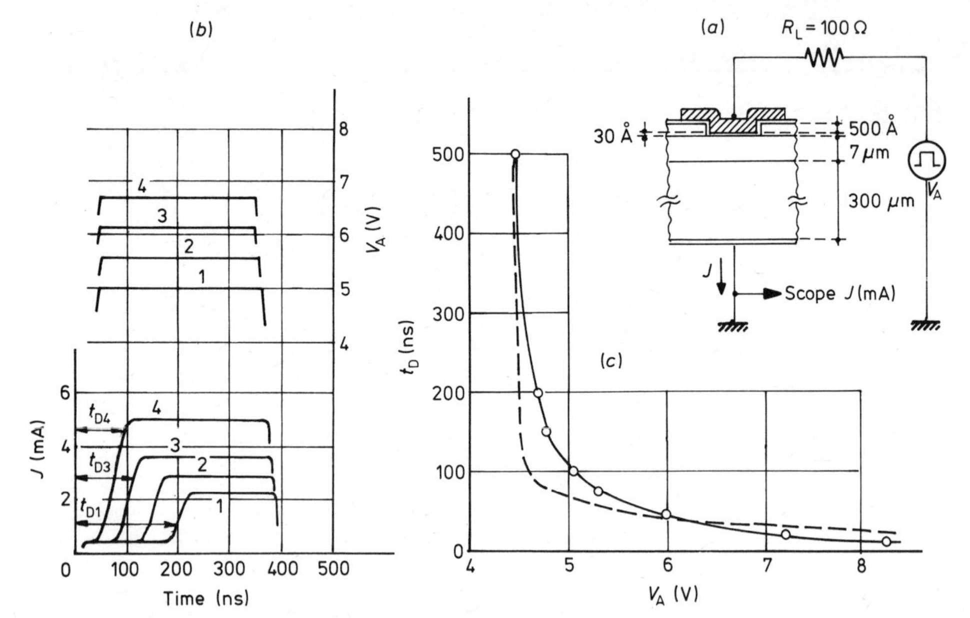

Figure 1. (*a*) Schematic illustration of the MISS structure and biasing circuit. (*b*) Experimental voltage and current pulses at switching, showing the dependence of the delay time t_D on the height of the applied voltage pulse V_A. (*c*) The delay time as a function of applied voltage. —O— experiment; – – – theory.

mately 800 °C; the oxide thickness was less than 50 Å. A gold contact was made to the p^+-substrate and an aluminium contact was evaporated onto the oxide.

2.2. *Experimental results*

When a voltage pulse V_A ($>V_S$) is applied to the switch (see figure 1*a*), the current remains low (corresponding to the OFF-state current) for a period t_D, after which it rises rapidly to the ON-state value. The period t_D is called the delay time. Figure 1(*b*) illustrates typical behaviour and shows that t_D is a function of the pulse height V_A (see Buxo *et al* 1978, Sarrabayrouse 1978, Sarrabayrouse *et al* 1979). Experimental data on the dependence of t_D on V_A are plotted in figure 1(*c*).

2.3. *Discussion*

In the framework of the model developed by the authors (Sarrabayrouse *et al* 1979), switching occurs when hole injection into the n-layer is large enough to store an inversion layer under the thin SiO_2 layer. If a thicker oxide layer is used, the build-up of holes is very rapid because the associated hole-tunnelling current through the insulator is low. It is assumed here that a fraction of the minority-carrier current injected from the p^+–n junction towards the Si–SiO_2 interface tunnels through the insulator, while the rest is stored and will eventually lead to an inverted charge Σ_{ss}. Thus

$$d\Sigma_{ss}/dt = J'_p - J_p \tag{1}$$

where J'_p is the flux of minority carriers injected by the p^+–n junction and J_p is the tunnelling flux. A schematic band diagram, indicating J_p and J'_p as well as other relevant

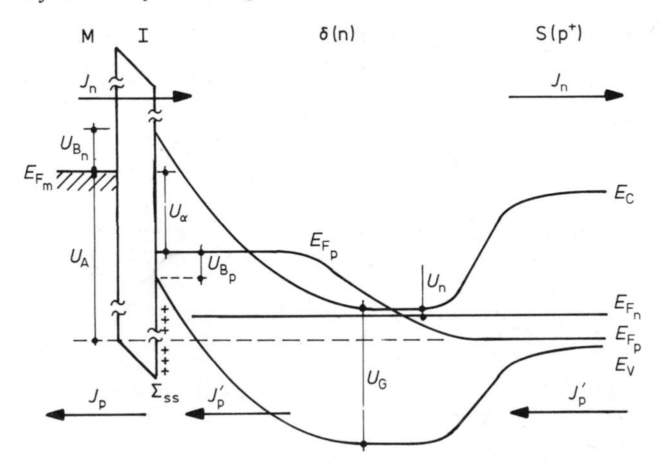

Figure 2. A band diagram of the reverse-biased MISS device. The parameters U_G, U_n, U_{B_p}, U_{B_n}, U_α and U_A are all normalised to U_T (see text).

parameters, is shown in figure 2. The flux J_p' is related to the reverse flux J_n through the MIS Schottky barrier, i.e.

$$J_p' = \gamma J_n \tag{2}$$

where γ is the injection efficiency of the p^+–n junction and where it is assumed that recombination across the n-layer is negligible. The charge Σ_{ss} can be expressed as

$$\Sigma_{ss} = qN_v u_T E_{Si}^{-1} \exp(-U_{B_p}(t)) \tag{3}$$

in which a linear approximation is made to the potential in the semiconductor near the Si–SiO_2 interface. In equation (3), E_{Si} is the electric field in the silicon near the interface, q is the electronic charge, $u_T = kT/q$ (T is the temperature and k the Boltzmann constant), N_v is the density of states in the silicon valence band and U_{B_p} is depicted in figure 2. The fluxes J_p and J_n can be expressed by (Sarrabayrouse 1978; see also figure 2)

$$\begin{aligned} J_n &= J_{n_0} \exp(-U_{B_n}) (\exp(U_A) - 1) \\ J_p &= J_{p_0} \exp(-U_{B_p}(t)) (\exp(U_\alpha) - 1). \end{aligned} \tag{4}$$

Solution of equation (1) with the condition $U_{B_p}(t = 0) = (U_G - U_{B_n})$ gives the time dependence of $U_{B_p}(t)$.

The onset of the inverted layer *first* occurs when the hole density at the interface equals the doping level of the epitaxial layer N_D, i.e.

$$U_{B_p}(t_D) = U_n \tag{5}$$

where U_n is defined in figure 2. Equation (5) determines t_D with the result

$$t_D = -qN_v u_T E_{Si}^{-1} J_{p_0} \ln \left(1 - \frac{\exp(U_{B_n}(u_S))(\exp(u_S) - 1)}{\exp(U_{B_n}(u_A))(\exp(u_A) - 1)}\right) \tag{6}$$

where $U_{B_n}(u_S)$ and $U_{B_n}(u_A)$ are the barrier heights U_{B_n} when the normalised voltages are $u_S = (qV_S/kT)$ and $u_A = (qV_A/kT)$ respectively. Equation (6) is plotted as the broken curve (theory) in figure 1(*c*) using the physical parameters that give the best fit to the experimental values of the switching voltage V_S (Sarrabayrouse 1978).

3. Dynamics of the turn-on mechanism of two adjacent cells

The possibility of using MISS switches in an analogous way to charge-transfer devices has been demonstrated by Yamamoto *et al* 1976). This section considers the electrical characteristics of a basic two-cell system and useful design information for charge-transfer-type devices is derived.

Figure 3(*a*) illustrates the two-cell system and defines some of the important symbols used subsequently. The main features of the model are as follows. Cell 1 is in the ON-state with I_1 its associated current. A minority-carrier density p_1 is, therefore, stored under the thin oxide layer (at $y = 0$) of cell 1 but the stored charge tends to diffuse laterally towards cell 2. The voltage V_2 cannot by itself make cell 2 switch ($V_2 < V_S$). Nevertheless, when enough of the laterally diffusing minority carriers reach cell 2 it will turn on; this occurs at a time t_{DT} after cell 1 has been switched on. To evaluate t_{DT} three simplifying assumptions are made:

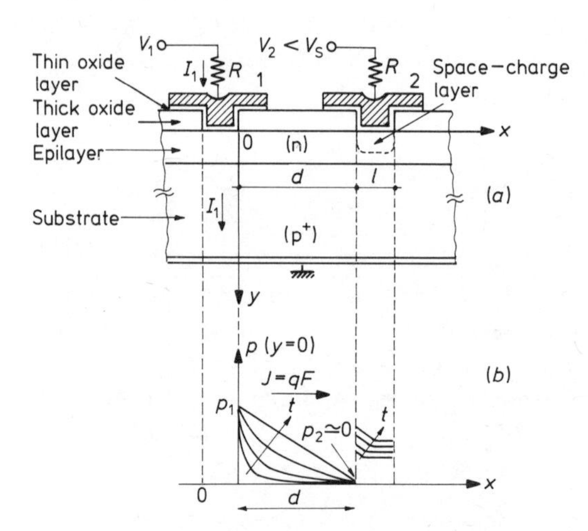

Figure 3. (*a*) A two-cell MISS device. (*b*) Illustration of the build-up of minority carriers under the oxide in both the inter-cell space and beneath cell 2.

(i) As long as cell 1 remains in the OFF-state, there is no out-diffusion of minority carriers from under the oxide ($y = 0$) within the inter-cell space of the n-layer.
(ii) Once cell 1 has been switched on, p_1 remains constant and depends only on I_1. The relationship between p_1 and I_1 has been given elsewhere (Sarrabayrouse 1978).
(iii) When a voltage $V < V_S$ is applied to cell 2, the depletion region extends into the n-layer with the result that the value of p at $x = d$ is $p_2 \approx (n_i^2/N_D) \ll p_1$ (n_i is the intrinsic carrier density).

Under these circumstances the time-dependent built-up of minority carriers within the inter-cell n-layer is governed by the following solution of the diffusion equation (Carslaw and Jaeger 1959):

$$p = p_1\left(1 - \frac{x}{d}\right) - \frac{2}{\pi}\sum_{n=1}^{\infty}\frac{p_1}{n}\sin\left(\frac{n\pi x}{d}\right)\exp\left(-\frac{n^2\pi^2 Dt}{d^2}\right) \tag{7}$$

where D is the diffusion coefficient of minority carriers. Recombination effects have been neglected in writing this solution, implying that $(d^2/\pi^2 D) < \tau$, where τ is the minority-carrier lifetime. Equation (7) shows that the largest time constant involved before a linear, steady-state regime is attained is of the order of $(d^2/\pi^2 D)$.

The induced switching of cell 2 is now considered and it will be assumed that the out-diffusing holes from cell 1, collected by the space-charge layer of cell 2, are stored under the oxide layer of cell 2. These holes will diffuse in the x direction, within the limits of cell 2 $(d < x < l + d)$ and meanwhile some of them will be lost by recombination or by tunnelling through the thin oxide. The hole current supplied by the p^+–n junction of cell 2 will make only a small contribution during the period $0 < t < t_{DT}$ and hence is neglected. The equation to be solved is

$$\partial p/\partial t = D(\partial^2 p/\partial x^2) - (p/\tau) \qquad (d < x < l + d). \tag{8}$$

The boundary conditions are

$$\partial p/\partial x|_{x=d} = F \tag{9}$$

where F denotes the gradient of the minority carriers flowing from cell 1 and will be considered to have attained its steady-state value for $t > (d^2/\pi^2 D)$, and also

$$\partial p/\partial x|_{x=l+d} = 0 \tag{10}$$

which implies that the hole flux ceases within $d < x < l + d$.

The following procedure has been used to solve equation (8). A solution U with $\tau = \infty$ is first sought and a solution, Z for finite τ, can then be obtained by Danckwert's method (Carslaw and Jaeger 1959):

$$Z = \frac{1}{\tau}\int_0^t \exp\left(-\frac{t'}{\tau}\right) U(t')\,dt' + U\exp\left(-\frac{t}{\tau}\right). \tag{11}$$

The functions U and Z are now referred to a new system the origin of which is taken at $x = d$. The solution U is then (Carslaw and Jaeger 1959)

$$U = Fl\left\{\theta + \tfrac{1}{2}\left(\frac{x}{l} - 1\right)^2 - \tfrac{1}{6} - \frac{2}{\pi^2}\sum_{n=1}^{\infty}\frac{(-1)^n}{n^2}\cos\left[n\pi\left(1 - \frac{x}{l}\right)\right]\exp(-n^2\pi^2\theta)\right\} \tag{12}$$

where θ is the reduced time variable, $\theta = (D/l^2)t$. Using equation (11) we obtain

$$Z = p(x, t)$$

$$= Fl\left\{\theta(\tau)\left[1 - \exp\left(-\frac{t}{\tau}\right)\right] + \left[\frac{1}{2}\left(\frac{x}{l} - 1\right)^2 - \frac{1}{6}\right] - \frac{2}{\pi^2}\sum_{n=1}^{\infty}\frac{(-1)^n}{n^2}\right.$$

$$\left.\times\cos\left[n\pi\left(1 - \frac{x}{l}\right)\right]\left[\exp\left(-n^2\pi^2\theta - \frac{t}{\tau}\right)\left(\frac{n^2\pi^2\theta(\tau)}{1 + n^2\pi^2\theta(\tau)}\right) + \left(\frac{1}{1 + n^2\pi^2\theta(\tau)}\right)\right]\right\} \tag{13}$$

with $\theta(\tau) = (D/l^2)$. The inversion layer will be first induced at $x = 0$ (in the new reference system) and only this point is considered in the following.

Much physical insight may be gained from equations (11), (12) and (13) and two particular cases are considered.

3.1. $\tau = \infty$

This implies that recombination is negligible and hence

$$p(0, t) = Fl\left(\theta + \frac{1}{3} - \frac{2}{\pi^2} \sum_{n=1}^{\infty} \frac{1}{n^2} \exp(-n^2\pi^2\theta)\right) \tag{14}$$

Thus, the largest time constant involved, before the linear regime ($p/Fl = \theta + \frac{1}{3}$) is attained, is of the order of $(l^2/\pi^2 D)$. This relationship is shown in figure 4. At shorter times, such that $\theta < (1/\pi^2)$, a larger number of terms in the series must be considered. It is also implicit in figure 4 that a necessary condition for rapid attainment of inversion ($p_{\text{inv}} \approx N_D$) is that

$$N_D/Fl \leqslant \tfrac{1}{3} \tag{15}$$

where N_D is the doping concentration of the n-layer.

3.2. $\tau \neq \infty$

Consideration of equation (13) with $x = 0$ suggests that the largest time constant involved is now of the order of $[(D\pi^2/l^2) + (1/\tau)]^{-1}$, so far as the contribution of the series is concerned. Furthermore, as $t \to \infty$, $p(0, t)$ tends to a constant value given by

$$p(0, \infty) = Fl\left(\theta(\tau) + \frac{1}{3} - \frac{2}{\pi^2} \sum_{n=1}^{\infty} \frac{1}{n^2(1 + n^2\pi^2\theta(\tau))}\right) \tag{16}$$

which is approached asymptotically with a time constant τ. A necessary condition for switching is therefore

$$\frac{N_D}{Fl} < \theta(\tau) + \frac{1}{3} - \frac{2}{\pi^2} \sum_{n=1}^{\infty} \frac{1}{n^2(1 + n^2\pi^2\theta(\tau))}. \tag{17}$$

Two cases have to be considered, as shown in figure 4. Firstly

$\tau > (l^2/\pi^2 D)$.

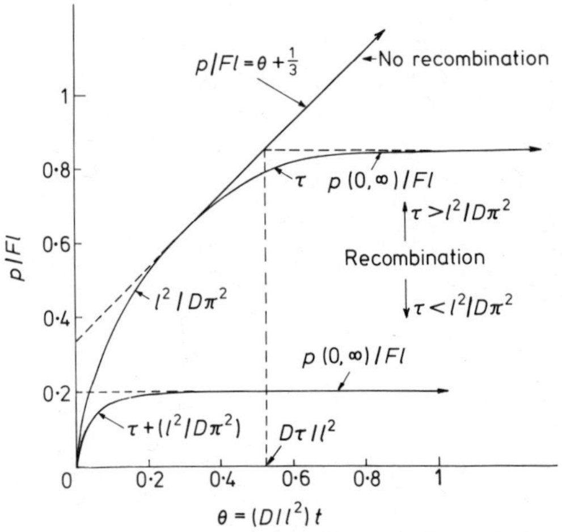

Figure 4. The dynamics of the build-up of carriers at the edge of cell 2 plotted in reduced variables as (P/Fl) against θ. The situations corresponding to the inclusion and omission of recombination are shown (see text).

During the period $0<t<\tau$, charge build-up follows equation (14) until $t \geqslant \tau$, when saturation occurs. Secondly

$$\tau < (l^2/\pi^2 D).$$

The magnitude of τ uniquely governs the dynamics of the build-up of charge and saturation is attained before the intermediate linear region occurs.

The foregoing analysis shows that there is a succession of regimes for the charge-transfer process from cell 1 to cell 2 with contributions from the charge dynamics of the inter-cell region and the dynamics of the build-up of carriers under the oxide of cell 2.

The analytical results provide simple criteria for rapid switching and they also allow evaluation of the successive time constants involved in the switching process.

References

Buxo J, Owen A E, Sarrabayrouse G and Sabaa J-P 1978 *Rev. Physique Appl.* **13** 767
Carslaw H S and Jaeger J C 1959 *Conduction of Heat in Solids* 2nd edn (Oxford: Clarendon)
Kroger H and Wegener H A R 1973 *Appl. Phys. Lett.* **23** 397
Sarrabayrouse G 1978 *Thèse de Doctorat d'Etat* Université Paul Sabatier, Toulouse, France
Sarrabayrouse G, Buxó J, Owen A E, Muñoz-Yagüe A and Sebaa J-P 1979 *IEE J. Solid St. Electron Dev.* submitted for publication
Simmons J G and El Badry A 1977 *Solid St. Electron.* **20** 955
Yamamoto T, Kawamura K and Shimizu H 1976 *Solid St. Electron.* **19** 701
Yamamoto T and Morimoto H 1972 *Appl. Phys. Lett.* **20** 269

Analysis of p-type CdTe-Langmuir film interface

M C Petty and G G Roberts

Department of Applied Physics and Electronics, University of Durham, South Road, Durham DH1 3LE, UK

Abstract. Langmuir films based on cadmium stearate have been successfully deposited on to the surface of p-type CdTe. Stringent control of the substrate preparation has been found necessary to obtain reproducible results. The measured *C–V* curves at high frequencies show, for the first time, the accumulation, depletion and inversion regions typical of an ideal device. The *G–V* curves show a sharp peak in the depletion region which is thought to be due to losses at interface states. An initial estimate of the density yields a value of 9×10^{11} states $cm^{-2} eV^{-1}$ at an energy level in the band gap 0·2 eV above the valence-band edge.

1. Introduction

The surface characteristics of semiconductors may be investigated by measuring the electrical properties of metal–insulator–semiconductor (MIS) structures. However, the lack of suitable insulating layers has confined this study almost exclusively to silicon. This means that there is still little information available describing the surfaces of many other important semiconductors. A great deal of effort is currently being devoted to studying insulating films on III–V compounds. For example, we recently reported encouraging results for MIS structures based on semiconducting InP and Langmuir films (Roberts *et al* 1978a,b). We show in this paper that cadmium stearate films can be deposited successfully on to CdTe single crystals; our preliminary results indicate that useful MIS devices can be constructed with this particular II–VI semiconductor.

2. Langmuir–Blodgett films

Langmuir–Blodgett films are an interesting example of ordered organic systems. They can be assembled one monolayer at a time to form a planar two-dimensional sheet of accurately controlled thickness. The preparation of such films was suggested by Langmuir (1920) and his ideas were extensively applied by Blodgett (1935). The technique consists of transferring monolayers of amphipathic molecules on to solid substrates by dipping and raising the latter through a compact monolayer floating on the surface of purified water. If the deposition conditions are carefully controlled, a single monolayer is transferred on to the substrate during each traversal of the water surface. The thickness of a Langmuir film is determined by the number of monolayers deposited and the molecular size of the material used. Various organic materials have been deposited using this technique, but most data have been obtained for fatty acids or their salts. These substances possess molecular structures which consist of a hydrophilic carboxyl group (–COOH) at one end of a straight carbon chain and a hydrophobic group ($-CH_3$) at the other end.

0305-2346/80/0050-0186$01.00

Under normal deposition conditions the carbon chains are aligned normal to the substrate surface with successive layers oriented in opposite directions. The molecular size (half the distance separating the carboxyl groups) is approximately 2·5 nm for stearic acid ($C_{17}H_{35}COOH$) and 2·75 nm for arachidic acid ($C_{19}H_{39}COOH$). Langmuir films formed from the cadmium salts of these two materials have been shown to be good insulators (Roberts *et al* 1978a, b) with high breakdown fields (Agarwal and Srivastava 1975).

3. Experimental details

3.1. Preparation of semiconductor substrates

The experiments were carried out using p-type CdTe crystals (⟨111⟩ orientation) with carrier concentrations in the range 10^{15}–10^{16} cm^{-3}. The single crystals were grown at RSRE, Malvern, using a modification of the solvent evaporation technique reported by Lunn and Bettridge (1977). The crystal slices (typically 300 μm thick and 1 cm^2 in area) were mechanically polished to a finish of 0·25 μm using diamond paste and then chemically polished in a solution of 1% bromine in methanol. Immediately before evaporation of gold back contacts, the samples were etched in 40% hydrofluoric acid for 1 min. This was found necessary to give reproducible Ohmic contacts. Treatment applied to the semiconductor surface immediately before Langmuir film deposition was found to be of vital importance in achieving reproducible MIS data. The technique that produced the results described in this paper comprised leaving the semiconductor for several days in air after HF etching/back contacting and then refluxing the substrate in isopropyl alcohol vapour just before depositing the film.

3.2. Langmuir–Blodgett film deposition

Cadmium stearate ($CdSt_2$) films were deposited on to the CdTe surface using the technique established by Blodgett (1935).

A micrometer syringe was used to spread a few drops of stearic acid (Sigma grade 1) dissolved in chloroform (BDH Aristar) on to the surface of water obtained from a Millipore purification system. This aqueous subphase contained $2{\cdot}5 \times 10^{-4}$ molar cadmium chloride (BDH Analar) and its pH was adjusted to a value of 5·6–5·8 by the addition of ammonia (BDH Aristar) or hydrochloric acid (BDH Analar). The surface area and hence the surface pressure of the monolayer could be varied using a motor-driven PTFE-coated glass fibre barrier. The surface pressure was measured with a Wilhelmy plate suspended from a sensitive microbalance (Gaines 1966). An electronic feedback system connected to the barrier motor was used to maintain a constant monolayer surface pressure of $2{\cdot}5 \times 10^{-2}$ N m^{-1} during the deposition process. In fact, both the surface area and the surface pressure were monitored continuously during the deposition process.

The CdTe substrates were dipped and raised through the compressed monolayer using a variable-speed motor attached to a micrometer. When the CdTe single crystal was first lowered through the monolayer/water interface, no material was deposited, indicating that the semiconductor surface was hydrophilic. The first monolayer was transferred on to the substrate as it was raised through the interface. A rate of about 2 mm min^{-1} ensured that this first layer emerged completely dry from the subphase. Subsequent layers could then be transferred to the CdTe at dipping rates of about 1 cm min^{-1}. The films were stored in a desiccator under a low pressure of nitrogen for at least 3 days before a series of top electrodes ~ 1 mm in diameter (Au or Au/Ge eutectic alloy) were deposited by thermal evaporation. The substrates were maintained at a temperature of approximately –100 °C during this process.

3.3. Admittance measurements

All measurements were performed at room temperature with the devices in the dark under a low pressure of nitrogen. Pressure top contacts were made to the top electrodes using 1 mm diameter gold balls attached to micromanipulator controls. The admittance of the MIS structures was measured as a function of gate bias using an Ortholoc 9502 two-channel phase-sensitive detector. The voltage scanning rate was approximately 10 mV s^{-1}. The amplitude of the AC test signal superimposed upon the ramp voltage was less than 25 mV to avoid non-linear effects.

4. Results and discussion

4.1. Admittance data

Figure 1 shows an example of the conductance and capacitance versus gate bias curves measured at a frequency of 100 kHz for a Au/Ge–$CdSt_2$–CdTe MIS device. The dielectric thickness was 80 nm, corresponding to 31 layers of $CdSt_2$. At a voltage scanning rate of 10 mV s^{-1}, this particular device showed no hysteresis.

The capacitance curve shows accumulation, depletion and inversion characteristics similar to those reported for many Si–SiO_2 devices at high frequencies. The value of the accumulation capacitance is in good agreement with that expected from the known values of top electrode area, dielectric constant and thickness of the Langmuir film. At frequencies below 100 Hz, this measured accumulation capacitance was about 5% greater than the high-frequency value. The increase in capacitance due to minority-carrier response in the inversion region has not been observed in any of our devices, even at the

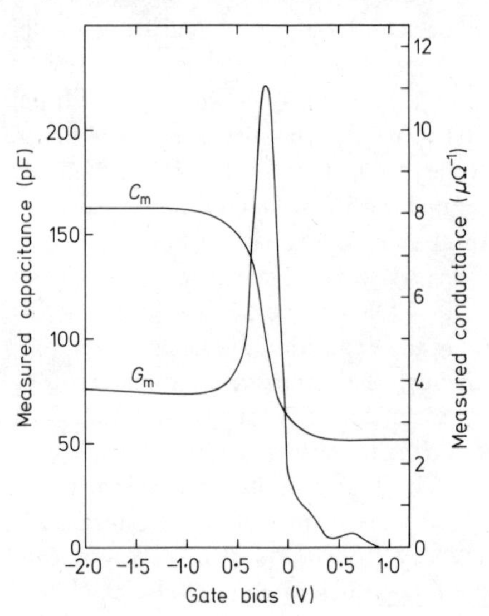

Figure 1. Measured values of capacitance and conductance at 100 kHz for a p-type CdTe/$CdSt_2$ capacitor. Dielectric thickness 80 nm; device area $5{\cdot}5 \times 10^{-7}\,m^2$.

lowest measurement frequency used (8 Hz). This effect, however, was seen when light was focused on to the sample.

The conductance peak in the depletion region is thought to be due to losses at interface states. The finite value of conductance at negative biases indicates that the equivalent circuit of the device in the accumulation region is more complicated than a single capacitor representing the insulator. At frequencies above 10 kHz, the magnitude of this accumulation conductance is consistent with that expected from the series resistance of the substrate (~300 Ω). However, at lower frequencies this conductance becomes much higher than that expected from the substrate resistance alone. We are at present unsure of the origin of the small subsidiary conductance peaks in the inversion region. This effect is only observed in certain devices at frequencies above 10 kHz. Figure 2 shows the effect of frequency on the conductance curves. As the measurement frequency is reduced, the height of the G_m/ω peak increases and its position shifts towards the weak-inversion region.

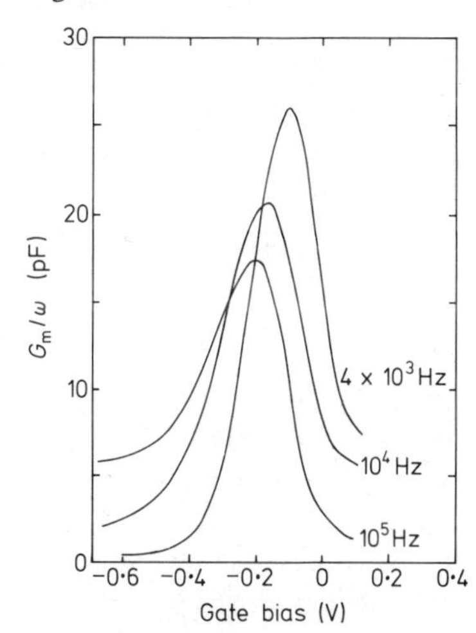

Figure 2. Conductance voltage characteristics at different frequencies for the MIS device the high-frequency characteristics of which are shown in figure 1.

4.2. Device reproducibility

Not all of our devices displayed the *C–V*, *G–V* characteristics as shown in figure 1. In many cases the constant capacitance in the accumulation and inversion regions was not observed and the devices displayed large hysteresis effects. Systematic experiments using the same substrate slice seemed to indicate that the surface treatment of the CdTe immediately before Langmuir film deposition was the stage in the device preparation that had most influence on the measured *C–V*, *G–V* curves. Such experiments led to the substrate preparation routine outlined in §3.1, i.e. leaving the substrate in air for some time after the etching/back contacting stage before depositing the film. It is thought that this procedure allows the growth of a passivating surface layer on the CdTe. The measured conductance peak would then probably originate from losses at interface states between the bulk CdTe and this natural 'oxide' layer. Thus it would seem likely that our devices are double-dielectric-type structures. However, even MIS structures prepared under seemingly identical conditions and using the same substrate slice showed some inconsistencies in their *C–V* and *G–V* characteristics. In particular, the magnitude of the measured conductance peak varied quite markedly from one device to another and in some cases resulted in difficulties in using the conductance technique for surface-state analysis discussed in the next section.

4.3. *Evaluation of interface state parameters*

From the ratio of the maximum to the minimum capacitance, the impurity concentration of the CdTe under the insulator was determined using the method of Gupta and Anantha (1967). The value obtained for the device discussed in §4.1 was 3×10^{15} cm^{-3}. This represents an average over the maximum depletion region and takes no account of any impurity concentration profile that might exist at the surface of the semiconductor. Using this value the flat-band capacitance was calculated to be 133 pF (Sze 1969). From figure 1 the flat-band voltage is therefore $-0{\cdot}32$ V.

An initial estimate was made of the average density of surface states by calculating the theoretical flat-band voltage from a knowledge of the workfunction of the top electrode and the semiconductor electron affinity. Taking into account the slight variations in the flat-band voltage observed in different devices, an average value of 10^{11} states cm^{-2} was obtained.

We have also attempted to use the conductance technique introduced by Nicollian and Goetzberger (1965) to provide more accurate information on surface states. This method involves evaluating the equivalent parallel conductance (G_p) of the semiconductor depletion capacitance (C_D) in parallel with the series $R_s C_s$ network of the surface states. This was accomplished by converting the measured admittance into an impedance, subtracting the impedance in the accumulation region and finally converting the resulting impedance back into an admittance. Figure 3 compares the parallel and measured conductances as a function of bias voltage for an MIS device at 120 kHz. This structure was fabricated using the same substrate as the device for which the characteristics are shown in figures 1 and 2 and for which no distinct peak was observed in the G_p/ω versus bias curve. The errors in the accumulation region for the G_p/ω curve reflect the effect of a ±2% uncertainty in the measured values of the conductance and

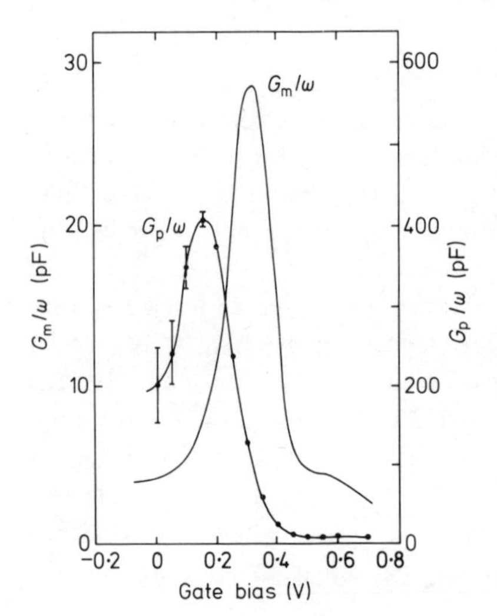

Figure 3. Measured conductance and calculated parallel conductance versus gate bias curves for a p-type CdTe/$CdSt_2$ capacitor at 120 kHz. Dielectric thickness 80 nm; device area $6{\cdot}5 \times 10^{-7}$ m^2.

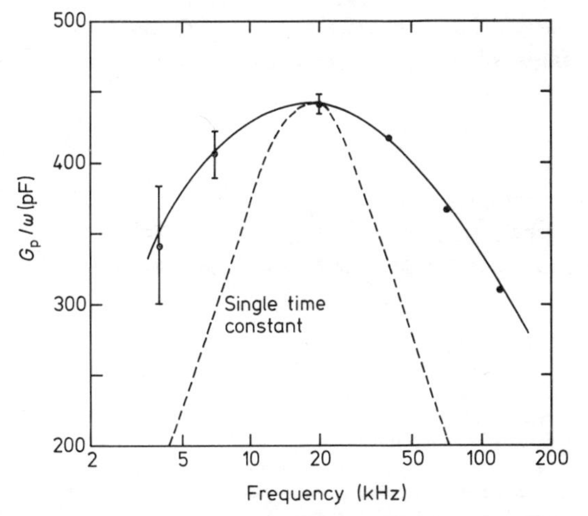

Figure 4. Parallel conductance data versus frequency at a gate bias (V_G = 0·225 V) corresponding to flat-band conditions for the device the high-frequency conductance of which is shown in figure 3.

capacitance. Because of the relatively narrow frequency range over which we were able to make reliable conductance measurements (4–120 kHz), we could only obtain distinct maxima in the G_p/ω versus frequency curve over a very small voltage range. Figure 4 shows such a curve corresponding to flat-band conditions for the device shown in figure 3. The curve is broader than is predicted by a single time-constant model. Time-constant dispersion can arise from statistical fluctuations of the surface potential (Goetzberger *et al* 1976) or from tunnelling of carriers from the semiconductor surface to traps located in the insulator (Preier 1967). Complex data processing and curve fitting are required to extract accurate parameters. However, using the single time-constant model to give an order-of-magnitude estimate for the density of surface states, we obtain $N_{ss} = 9 \times 10^{11}$ cm^{-2} eV^{-1} at flat-band conditions, i.e. approximately 0·2 eV above the valence-band edge for this particular device.

5. Conclusion

The results presented demonstrate that cadmium stearate films may be used to help characterise the surface of CdTe. Stringent control of the device preparation has been found necessary to obtain reproducible results.

Although we have reported data for relatively thick insulating layers, we feel that useful devices might be fabricated by exploiting tunnelling structures based on II–VI semiconductors. For example, it has been shown by Singh and Sewchun (1976) that the conversion efficiency of a photovoltaic Schottky-barrier device can be enhanced considerably by the presence of an interfacial layer. Theoretical studies emphasise the important influence of surface states and the critical control of insulator thickness required to optimise solar-cell performance. Over the years there has also been considerable interest in the possibility of using tunnel injection of minority carriers into direct band-gap semiconductors as an alternative to p–n junctions in electroluminescent devices (Clark *et al* 1976). In principle the Langmuir–Blodgett technique is ideally suited for the production of such devices, because insulating films one monolayer thick may be readily deposited. The particular dielectrics used in our investigations have quite low melting points and are, therefore, unlikely to form the basis of these devices. However, novel

Langmuir–Blodgett multilayer films that may have greater thermal stability than the simple fatty acid salts are at present being studied and we hope to report the results in due course.

Acknowledgments

The authors wish to thank J B Mullin (RSRE, Malvern) for supplying the CdTe crystals and for useful discussions. They also wish to acknowledge the technical assistance of W A Barlow and J A Finney (ICI Ltd).

References

Agarwal D K and Srivastava V K 1975 *Thin Solid Films* **27** 49–62
Blodgett K B 1935 *J. Am. Chem. Soc.* **57** 1007–22
Clark M D, Baidyaroy S, Ryan F and Ballantyne J M 1976 *Appl. Phys. Lett.* **28** 36–8
Gaines G L Jr 1966 *Insoluble Monolayers at Liquid–Gas Interfaces* (New York: Wiley Interscience)
Goetzberger A, Klausmann E and Schulz M J 1976 *CRC Crit. Rev. Solid St. Sci.* **6** 1–43
Gupta D C and Anantha N G 1967 *Proc. IEEE* **55** 1108
Langmuir I 1920 *Trans. Faraday Soc.* **15** 62
Lunn B and Bettridge V 1977 *Rev. Physique Appl.* **12** 151–4
Nicollian E H and Goetzberger A 1965 *Appl. Phys. Lett.* **7** 216–9
Preier H 1967 *Appl. Phys. Lett.* **10** 361–3
Roberts G G, Pande K P and Barlow W A 1978a *IEE J. Solid St. Electron Dev.* **2** 169–75
Roberts G G, Vincett P S and Barlow W A 1978b *J. Phys. C: Solid St. Phys.* **11** 2077–85
Singh R and Sewchun J 1976 *Appl. Phys. Lett.* **28** 512–4
Sze S M 1969 *Physics of Semiconductor Devices* (New York: Wiley) ch 9 pp425–504

MNOS memory transistors

M Pepper
Cavendish Laboratory, Madingley Road, Cambridge CB3 0HE, UK

Abstract. In this article various properties and problems of the MNOS memory transistor are summarised. The different models for the tunnelling process are discussed, as is the evidence in favour of band-to-band tunnelling followed by trapping in the nitride rather than direct tunnelling. Some of the problems encountered during device use, such as collapse of the memory window, are presented and results are summarised on their possible relation to processing variables.

1. Introduction

The study of the properties and potential of MNOS memory transistors cuts across many areas. It embraces carrier transport and trapping in dielectrics, the nature of the silicon–dielectric interface, quantum-mechanical tunnelling through potential wells of different types and the most useful configuration of memory devices on the final chip. As a result of a considerable amount of effort, the memory transistor now has a definite role in the hierarchy of electronic devices; in turn this has led to the issue of an IEEE standard (581-1978) on the device. In addition to discussing device characterisation this states '... the ability of the device to withstand continuously repeated writing pulses of alternating polarity without changing its device characteristics ...'. It is perhaps this problem of changes with successive write–erase cycles that is the major problem of the device and is the subject of considerable work. Nevertheless, there are chips available containing memory devices offering 1 year's storage, stable to 10^6 cycles (Greenwood 1979a, b), and 24 h storage, stable to 10^{11} cycles (Marraffino *et al* 1978). Naturally, there are possible alternatives to the conventional MNOS structure, for example the use of tungsten to provide traps at the SiO_2–Al_2O_3 interface (Kahng *et al* 1974) and the use of a silicon buried gate, SIMOS, an electrically alterable FAMOS (Scheibe and Schulte 1977). The general trapping phenomenon in insulators renders other semiconductors, suitable in principle, e.g. CdSe, but the basic MNOS transistor is occupying an area of application despite the problem with repeated cycling. In this short article some of the more striking features of MNOS operation will be discussed.

2. Direct tunnelling models

This approach was first introduced by Wallmark and Scott (1969) and Ross and Wallmark (1969) and has been extended subsequently. The principal assumptions are (see figures 1 and 2):

† On leave of absence from the Plessey Company, Allen Clark Research Centre, Caswell, Towcester, Northamptonshire.

0305-2346/80/0050-0193$02.00

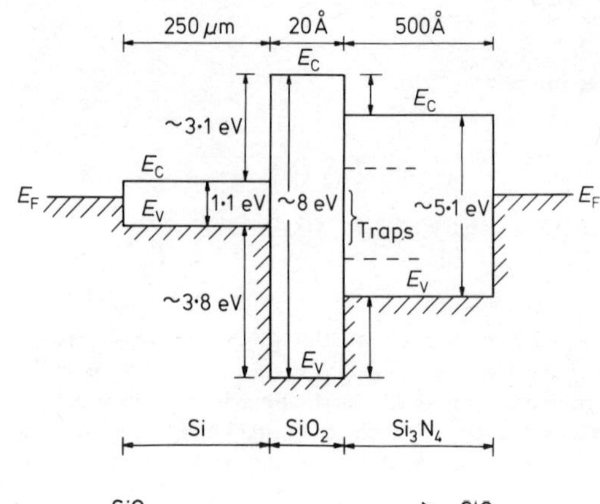

Figure 1. Schematic MNOS energy-level diagram.

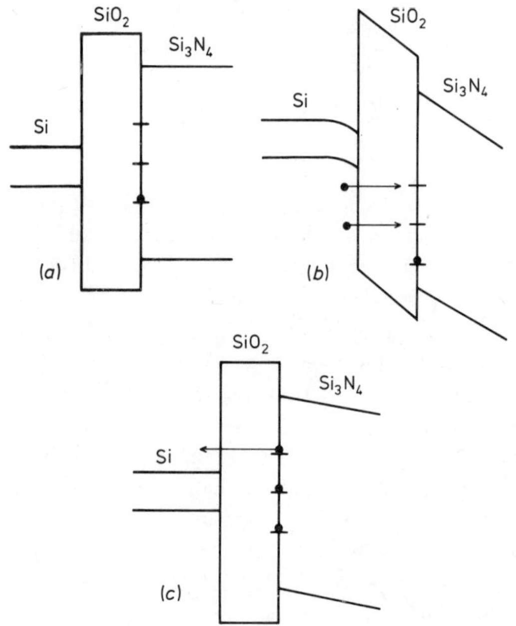

Figure 2. Charge exchange in the direct tunnelling model illustrating (*a*) the flat-band condition, (*b*) charging and (*c*) discharging on return to flat-band.

(i) Traps are distributed uniformly within the Si_3N_4 but at a single energy below the conduction-band edge.
(ii) Charging is by the direct tunnelling of electrons from the Si valence band to the traps. Discharging is by the tunnelling of these electrons from the traps to the Si conduction band.
(iii) The tunnelling is time-independent, i.e. the change in the nitride charge is neglected.

The effect of surface states at the Si–SiO_2 interface is generally omitted, but if necessary their effect can be included by a simple modification of the tunnelling probability. The result of the analysis in the form proposed by Ferris-Prabhu (1972a, b) is that, after a short initial period, the time dependence of trap charging $N(t)$ is given by

$$N(t) \approx N_{\mathrm{TOT}}(\lambda/x_{\mathrm{m}})\ [\ln \omega_0 t - (t/t_{\mathrm{m}})] \qquad (1)$$

Here λ is the constant in the tunnelling exponent $(\exp(-x/\lambda))$, x_m is the maximum tunnelling distance introduced by Ferris-Prabhu (1972a,b) to simplify the algebra without loss of physical realism, ω_0 is the attempt frequency, and $t_m = \exp(x_m/\lambda)\omega_0$.

This simple model satisfactorily explains two observations, the logarithmic increase in nitride charge and the eventual saturation. In a similar manner, the discharging is found to be of the form

$$N_t \sim N_{TOT} \{1 - (\lambda/x_m) [\ln \omega_0 t - (t/t_m)]\}. \qquad (2)$$

The model was amended by Lundqvist *et al* (1973) who took account of the change in tunnelling probability caused by the change in nitride charge. It was found that the modified expression did not differ greatly from the original Ross and Wallmark model.

A different and in some respects more realistic model was suggested by Dorda and Pulver (1970). These authors proposed that the nitride traps are unformly distributed in energy but are spatially localised. This latter assumption is not as drastic as it may initially appear; the exponential decrease in tunnelling probability with distance ensures that the dominant role is played by those traps close to the Si. The expressions derived from this model were formally similar to equations (1) and (2), but as pointed out by Ferris-Prabhu (1977), because of the use of a model-dependent decay constant, the model lacks predictive ability.

A further treatment by White and Cricchi (1972) introduces the concept of tunnelling from the nitride traps to states at the Si–SiO_2 interface with the same energy. It is assumed that the traps are at a single level and are localised at the oxide–nitride interface. The novelty of this model is that, for the discharging mode, as the trapped charge decreases, the voltage drop across the oxide decreases and the traps are located opposite the middle of the Si energy gap. Here the density of interface states is at a minimum and consequently the tunnel current decreases. The direct tunnelling treatments have recently been reviewed by Ferris-Prabhu (1977), who points out that attempts to refine them further may be fruitless in view of the uncertainty surrounding the values of the parameters. Thus, values of ω_0 between 10^{11} and $10^{14}\,s^{-1}$ have been reported and, in addition, it is clear that the device switching characteristics are dependent on nitride deposition conditions. However, more seriously, evidence is available to show that charge is distributed throughout the nitride and that the dependence of charge trapping on write voltage cannot therefore be explained by direct tunnelling models.

3. Band-to-band tunnelling

There is a considerable body of evidence against the direct tunnelling models. Briefly, the main points are:

(i) The work of Yun (1974), extended by Arnett and Yun (1974) (and also Lehovec *et al* 1978) shows that the memory charge is distributed throughout the Si_3N_4.
(ii) Gordon and Johnson (1973) found that the discharging of n-type MNOS capacitors, i.e. negative gate bias, coincides with the onset of deep depletion in the Si. The inability of the system to sustain an inversion layer is due either to holes tunnelling into the dielectric or to holes in the Si recombining with electrons tunnelling back. The authors showed that the recombination is not sufficient and that the tunnelling of holes from

the Si valence band into the nitride valence band is occurring. The voltage dependence suggests that holes are tunnelling through a region of the nitride as well as the oxide.
(iii) If direct tunnelling is occurring, then charging and discharging are achieved by varying the voltage drop across the SiO_2 by a value less than or equal to 1·1 eV. In practice, it is found that the voltage drop across the oxide is greater than ~2 eV.

It thus appears that the memory process arises from tunnelling from the Si into the energy bands of the nitride, followed by carrier transport and subsequent trapping (see figure 3). The first treatment of this mechanism by Lundstrom and Svensson (1972) did not consider in detail the change in charge and hence voltage across the oxide with time. A more complete theory was developed by Chang (1977), who assumed that modified Fowler–Nordheim tunnelling was occurring through the oxide and a small region of the nitride. It was assumed that once carriers tunnelled into the nitride, the subsequent trapping did not limit the process, but merely altered the field across the oxide and hence the tunnelling rate.

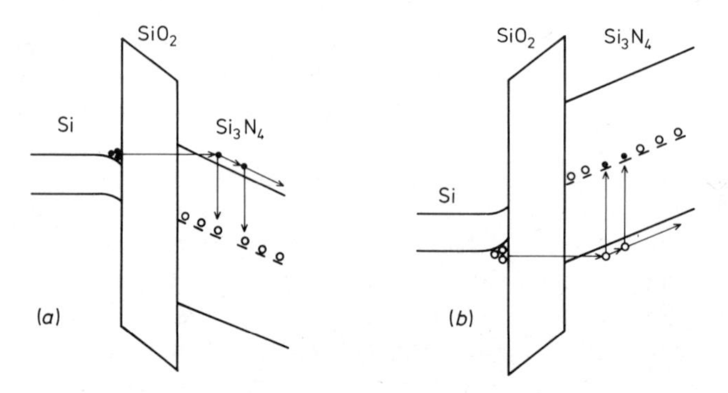

Figure 3. Charge exchange in the band-to-band tunnelling model (*a*) positive gate voltage and (*b*) negative gate voltage.

Beguwala and Gunckel (1978) have performed a simulation of the charging process that incorporates both tunnelling and conduction in the nitride. The change in the flat-band voltage $\Delta V_{FB}(\tau)$ for a pulse of duration τ and amplitude V_g can be simply derived from the relation

$$-dQ/dt = J_2 - J_1$$

where Q is the charge in the nitride, J_2 is the current flowing through the nitride to the gate and J_1 is the current flowing into the SiO_2. Omitting the algebra, ΔV_{FB} can be written as

$$\Delta V_{FB}(\tau) = (1 - \bar{x}/d_N)(d_N/\epsilon_N)\int_0^{\tau} J_2(V_g, t) - J_1(V_g, t)\, dt$$

where $\bar{x}$ is the centroid of the memory charge in the nitride and d_N and ϵ_N are the thickness and dielectric constant of the nitride. The oxide current J_1 can be calculated from the generalised Fowler–Nordheim expression given by Murphy and Good (1956):

$$J_1 = (ekT/2h^2c \sin \pi ckT) \exp(-b)$$

where b and c are complex expressions dependent on the shape of the potential barrier through which tunnelling is occurring. Chang (1977) has given suitable expressions for b and c.

The nitride current J_2 can be written as

$$J_2(V_g, t) = AF_N(V_g, t)\exp(-\Delta E/kT) + BF_N^2(V_g, t)\exp(-W/F_N) + DF_N(V_g, t)\exp\{-e[\phi_B - (eF_N(V_g, t))^{1/2}/\pi\epsilon_N]/kT\}.$$

The first term arises from Ohmic hopping conductivity, $F_N(V_g, t)$ is the field across the nitride, W is an activation energy for the process and A is a constant. The second term represents field emission or tunnelling of electrons into the nitride conduction band; B and C are empirically determined. The third term arises from Poole–Frenkel conduction, the field-assisted hopping of electrons from the traps into the conduction band. Here D is empirically determined and ϕ_B is the trap depth.

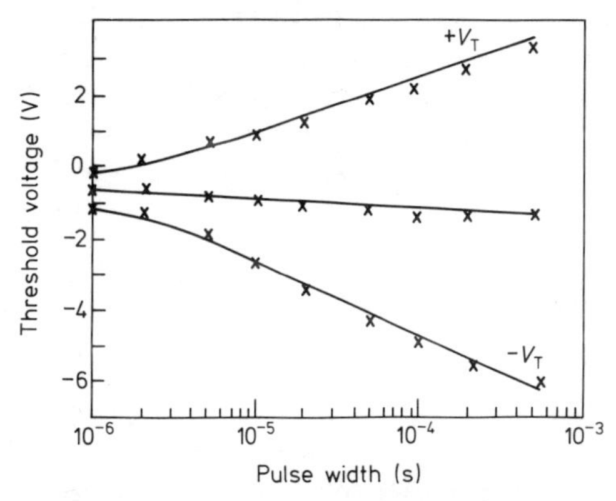

Figure 4. The variation of threshold voltage as a function of pulse width (Beguwala and Gunckel 1978). The crosses are experimental points and the curves are the result of computer calculations; the central line is the window centre. The applied voltage is (±) 22 V, the SiO_2 thickness 19 Å and the Si_3N_4 thickness 340 Å; the values required to fit the experimental points are slightly greater.

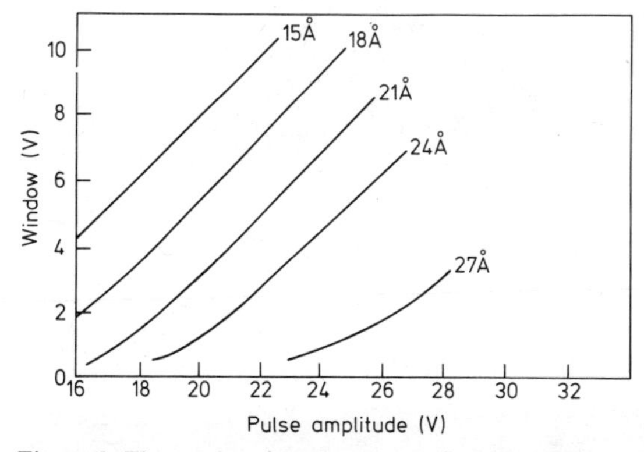

Figure 5. The results of Beguwala and Gunckel (1978) showing the calculated oxide thickness necessary to account for the window width plotted as a function of applied voltage. The Si_3N_4 thickness is ~400 Å; the SiO_2 thickness 19 Å and the pulse width 100 μs; these results agree with experiment to within experimental error.

By obtaining values for the constants from curve fitting and published data, Beguwala and Gunckel were able to fit their expressions to experimental data. Figures 4 and 5 show the agreement that they obtained. Values of the more meaningful constants are as follows: $W = 0{\cdot}1$ eV; $\phi_B = 1{\cdot}3$ eV; $A = 5{\cdot}1 \times 10^{-14}$ A V^{-1} cm^{-1}; $B = 3{\cdot}5 \times 10^{-10}$ A V^{-1}; $D = 3 \times 10^{-9}$ A V^{-1} cm^{-1}; $C = 1{\cdot}2 \times 10^{8}$ V cm^{-1}.

4. Avalanche punch-through erase

One of the problems encountered in circuit use of the MNOS memory is that the processes of writing and erase require the availability of pulses of both polarities. Uchida (1973) and Uchida *et al* (1978) proposed a method of circumventing this difficulty; by applying the erase pulse to the source and drain, the memory charge can be erased in the normal way. It was suggested that the high source (and drain) to substrate voltage resulted in an avalanche and the injection of hot carriers into the dielectric. However, Kahng *et al* (1977) showed that the erase process has the same efficiency for both n- and p-channel devices, i.e. electron and hole injection. In view of the greater barrier height for hole injection, it is unlikely that hot-carrier emission over the barrier is responsible for the erase. The authors have produced evidence to show that both punch-through and avalanche are required for the erase. When punch-through occurs, a potential minimum is created at the silicon surface and this acts as a well to trap majority carriers created by the avalanche. Figure 6 is a schematic illustration of the potential distribution in an n-channel device with the gate grounded. Carrier trapping in the pocket causes an increase in the surface potential until the supply of carriers is equalled by leakage into the substrate. It can be seen from figures 6(*a*) and (*b*) that before avalanching the surface potential is not significant, but after avalanche the enhancement is such as to allow appreciable tunnelling between the dielectric and the Si. Figure 7 shows the computed potential as a function of distance into the substrate for a line that is central between source and drain. The source and drain voltages are +35 V and the plots are shown for different values of the potential at the Si–SiO_2 interface, determined principally by the

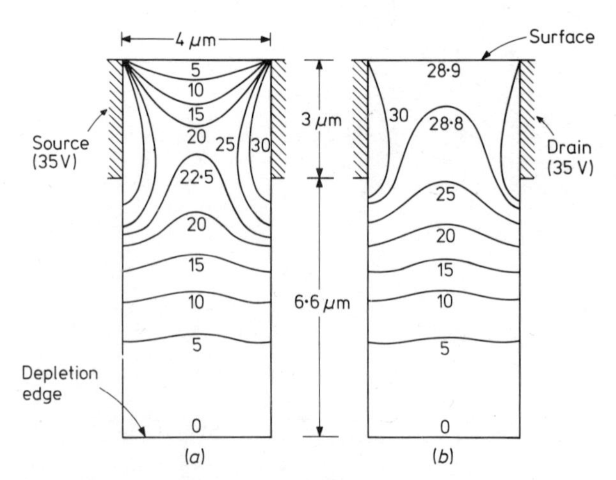

Figure 6. The equipotential lines for a device in punch-through condition, after Kahng *et al* (1977). (*a*) Before avalanche; (*b*) after avalanche.

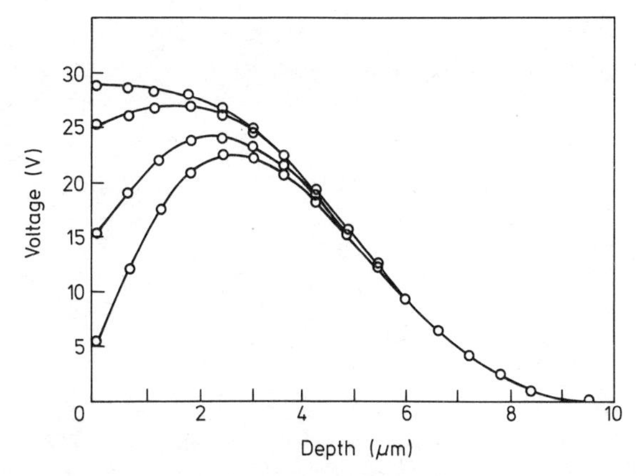

Figure 7. The computed potential plotted against distance into the bulk for a line midway between source and drain (Kahng *et al* 1977). The curves are for different values of surface potential, i.e. different values of majority-carrier concentration at the Si surface. The source and drain voltages are +35 V, source–drain separation = 4 μm, junction depth = 3 μm and $N_A = 10^{15}\ cm^{-3}$.

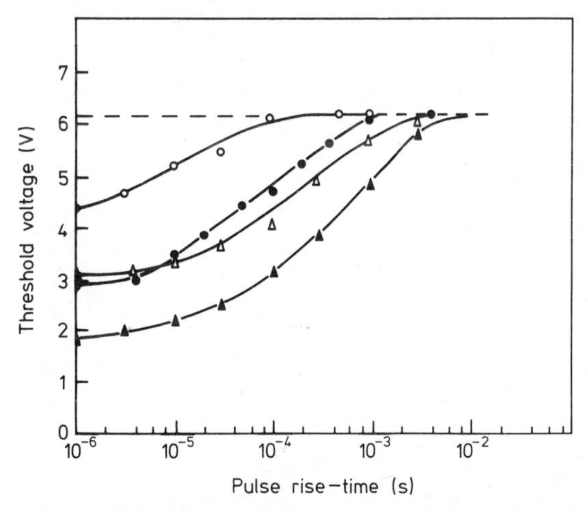

Figure 8. The effects of pulse rise-time on threshold voltage in the avalanche punch-through erase mode. The pulse rise is followed by 10 ms application. Points ○ and ● are +32 and +35 V respectively for a device with a channel width of 25 μm and a channel length of 4 μm. Points △ and ▲ are +27 and +29 V for a device with a channel width of 75 μm and a channel length of 4 μm. The broken line is the initial written state.

gate voltage. There is a maximum value of surface potential above which the carriers leak into the bulk; for the device dimensions used in the derivation of figure 7, this is ~29 V, a value that is sufficiently high to allow appreciable tunnelling. As expected with such a mode of operation, the surface potential is sensitive to device dimensions and substrate doping. The pulse rise-time also plays a significant role; a long rise-time results in an inefficient erase, presumably because of leakage of carriers from the well or an insufficient supply (see figure 8).

With this technique, erase can be minimised and prevented by the application of a voltage of the same polarity to the gate. Kahng *et al* (1977) showed the advantages of the technique for devices with SiO_2 60 Å thick, Al_2O_3 450 Å thick and tungsten present at the Al_2O_3–SiO_2 interface to increase the trapping efficiency.

In a similar vein, devices have been proposed where both electrons and holes can be injected onto a floating buried gate from p^+–n and n^+–p diodes (Tarui *et al* 1972 and the DIFMOS described by Gosney 1977). Conceptually these devices are an advance on the original FAMOS but are also slow; a typical time to charge the gate is of the order of tens of milliseconds.

5. Memory charge decay and fatigue

There are two main factors affecting decay of the memory charge. The first is the movement of charge deep into the nitride film with a consequent increase in the erase time. The second is the presence of states in the gap at the Si–SiO_2 interface. These states can receive charge tunnelling back from the nitride which then flows into the substrate, either by recombination or emission into a band. This can be a problem in view of the creation of interface states by repeated cycling. Investigation of the creation of interface states by conventional C–V techniques is not unambiguous because of current flow in the nitride and possible non-uniformities in the lateral charge distribution. Schauer *et al* (1978) have investigated interface-state creation as a function of cycling and a typical result is shown in figure 9. These authors found that if the memory charge was positive, which implies trapped holes, the decay rate varied linearly with the logarithm of the number of interface states N_{ss}. This is of course expected on any simple theory as the tunnelling rate will vary as N_{ss}. When the nitride charge was negative, the decay rate did not vary with write–erase cycling. However, as the density of interface states did increase, it was concluded that in this case direct tunnelling between the traps and the interface states is not significant, in contrast to the results obtained by White and Cricchi (1972).

The problem of fatigue is summarised in figure 10, where the change in the memory window for write–erase pulses of ±40 V and 5 ms width is shown. Beyond 10^4 cycles, the window starts to close and also shifts in the negative direction.

Zirinsky (1975) has studied the dependence of fatigue on nitride and oxide processing

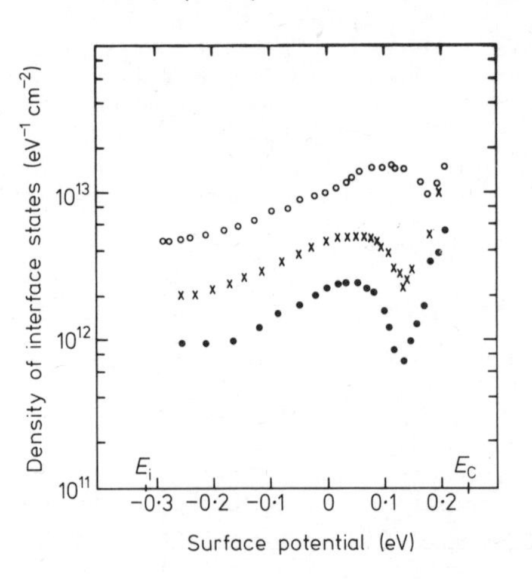

Figure 9. The results of Schauer *et al* (1978) showing the creation of interface states with increasing number of cycles. Points •, × and ○ are for zero, 10^6 and 10^7 cycles respectively. The applied voltage is ± 30 V and pulse width is 5 ms. The dip in the density of interface states near the conduction-band edge E_C may be due to interface-state capacitance.

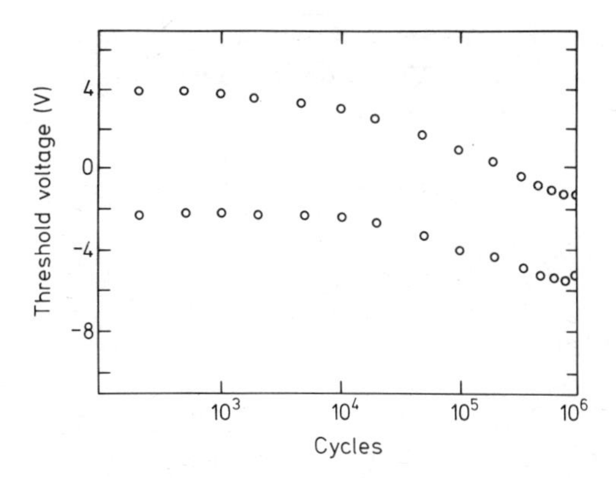

Figure 10. Results of Tanaka and Nishi (1975) showing the change in the memory window (fatigue) as a function of the number of pulses. The applied voltage is ±40 V, pulse width = 5 μs.

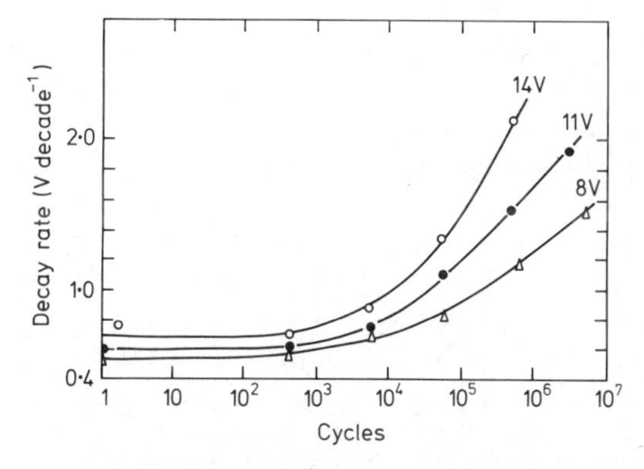

Figure 11. Results of Neugebauer and Burgess (1976) showing the rate of decay of memory charge as a function of the number of cycles and initial value of the window, this is marked on the curves. d_N = 743 Å.

variables. He found that if the thin SiO_2 layer was grown in oxygen, either pure or mixed with nitrogen, failure occurred between 10^4 and 10^7 cycles for voltages in the range 35–40 V and pulse widths of about 1 ms. The failure mechanism was a collapse of the memory window accompanied by, and possibly partly attributable to, a considerable increase in the density of interface states (see figure 11). On the other hand, if the oxide was grown in NO in the temperature range 900–1100 °C, window stability was achieved up to at least 10^8 cycles. In general, window stability can be sustained by decreasing the pulse width.

White *et al* (1977) have also investigated fatigue for different oxide preparation conditions. They suggested that hydrogen renders the oxide more sensitive to prolonged cycling but point out that oxides can be prepared offering window stability up to 10^{10} cycles. It is interesting to note that exposing the untreated Si surface to HCl gas increases fatigue in the completed device, but can increase the charge retention time (McLouski *et al* 1978). This may be related to the well-known effect of HCl on the reduction of the density of interface states and surface defects.

The decay of the memory charge with time can be divided into three regimes. Initially the charge is stable for a period of time up to ~10^{-4} s; this is followed by a period during which the charge decays logarithmically with time. The final regime is also logarithmic decay but at an increased rate. Both the logarithmic decay rates are enhanced by repeated cycling and increase with the amplitude and duration of the pulses. At present the cause of the collapse of the window is not definitely established.

It is often not possible to investigate the window or charge storage characteristics directly when the memory transistor is in circuit configuration. Methods have, however, been developed to overcome this problem (see Schuermeyer 1977, Hsia 1977). For example, write pulses can be applied for increasing times and the time necessary for destruction of the information can be obtained. The state of charge storage and the window amplitude can then be obtained from the retention characteristics of the device. At present, circuits are appearing in which the problem of degradation is circumvented by appropriate word organisation.

6. Relevant dielectric properties

The determination of the charge distribution in the nitride film by Yun (1974) was of major significance in understanding device behaviour. The principal feature of the method is that if the device is pulsed from flat-band, so changing the nitride charge, then the total change in the charge on the metal gate on restoring the flat-band condition is equal to the change in the nitride charge. Measurements of the charge flow into the metal gate between successive flat-band conditions and the flat-band voltage allow determination of the centroid of the nitride charge. Provided that there is no significant charge injection from the gate and carriers injected from the Si have not left the nitride, the centroid $\bar{x}$ of the change in the nitride charge is given by

$$\bar{x} = d_N [1 - (C_N \Delta V_{FB}/Q)]$$

where Q is the change in nitride charge and C_N is the capacitance. It is assumed that the surface-state and oxide charges have not changed, Q is obtained by electronic integration. In practice a feedback circuit is often used to maintain the flat-band condition and the response time of the circuit is greater than the time of the charging pulse. The initial work by Yun showed that injected electrons are present throughout the nitride film, the centroid being about one-third of the film thickness.

A slight modification of Yun's method was proposed by Lehovec *et al* (1978). Instead of returning the device to the charge-free state before the application of the next charging pulse, these authors applied pulses continuously, in a staircase pattern, and measured the flat-band charge between the pulses. It was found that when the nitride charge is greater than ~10^{13} cm^{-2}, the charge distribution differs depending on whether a single charging pulse is used (as in Yun's method), or successive pulses, where the total time of application is equal to that of the single pulse. In the latter case (the staircase method), the charge is much deeper into the nitride. The increasing depth of the centroid with magnitude and time of the charging pulse is clearly evident in Yun's work. Arnett and Yun (1975) used the charge-centroid method to investigate the distribution of traps within the nitride. They found that if $\bar{x}$ is less than about 20% of the film thickness, it increases uniformly with Q and is independent of the applied voltage. The value of $\bar{x}$ then increases

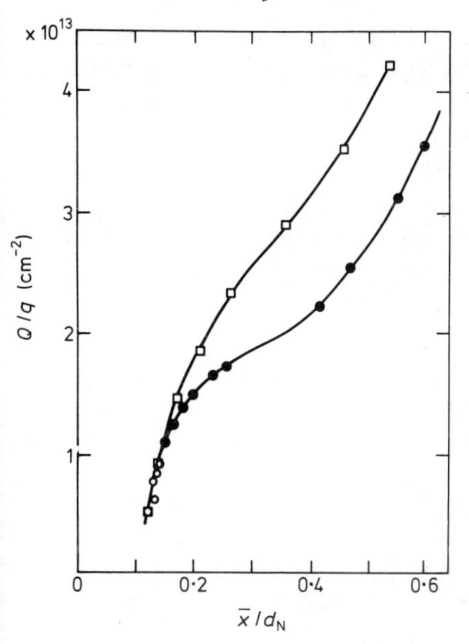

Figure 12. Results of Lehovec *et al* (1978) comparing the charge centroid, as a function of nitride thickness, with total nitride charge. Points □ were obtained by the staircase method and points • by Yun's (1974) method.

rapidly and becomes voltage-dependent as shown in figure 12. If the space-charge density is sufficiently great, the high field produced can enhance detrapping at the edge of the space-charge region. The analysis of Arnett and Yun shows that the total trap concentration is about 6×10^{18} cm^{-3} with a capture cross section of $\sim 4{\cdot}6 \times 10^{-13}$ cm^{-3}, consistent with a positively charged centre as suggested by the observation of Poole–Frenkel conduction in nitride films. Work such as this has revealed the existence of back tunnelling from the dielectric to the silicon in the earliest stage of charging and also carrier injection from the metal gate. However, neither of these effects is sufficiently great to modify the analysis seriously.

The conductivity of the nitride can also be detected from more conventional charge-storage measurements. Thus Beguwala and Gunckel (1978) incorporated the nitride conductivity into their model of MNOS operation and Williams and Beguwala (1978) investigated the effect of temperature on the decay of the memory charge. In this work it was found that the rate of decay up to 1 s was identical at both 23 and 244 °C, as expected for tunnelling. However, after 1 s the higher temperature produced a more rapid rate, indicative of enhanced nitride conduction assisting the decay.

The role of the deposition conditions in determining device properties has been investigated by many workers, initial studies were performed by Ross *et al* (1969). Using conventional devices with 20 Å thick layers of SiO_2, they varied the Si_3N_4 deposition temperature between 650 and 1100 °C. Increasing the temperature of deposition increased the rate of decay of the memory charge and also decreased the positive dielectric charge prior to electrical operation. The dependence of the DC conductivity on the nitride deposition conditions has been investigated by, for example, Brown *et al* (1968). Work on MNS structures has shown that increasing the silane–ammonia ratio increased the conductivity of the film by lowering the activation energy in the Poole–Frenkel exponent. Brown *et al* (1968) suggested that in some way this is associated with the inclusion of excess Si (see also Zirinsky 1975). The structure of the nitride is pertinent here: if the deposition temperature is greater than 900 °C and a high ammonia: silane ratio is used, small crystallites ~10 Å across are formed (Topich and Yon 1976). The high density of defect sites at the 'grain boundaries' may enhance the Poole–Frenkel conduction.

There is discussion in the literature on the role of hydrogen in the dielectrics. Multiple internal reflection techniques (Stein *et al* 1978) have shown that the amount of hydrogen in the oxide is very dependent on the nitride deposition conditions. Johannessen *et al* 1977) have suggested that, after nitride deposition the oxide is converted to an oxy-nitride.

It is clear that there is now considerable scope for correlating chemical and electrical properties. Thus Zirinsky (1975) has found that the presence of hydrogen in the oxide shifts the memory window to the negative side, although the presence of hydrogen and HCl can result in faster switching. Considerable progress in understanding these effects is to be expected in the future.

Acknowledgments

I thank the Warren Research Fund Committee of the Royal Society for the provision of a Warren Research Fellowship. I am grateful to V A Browne, G Greenwood and R E Oakley for discussions on this topic. The author's research has been partially supported by the US Army through its European Research Office.

References

Arnett P C and Yun B H 1974 *Appl. Phys. Lett.* **25** 340–2
Beguwala M M E and Gunckel T L 1978 *IEEE Trans. Electron Devices* **ED-25** 1023–30
Brown G A, Robinette W C and Carlson H G 1968 *J. Electrochem. Soc.* **115** 948–55
Chang J J 1977 *IEEE Trans. Electron Devices* **ED-24** 511–8
Dorda G and Pulver M 1970 *Phys. Stat. Solidi* **A** 71–9
Ferris-Prabhu A V 1972a *Appl. Phys. Lett.* **20** 149–50
—— 1972b *Phys. Stat. Solidi* **A11** 81–6
—— 1977 *IEEE Trans. Electron Devices* **ED-24** 524–30
Gordon N and Johnson W C 1973 *IEEE Trans. Electron Devices* **ED-20** 253–6
Gosney W M 1977 *IEEE Trans. Electron Devices* **ED-24** 594–9
Greenwood G 1979a *Plessey Data Sheet* MN9105
—— 1979b *Plessey Data Sheet* MN9102
Hsia Y 1977 *IEEE Trans. Electron Devices* **ED-24** 568–77
Johannessen J S, Helms C R, Spicer W E and Strausser Y E 1977 *IEEE Trans. Electron Devices* **ED-24** 547–51
Kahng D, Brews J R and Sundburg W J 1977 *IEEE Trans. Electron Devices* **ED-24** 531–5
Kahng D, Sundburg W J, Boulin D M and Ligenza J R 1974 *Bell Syst. Tech. J.* **53** 1723–39
Lehovec K, Chen C H and Fedotowsky A 1978 *IEEE Trans. Electron Devices* **ED-25** 1030–6
Lundqvist L, Lundstrom K and Svensson C 1973 *Solid St. Electron.* **16** 811–23
Lundstrom K and Svensson C 1972 *IEEE Trans. Electron Devices* **ED-19** 826–36
Marraffino P, Newman R, Wegener R, Borovick M B, Lewis E T and Lodi R J 1978 *IEEE Trans. Electron Devices* **ED-25** 1054–60
McLouski R M, Peckerar M C and Schreurs J J 1978 *J. Electrochem. Soc.* **125** 985–8
Murphy E L and Good R H 1956 *Phys. Rev.* **102** 1464
Neugebauer C A and Burgess J F 1976 *J. Appl. Phys.* **47** 3182–91
Ross E C, Duffy M T and Goodman C 1969 *Appl. Phys. Lett.* **15** 408–9
Ross E C and Wallmark J T 1969 *RCA Rev.* **30** 366–81

Schauer H, Arnold E and Murau P C 1978 *IEEE Trans. Electron Devices* **ED-25** 1037–41
Scheibe A and Schulte H 1977 *IEEE Trans. Electron Devices* **ED-24** 600–6
Schuermeyer F L 1977 *IEEE Trans. Electron Devices* **ED-24** 564–8
Stein H J, Picraux J T and Holloway P H 1978 *IEEE Trans. Electron Devices* **ED-25** 1008–14
Tanaka T and Nishi Y 1975 *Japan J. Appl. Phys.* Suppl. **44** 203–9
Tarui Y, Hayashi Y and Nagai K 1972 *IEEE J. Solid St. Circuits* **SC-7** 369–75
Topich A and Yon E T 1976 *J. Electrochem. Soc.* **123** 535–9
Uchida Y 1973 *Japan. J. Appl. Phys.* **42** 151
Uchida Y, Saito S, Nakane M, Endo N, Matsuo T and Nishi Y 1978 *IEEE Trans. Electron Devices* **ED-25** 1066–70
Wallmark J T and Scott J H 1969 *RCA Rev.* **30** 335–65
White M H and Cricchi J R 1972 *IEEE Trans. Electron Devices* **ED-19** 1280–8
White M H, Dzimianski J W and Peckerar M C 1977 *IEEE Trans. Electron Devices* **ED-24** 577–80
Williams R A and Beguwala M M E 1978 *IEEE Trans. Electron Devices* **ED-25** 1019–23
Yun B H 1974 *Appl. Phys. Lett.* **25** 340–2
Zirinsky S 1975 *J. Electron Mater.* **3** 591-624

Chemical composition and electronic states of MNOS structures studied by Auger electron spectroscopy and electron energy-loss spectroscopy

N Lieske and R Hezel

Institut für Werkstoffwissenschaften VI, Universität Erlangen-Nürnberg, 8520 Erlangen, West Germany

Abstract. For the first time a depth profile of a MNOS structure is presented, based upon a combination of results obtained by Auger electron spectroscopy (AES) and low-energy electron loss spectroscopy (ELS) with low-energy argon-ion sputtering. Results are presented for a MNOS structure made up of a 3·4 nm Si oxide layer and a 45 nm low-pressure CVD Si nitride layer on p-Si (100). The chemical composition as a function of depth is characterised in detail by the core- and valence-electron excitation spectra obtained with ELS. A major result is that the thin Si oxide has changed during nitride deposition into a Si oxynitride layer made from Si–O, Si–N, Si–Si bonds with broken Si–O and Si–N bonds. The electron states of this interfacial Si oxynitride depend strongly on thermal annealing treatments. At the Si surface, electron states are detected which are not visible at the crystalline or amorphous surface.

1. Introduction

The electrical behaviour of MNOS transistors depends strongly on the bulk properties of the thin amorphous Si oxide and Si nitride dielectric layers and their interface regions, particularly the chemical composition and electronic states, which are closely related to each other. The aim of the present work is to obtain information about these properties as a function of depth in a MNOS system, with a depth resolution of about 10 Å. This can be achieved by low-energy electron spectroscopy, because of the short mean-free path (5–30 Å) of electrons in the energy range 0–2000 eV (Morrison 1977, Kane and Larrabee 1974).

In recent papers (Lieske and Hezel 1979a,b,c) it was shown that a combination of Auger electron spectroscopy (AES) and low-energy electron loss spectroscopy (ELS) is a suitable method for obtaining basic information about the elemental composition and chemical bonding states of thin amorphous Si oxide and Si nitride films. Valence- and core-electron excitation spectra measured by ELS provide electron energy-level schemes for ground and excited states. An important result, derived from the excellent agreement between ELS spectra, optical and x-ray experiments and theoretical calculations, is that the electronic states in amorphous SiO_2 and Si_3N_4 can be described by localised molecular states. Based upon this method, the formation of basic tetrahedra [SiO_4] and [SiN_4], and mixed tetrahedra with excess silicon and the presence of broken Si–O and Si–N bonds can be detected.

0305-2346/80/0050-0206$02.00

The principal mechanisms of AES and ELS have been described elsewhere (see for example Sevier 1972, Ludeke and Esaki 1975, Rowe *et al* 1975, Bauer 1969, Lieske and Hezel 1979a, b, c, d), but two important features should be pointed out:

(i) Since the energy of an Auger electron is due to a three-particle process, the interpretation of the AES spectra, apart from the elemental composition, is complicated. On the other hand, the value of a characteristic energy loss depends only on two energy levels and thus the interpretation of the ELS spectra is more straightforward.
(ii) Whereas the Auger process only gives information about the filled electron states, both the ground and excited states of a system can be characterised by ELS.

In this work AES was used mainly to determine the quantitative elemental composition; ELS provided detailed information about the chemical bonding states. By combining AES and ELS with low-energy Argon-ion bombardment, the information could be obtained as a function of depth, with a depth resolution of ~10 Å.

For the first time, AES and ELS have been used to determine the depth profile for a thin-oxide MNOS structure. In contrast to the methods used for depth profiling until now (AES, SIMS, XPS and RIBS), only ELS is able to give information about the excited electronic states, which essentially determine the electronic behaviour of a system exposed to external fields.

2. Experiments

The experiments were performed in a stainless steel UHV chamber equipped with a four-grid LEED system, a single-pass cylindrical mirror analyser and a quadrupole mass analyser. A special flat-beam-profile ion gun provided noble-gas ion beams with energy that could be varied between 200 eV and 6 keV in order to clean or sputter the sample. The AES and ELS spectra were measured with the CMA and monitored in the first and second derivative mode using a modulation technique.

The MNOS structure was made up of a 3·4 nm thick layer of Si oxide and a 45 nm thick layer of low-pressure CVD Si nitride on p-Si(100). The index of refraction and the thickness of the dielectric layers were determined using an optical ellipsometer ($\lambda = 632{\cdot}8$ nm). For further experimental details see Lieske and Hezel (1979b) and Lieske (1979).

3. Results and discussion

3.1. AES and ELS spectra of Si, Si oxide and Si nitride

In this section the results derived from the AES and ELS spectra of elemental silicon and amorphous Si oxide and Si nitride are summarised. These basic spectra are required for the interpretation of the spectra obtained during depth profiling of an MNOS structure. For further details see Lieske and Hezel (1979a, b, c, d) and Lieske (1979).

3.1.1. Auger spectra. In figure 1 the AES spectra of Si, SiO_2 and Si_3N_4 are shown. There are characteristic differences between the three samples regarding the energy positions and lineshapes of the Auger signals. The quantitative elemental composition can be deduced from the peak-to-peak heights of the main Auger transitions in the N_{KLL},

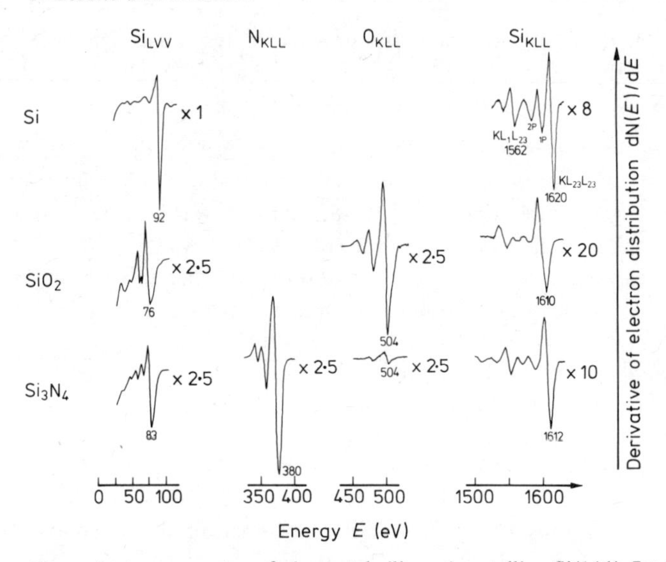

Figure 1. Auger spectra of elemental silicon (crystalline Si(111) 7 × 7 surface), amorphous Si oxide (produced by *in situ* thermal oxidation of the crystalline Si surface) and Si nitride (CVD Si nitride produced by NH_3/SiH_4 reaction, optical index of refraction $n = 1{\cdot}98$). AES parameters: primary beam at normal incidence $E_{pr} = 2$ keV/3 keV for Auger energies $E < 1000$ eV/>1000 eV; $I_{pr} = 5$ μA; modulation frequency = 17 kHz; modulation voltage $U_m = 2$ V/5 V for Auger energies $E < 1000$ eV/>1000 eV.

O_{KLL} and Si_{KLL} spectra using an approximation method for quantitative Auger analysis (Palmberg *et al* 1972, Palmberg 1976, Lieske 1979). For the samples of Si oxide and Si nitride shown in figure 1, the elemental compositions $Si_{35}O_{65}$ and $Si_{39}N_{58}O_{03}$ were found, which are very near to the stoichiometric values of SiO_2 and Si_3N_4.

Information on whether the elements are chemically bonded can be obtained from the peak shifts of the KLL Auger signals and from the characteristic lineshapes of the Si_{LVV} spectra. Detailed information about the filled and empty electron states and thus the nature of the chemical bonding is derived from core- and valence-electron excitation spectra measured with energy-loss spectroscopy.

3.1.2. ELS spectra of elemental silicon. Figure 2 shows, for the Si(100) 2 × 1, Si(111) 7 × 7 and amorphous silicon surfaces the valence-electron excitation spectra measured at primary electron energies of 100 and 300 eV, the Si(2P) core-electron excitation spectra and LEED pictures of the two crystalline surfaces. The amorphous Si surface was obtained by 3 keV Ar-ion bombardment of the crystalline surfaces. Much experimental and theoretical work has been done on the interpretation of such spectra (Bauer 1969, Ibach and Rowe 1974, Koma and Ludeke 1975, 1976, Lieske and Hezel 1979a, b, c, d) and the results can be summarised as follows: the valence-electron spectra show surface-state transitions (S), bulk interband transitions (B) and the surface and bulk plasmon excitations $h\omega_S$ and $h\omega_B$. In the valence-electron spectra of Si(100) 2 × 1 and Si(111) 7 × 7 obtained at 100 eV, the energy loss peaks S_1 (1·7 eV) and S_1 (2·0 eV) correspond to transitions from filled dangling-bond surface states and the loss peaks S_2–S_4 to transitions from filled back-bond surface states. The surface-state loss peaks are more inten-

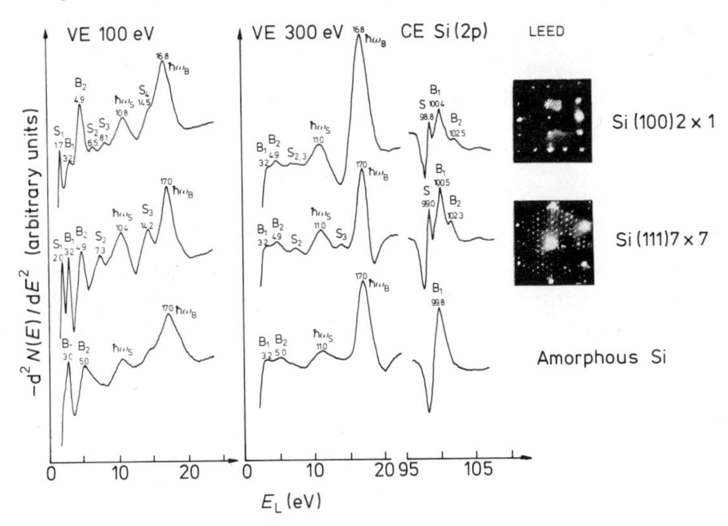

Figure 2. Valence-electron (VE) and Si(2p) core-electron (CE Si(2p)) spectra for crystalline Si(100) 2 × 1 and Si(111) 7 × 7 surfaces and the amorphous Si surface produced by argon-ion bombardment (ion energy 3 keV, ion-beam density ≈ 10 μA cm^{-2}, t = 10 min) of the Si(111) surface, together with LEED pictures (E_{pr} = 50 eV) for the crystalline surfaces. ELS parameters: 100 eV VE spectrum: E_{pr} = 100 eV, I_{pr} = 1 μA, U_m = 0·5 V; 300 eV VE spectrum: E_{pr} = 300 eV, I_{pr} = 4 μA, U_m = 1 V; CE Si(2p) spectrum: E_{pr} = 170 eV, I_{pr} = 2·5 μA, U_m = 1 V.

sive for the Si(111) than for the Si(100) surface, because of the higher density of surface states on the Si(111) surface. These peaks have almost vanished from the spectra for the amorphous surface. The bulk interband transitions and plasmon losses have very similar energies for the three samples. The Si(2p) core-electron spectra of the crystalline surfaces show transitions into empty dangling-bond surface states (S) and bulk states (B_1, B_2). For the amorphous Si surface only the B_1 transition can be seen, shifted to lower energies; this indicates the presence of electron states within the energy gap of Si.

3.1.3. ELS *spectra of Si oxide.* In figure 3 the valence-electron spectra, together with the Si(2p) and O(1s) core-electron spectra are shown for the thermal Si oxide of figure 1 in both the as-grown state and after Ar-ion bombardment. In addition, the electron energy-level scheme for the interpretation of the spectra is given. The basic molecules of Si oxide are [SiO_4] tetrahedra, which are linked by their oxygen atoms to form an amorphous network. The Si–O bond is formed by the Si(3s, 3p) and the O(2p) valence electrons. The results of the interpretation (Lieske and Hezel 1979a, b, c, Lieske 1979) can be summarised as follows: the valence-electron spectrum of nearly stoichiometric Si oxide is made up of interband transitions from ground into excited states of the Si–O bond (E_L = 10·2, 12·0, 13·8, 17·0, 20·5 eV), from the O(2s) level into Si–O excited states (E_L = 27·5, 30·0 eV) and the bulk plasmon excitation at 22·5 eV. The Si(2p) and O(1s) excitation spectra show transitions from these core levels into the excited states of the Si–O bond. The existence of Si–Si bonds is detected in the valence electron and core-electron Si(2p) spectrum with energy losses at 3·2, 6·5 and 101·0 eV. After Ar^+-ion bombardment, the existence of broken Si–O bonds is reflected by additional energy losses at 5·0 and 7·0 eV in the valence-electron spectrum and at 529·0 and 531·0 eV in the O(1s) spectrum. The corresponding new electron states are located at the oxygen atoms (marked O* in the energy-level scheme).

To summarise, the ELS spectra of Si oxide show the existence of Si—O bonds as in $[SiO_4]$ tetrahedra and of Si—Si bonds in mixed tetrahedra, indicating deviations from stoichiometry and the presence of broken Si—O bonds.

3.1.4. ELS spectra of Si nitride. Figure 4 shows the valence-electron spectrum and the Si(2p) and N(1s) core-electron spectra for the CVD Si nitride of figure 1, together with the electron energy-level scheme for the interpretation of these spectra (Lieske and Hezel 1979a, b, c, Lieske 1979). Si nitride is formed from $[SiN_4]$ tetrahedra where three tetrahedra are linked by one nitrogen atom. The Si—N bond is formed by the Si(3s, 3p) and N(2p) valence electrons. The valence-electron spectrum shows characteristic energy losses attributable to electron transitions between ground and excited states of the Si—N bond (9·3, 10·8, 12·5, 14·5, 16·0 eV) and transitions from the N(2s) level into the Si—N excited states (near 23—24 eV) and the bulk plasmon excitation at 18·6 eV. In the lower-energy range, loss peaks corresponding to the presence of Si—Si bonds (at 3·2 eV) and broken Si—N bonds (at 4·6 and 6·8 eV) can be detected. The broken Si—N bonds

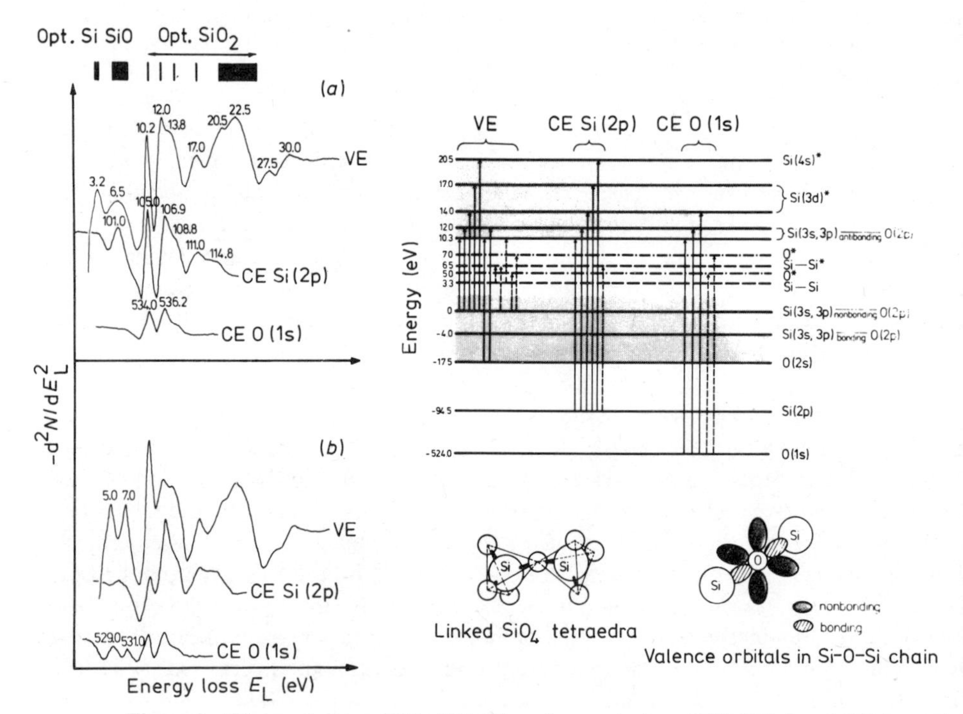

Figure 3. Valence-electron (VE, 100 eV) and core-electron (CE) Si(2p) and O(1s) spectra for thermal Si oxide in the as-grown state (*a*) and after argon-ion bombardment (*b*) ion energy = 500 eV, ion-beam density ≈ 1 μA cm^{-2}, t = 5 min), bonding models of the Si—O bond and electron energy-level scheme for the interpretation of the ELS spectra. ELS parameters for the VE and CE Si(2p) spectra are as for figure 2; for CE O(1s): E_{pr} = 670 eV, I_{pr} = 5 μA, U_m = 2 V. See text for details of the energy-level scheme.

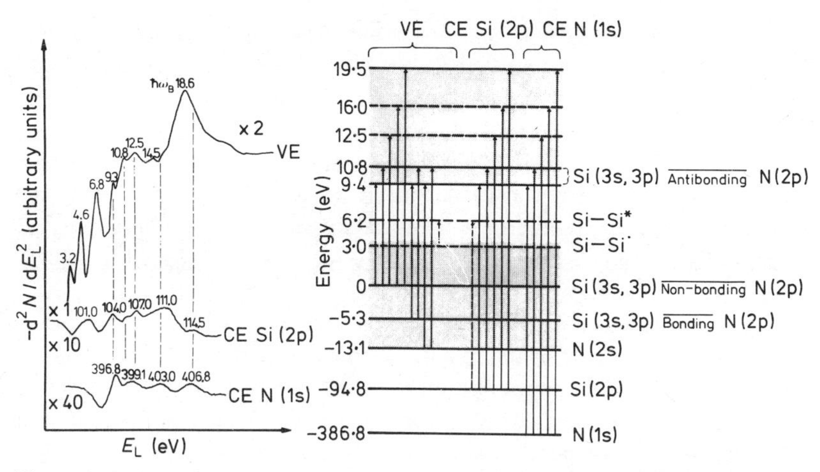

Figure 4. Valence-electron (VE, 100 eV) and core-electron (CE) Si(2p) and N(1s) spectra of CVD Si nitride and electron energy-level scheme for the interpretation of the ELS spectra. ELS parameters for the VE and CE Si(2p) spectra are as in figure 2; for CE N(1s): E_{pr} = 540 eV, I_{pr} = 5 μA, U_m = 2 V. See text for details of the energy-level scheme.

were produced by Ar^+-ion bombardment of the Si nitride. The Si(2p) and N(1s) core spectra exhibit transitions from these core-electron states into the excited states of the Si–N bond.

It may be concluded from these results that the combination of AES and ELS can be used to characterise Si oxide and Si nitride, particularly with regard to deviation from the ideal SiO_2 and Si_3N_4 structures. Such deviations are expected to be present at semiconductor–insulator or insulator–insulator interface regions in MIS systems and may be responsible for, or at least contribute to, the existence of interface states and charges (Cheng 1977).

3.2. AES/ELS depth profile of a thin-oxide MNOS system

To obtain a basic understanding of the electrical properties of interfaces in MIS systems, it is of great importance to know the chemical composition and electronic states as a function of depth, for which a high depth resolution, of the order of 10 Å, is required. For this purpose a number of depth profile studies with several techniques such as AES, SIMS (secondary-ion mass spectroscopy), XPS (x-ray photoelectron spectroscopy) and RIBS (Rutherford ion backscattering) have been performed for MOS and MNOS structures (see for example Pantelides 1978, Johannessen *et al* 1976, 1977). This paper describes, for the first time, the use of a combination of AES and ELS with noble-gas-ion sputtering for depth profiling and results are given for a thin-oxide MNOS structure.

3.2.1. Experimental conditions. A number of experimental artifacts must be considered when interpreting a depth profile experiment in which electron and ion beams are used. Since these artifacts cannot be avoided, the corresponding experimental errors induced must be established.

Artifacts arise from electron- and ion-beam effects and residual gas adsorption. The following argon-ion and electron-beam parameters were used: primary energy E_{pr}^{ion} = 500 eV, $E_{pr}^{el} \gtrsim 100$ eV; angles of incidence on the sample $\theta^{ion} = 40°$, $\theta^{el} = 90°$. From a number of experimental and theoretical results on argon-ion energies (Johannessen *et al* 1976, Ishitani and Shimizu 1975) and the mean-free paths of low-energy electrons (Johannessen *et al* 1976, Shelton 1974, Flitsch and Raider 1975, Akkerman and Chernov 1978, Koval *et al* 1978), the corresponding values for the argon-ion range δ_{ion} and information depth δ_{inf} could be estimated as $\delta_{ion} \lesssim 4$ Å and 5 Å $\lesssim \delta_{inf} \lesssim 20$ Å. Thus the critical condition $\delta_{inf} > \delta_{ion}$ could be fulfilled in the present work.

For the depth resolution, the experimental broadening caused by ion knock-on, the electron escape depth and sputtering inhomogeneities could be estimated with an upper limit of $\Delta W_{exp} \simeq 10$–30 Å, dependent on the thickness of the layer removed and on the energy of the electrons detected by AES and ELS. Further details concerning the experimental artifacts have been published elsewhere (Lieske 1979).

3.2.2. AES depth profile. Figure 5 shows the concentration depth profiles for the elements Si, N, O and C obtained with AES on a MNOS memory device. The MNOS structure was built up from a 40 nm thick low-pressure CVD layer of Si nitride on a 3·4 nm thick layer of Si oxide thermally grown at 800 °C on a p-Si(100) substrate. The profiles show four main regions.

Region A is characterised as a surface oxynitride layer with an elemental composition at the surface of $Si_{37}N_{34}O_{18}C_{11}$. With an experimental broadening of $\Delta W_{exp} \approx 10$–20 Å due to ion knock-on and the electron escape depth, the layer thickness is estimated to be $d_{ox} \approx 30$–40 Å.

Region B marks the bulk Si nitride with the average composition $Si_{42}N_{53}O_3C_2$. The Si content is near that of Si_3N_4, but the N content is diminished because of contamination with O and C. Small depth-dependent variations in the elemental composition can be seen.

Region C is the Si nitride–Si oxide–Si interface layer. The most striking feature is that

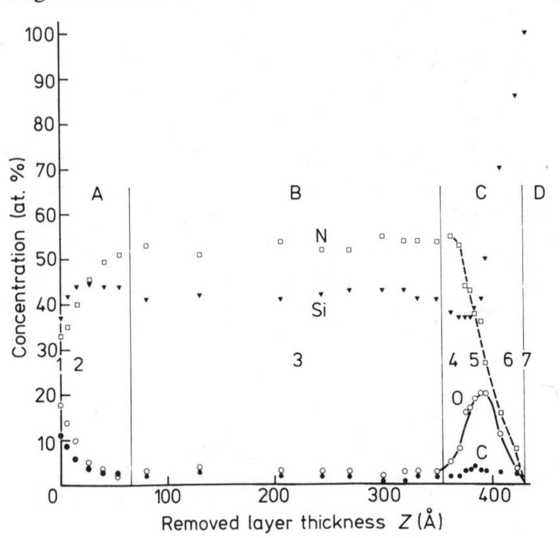

Figure 5. AES depth profile for the elemental concentrations of Si, N, O and C in a MNOS system: 40 nm low-pressure CVD Si nitride on 3·4 nm thermal oxide on p-Si(100) substrate. AES parameters as in figure 1.

this layer never has the stoichiometry of SiO_2, but is rather a Si oxynitride with depth-dependent elemental composition. At each point of the interface layer the oxygen to nitrogen atomic ratio is $O:N<1$; the maximum O content is in the middle of the layer at a composition of $Si_{41}N_{36}O_{20}C_3$. With an experimental broadening $\Delta W_{exp} \lesssim 30$ Å, the thickness of the interface layer is estimated to be 30–40 Å.
Region D marks the bulk silicon.

These AES results are in excellent agreement with those obtained by Johannessen *et al* (1977) for a thin-oxide MNOS structure.

3.2.3. ELS depth profile. ELS spectra were measured in order to obtain detailed information about the nature of chemical bonding and the electron states in the MNOS system. Valence-electron spectra at different primary energies (100, 300 eV) and Si(2p) core-electron spectra, measured at the most interesting points (1–7) of the AES concentration profile, are shown in figures 6 and 7. The broken curves were obtained for an equivalent sample but after annealing for 30 min at 1100 °C in the UHV chamber. This procedure left the AES concentration profile unchanged, to within experimental error.

The ELS spectra measured at the surface (figure 6, part 1) are very similar to those of Si oxide (for comparison see figure 3). This result is typical for CVD Si oxynitride with $O/N \gtrsim 0{\cdot}2$ (Lieske and Hezel 1979a). At a depth $Z = 16$ Å (figure 6, part 2) the ELS spectra clearly show the transition to Si nitride, which is best seen from the energy shifts of the bulk plasmon loss in the valence-electron spectra (22·5 eV → 19·6 eV) and of the Si(2p) excitation (104·7 eV → 104·0 eV; for comparison see figures 3 and 4).

Within the Si nitride bulk layer (figure 6, part 3), the valence-electron excitations of the sample without annealing are similar to those of plasma-deposited Si nitride (Lieske 1979, Lieske and Hezel 1979e). There is a new excitation at 8·5 eV, the 6·8 eV energy

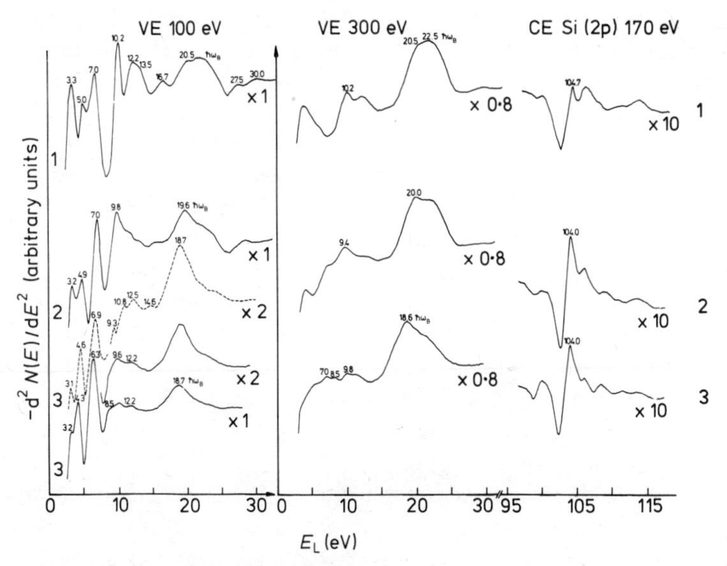

Figure 6. ELS depth profile for the MNOS system shown in figure 5: valence-electron (VE) and core-electron (CE) spectra measured at points 1, 2 and 3 of figure 5. ELS parameters as in figures 2, 3 and 4.

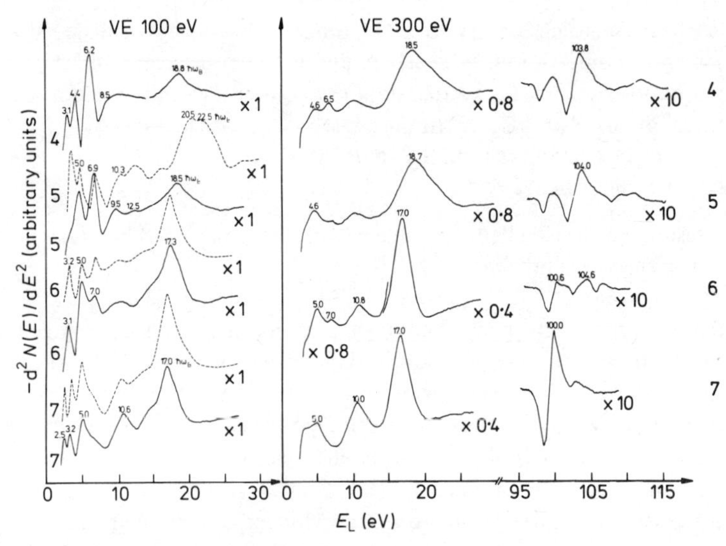

Figure 7. ELS depth profile for the MNOS system shown in figure 5: valence-electron (VE) and core-electron spectra measured at points 4–7 of figure 5. ELS parameters as in figures 2, 3 and 4.

loss has shifted to 6·3 eV and the interband excitations at $E_2 > 10$ eV have decreased in intensity. For the annealed sample (figure 6, part 3, broken curve), all of these effects have vanished and the valence-electron spectrum typical of Si nitride can be seen. These results are attributed to the presence of hydrogen in the initial sample and its disappearance by out-diffusion after thermal annealing. A similar result has been obtained by IR spectroscopy (Stein and Wells 1977).

At the edge of the interface layer (figure 7, part 4) the ELS spectra are nearly the same as in the bulk nitride. In the middle of the oxynitride interface layer (figure 7, part 5), the valence-electron excitations of the non-annealed sample are the same as for Si nitride, whereas after thermal annealing the characteristic energy losses of Si oxide begin to appear (for comparison see figures 3 and 4). This result leads to the following model for the influence of Si nitride deposition: fractions of SiN_4 tetrahedra penetrate into the Si oxide network, thereby breaking most of the Si–O bonds. These Si–O bonds are restored again during the annealing process.

10 Å away from the Si surface (figure 7, part 6), the ELS spectra show the increasing Si concentration: the bulk plasmon excitation $\hbar\omega_B$ is shifted towards 17 eV and the energy losses corresponding to Si–Si bonds (valence-electron, 3·2 eV; Si(2p), 100·6 eV) become more intense relative to the other signals. At the Si surface (figure 7, part 7) the valence- and core-electron spectra are equivalent to those of clean amorphous Si (see figure 2), but with an additional energy loss at 2·5 eV in the 100 eV valence-electron spectrum that increases in intensity after annealing.

Various annealing and adsorption experiments have shown that this energy loss at 2·5 eV corresponds to surface states that are characteristic of an intermediate state between the crystalline and the amorphous Si(100) surface. It is suggested that the energy position of these filled surface states may be within the upper part of the surface-state valence band or even within the energy gap of bulk Si.

4. Conclusions

For the first time, a depth profile based upon a combination of Auger spectroscopy and energy-loss spectroscopy with ion milling has been presented. The results for a thin-oxide MNOS structure show that additional information can be gained from ELS, especially concerning the properties of the Si nitride–Si oxide–Si interface region.

(i) During nitride deposition, the thin layer of Si oxide is changed into a Si oxynitride layer made up of Si–O, Si–N and Si–Si bonds and broken Si–O and Si–N bonds.
(ii) Thermal annealing of the sample in UHV above the nitride deposition temperature drastically changes the electron states of this interfacial oxynitride.
(iii) At the Si surface electron states are detected that are not visible at the crystalline or amorphous surface.

To correlate these results with the electrical interface properties of MNOS memory devices, further experiments based upon variation of the technological parameters combined with AES/ELS depth-profile measurements must be performed.

References

Akkerman A F and Chernov G Y 1978 *Phys. Stat. Solidi* **B89** 329
Bauer E 1969 *Z. Phys.* **224** 19
Cheng Y C 1977 *Prog. Surface Sci.* **8** 181
Davis L E, MacDonald N C, Palmberg P W, Riach G E and Weber R E 1972 *Handbook of Auger Electron Spectroscopy* (Minnesota: Physical Electronics Industries Inc.)
Flitsch R and Raider S J 1975 *J. Vac. Sci. Technol.* **12** 305
Ibach H and Rowe J E 1974 *Phys. Rev.* **B9** 1951
Ishitani T and Shimizu R 1975 *Appl. Phys.* **6** 241
Johannessen J S, Spicer W E and Strausser Y E 1976 *J. Appl. Phys.* **47** 3028
Johannessen J S, Helms C R, Spicer W E and Strausser Y E 1977 *IEEE Trans. Electron Devices* **ED-24** 547
Kane P F and Larrabee G B 1974 *Characterization of Solid Surfaces* (New York: Plenum)
Koma A and Ludeke R 1975 *Phys. Rev. Lett.* **35** 107
—— 1976 *Surface Sci.* **55** 735
Koval I P *et al* 1978 *Surface Sci.* **77** 40
Lieske N 1979 *Thesis* University of Erlangen–Nürnberg, West Germany
Lieske N and Hezel R 1979a *Proc. 4th Int. Congr. on Thin Film Properties in Relation to Structure, Loughborough, September 1978* (Lausanne: Elsevier Sequoia SA)
—— 1979b *Thin Solid Films* **61** 197
—— 1979c *Thin Solid Films* **61** 217
—— 1979d *Phys. Stat. Solidi* **B92** 159
—— 1979e to be published
Ludeke R and Esaki L *1975 Surface Sci.* **47** 132
Morrison S R 1977 *The Chemical Physics of Surfaces* (New York: Plenum)
Palmberg P W 1976 *J. Vac. Sci. Technol.* **13** 214
Pantelides S T (ed) 1978 *The Physics of SiO_2 and Its Interfaces* (New York: Pergamon)
Rowe J E *et al* 1975 *Surface Sci.* **48** 44
Sevier K 1972 *Low Energy Electron Spectrometry* (New York: Wiley Interscience)
Shelton J C 1974 *Surface Sci.* **44** 305
Stein H J 1976 *J. Electron. Mater.* **5** 161

Nature of interface states at III–V insulator interfaces†

W E Spicer‡, P W Chye, P R Skeath, C Y Su and I Lindau
Stanford Electronics Laboratories, Stanford University, Stanford, California 94305, USA.

Abstract. Synchrotron radiation has been used to make photoemission measurements of the energy levels induced on a clean (110) surface by adsorption of oxygen or various metals. Before such adsorption, there are no intrinsic surface states in the band gap; thus pinning of the Fermi level at the surface is produced by the adsorbate. The final pinning position gives the energy of the levels so induced. Oxygen and a number of metals varying from Au to Al and Cs are found to give almost identical pinning positions. These data are explained by defect levels produced by the interaction of the adsorbate with the semiconductor. Studies of core and valence levels give support for this explanation. For the insulator III–V interface, these levels are superimposed on a 'U'-shaped background of levels (similar to that found in Si) attributed to strain at the interface. The defect levels produce 'peaks' in the density of interface states due to missing column III or V elements. The following levels are found (measured from the valence-band minimum): GaAs, 0·5 and 0·75 eV; InP, 0·9 and 1·2 eV; GaSb, 0·1 eV. Good agreement is found between these levels and measurements on MIS structures.

1. Introduction

We report here models for the states (and their formation) at III–V oxide interfaces based on the application of some rather new experimental techniques. These depend strongly on the use of synchrotron radiation (Lindau and Spicer 1979) to determine the electronic structure (including the position of the surface Fermi level E_{Fs}) as well as chemical bonding and stoichiometry within a few molecular layers of the surface. Three III–V compounds, GaAs, InP and GaSb, have been studied.

The first step was to study and understand the free (110) surface formed by cleavage in ultrahigh vacuum (Gregory *et al* 1974, Spicer *et al* 1976, Pianetta *et al* 1978a, b, Chye *et al* 1975, Spicer *et al* 1977a). Photoemission, LEED and theory have been combined to give a consistent and quite detailed view of surface atomic arrangements and the electronic structure of these clean (110) surfaces (Spicer 1977a, Mark 1976, 1977, 1978). The results from the free surfaces form a background of information essential to the understanding of Schottky barrier formation and the interface states between the oxide and semiconductor.

The second step in these investigations involved determining the changes that take place when either oxygen or metals are placed on the free surface of the III–V com-

† Supported by the Advanced Research Projects Agency of the Department of Defense and monitored by the Office of Naval Research under Contract No DAAK 02-74-C-0069 and by the Office of Naval Research, ONR N00014-75-C-0289. Part of the work was performed at SSRL which is supported by the National Science Foundation, NSF DMR73-07692, in cooperation with the Stanford Linear Accelerator Center and the Department of Energy.
‡ Stanford W Ascherman Professor of Engineering.

pounds. It will become clear as we present the data that the comparison between results from oxygen and metals provides critical insight into the processes taking place. Before describing these experimental results, some perspective will be given on the use of synchrotron radiation within photoemission research.

2. Study of the last two molecular layers at the surface using photoemission and synchrotron radiation

Recent years have seen a rapid increase in the number of tools that are available for the studies of solids. One method that has developed rapidly since the late 1950's for the investigation of the electronic structure of solids and their surfaces is photoemission (Spicer 1958, Spicer and Simon 1962, Berglund and Spicer 1964, Spicer 1976, Feuerbacher *et al* 1978, Cardona and Ley 1978). A second tool that came into wide use somewhat later is synchrotron radiation. In this paper, we describe work in which these methods have been combined to gain information on the metal and oxide semiconductor interfaces. The use of other tools (Auger, sputter Auger, LEED etc) has also been essential to this work, but key results are due to the use of monochromatic synchrotron radiation in the photon energy range $20 < h\nu < 200$ eV (synchrotron radiation can provide a continuum from the visible to the x-ray region) to produce photoemission energy distribution curves (EDC). There are numerous important reasons for using a wide energy range. First, we are able to examine not only the valence states but also the highest-lying core (3d or 4d levels for In, Ga, As and Sb; 2p levels for P) levels. Studying the core levels allows the surface chemistry to be investigated on an atomic scale as oxygen or metals are added to the surface. It also provides a means of determining the density of a given atomic species near the surface since the amplitude of the intensity of the core levels of a given atom provides a direct measure of the concentration of that atom within the experimental sampling depth of the surface.

The sampling depth (Spicer 1976) is another critical parameter that can be varied by altering $h\nu$. The sampling depth is set by the escape depth of the photoelectrons. The probability of an electron escaping is given by $\exp(-X/L(E))$ where X is the depth beneath the surface at which the electron is excited and $L(E)$ is a characteristic escape depth that varies from one class of materials to another (Lindau and Spicer 1974). Figure 1 shows a plot of $L(E)$ for GaAs (Pianetta *et al* 1978a, b). The values in this figure are for the geometry appropriate to the average acceptance angle of the energy

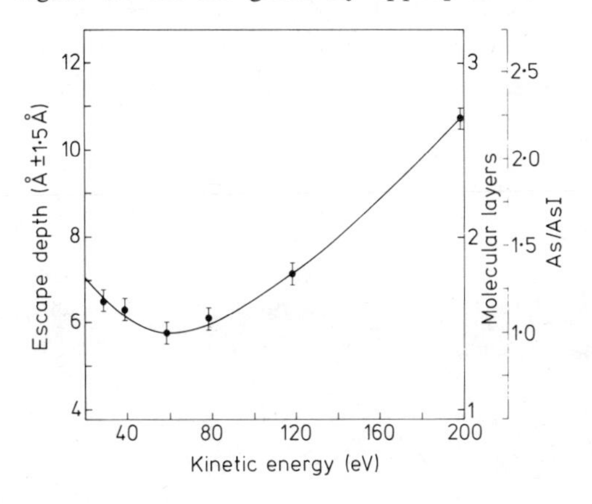

Figure 1. The escape depth for electrons photoexcited from GaAs. The energy of the electron is plotted on the abscissa.

analyser used in this work (see figure 2). To transform the escape depth to that appropriate for electrons emitted perpendicular to the surface ($L(E)$ is normally given in these terms), the values of $L(E)$ given in figure 1 should be multiplied by 1·4.

It can be seen from figure 1 that the escape depth has a fairly broad minimum near 60 eV. Most important, the escape depth near the minimum is quite small. Thus, by keeping the energy of the excited electron sufficiently close to 60 eV, the majority of the photoelectrons will come from the first two molecular layers of the surface, making the measurements highly surface-sensitive. To achieve this, the photon energy from the synchrotron source is tuned using monochromators. For example, for the 3d cores of Ga and As (located at -20 and -40 eV respectively), a photon energy of about 100 eV provides an optimum operation point, taking into account both $L(E)$ and the dependence of the matrix elements for the photoexcitation on photon energy (the matrix elements will be discussed below); for the 2p levels of P, however, that are located deeper than the 3d cores a higher value of $h\nu$ is needed.

Another consideration in choosing the photon energy for a given level or set of levels is the photon dependence of the matrix elements for photoexcitation of electrons from those levels. Details of the matrix elements involved can be found elsewhere (Lindau *et al* 1976). We will restrict ourselves here to a few important comments. The 3d and 2p matrix elements peak at about 50 eV above threshold and then decrease very slowly with increasing $h\nu$. The 4d elements also peak about 50 eV above threshold but decrease very rapidly with increasing $h\nu$. In contrast, the matrix elements for the valence electrons in the III–V compounds peak within about 5 eV of threshold and decrease fairly rapidly with increasing $h\nu$. Thus, an optimum photon energy to study the valence states near the surface is about 15–20 eV. The valence-band matrix elements for $h\nu \geqslant 100$ eV are so small that these states are barely visible at the optimum photon energy for studying the core levels. This has important consequences for studying the levels characteristic of an overlayer. For example, 10 or 20% of a monolayer of oxygen or Au on GaAs provides detectable oxygen 2p levels or Au 5d levels despite the fact that these levels are degenerate with the III–V valence levels. This is of importance to this work as it gives us the opportunity to examine in some detail the valence electronic structure of such overlayers at relatively low coverages.

In summary, the availability of a continuous distribution of radiation from the synchrotron source (the storage ring SPEAR) at the Stanford Linear Accelerator Center used in this work allowed us to study core and valence states characteristic of the last two molecular layers of III–V compounds. More important for the studies of interest here, we can study the changes induced by adlayers of oxygen from submonolayer to much higher coverages.

3. Experimental techniques

Since the experimental techniques used in these studies have been covered elsewhere (Pianetta 1976, Chye 1978, Garner 1978, Pianetta *et al* 1978a,b, Chye *et al* 1978b), we will not give a comprehensive review here; rather, we will consider two aspects that are particularly relevant to the work. Figure 2 is a simplified schematic diagram of the sample chamber (Pianetta 1976). Not shown are the LEED system, various metallic evaporators, a quartz oscillator to measure the rate of arrival of metal, systems for letting

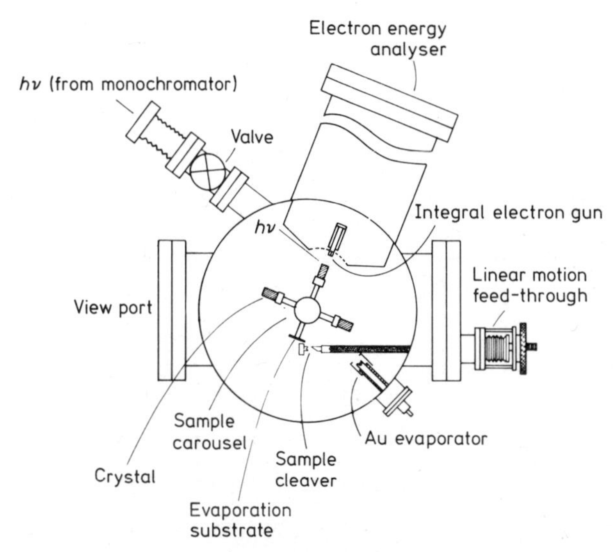

Figure 2. Experimental chamber. The sample carousel rotates through 360°. Not shown is the LEED apparatus. The integral electron gun allows Auger measurements.

gases into the chamber, a gate valve to isolate the pump from the system during gas exposures and many other accessories. Of particular interest is the rotatable sample holder that holds three crystals and an 'evaporation' substrate. Also shown is a Au evaporator from which a clean metallic film may be evaporated onto the substrate to be used as a reference sample in determining Fermi-level positions. We will return presently to the importance of this reference standard; however, first we should mention the importance of having three crystals available in the same chamber.

Early in this work, it was found that n-type crystals pinned on cleaving (Gregory *et al* 1974). However, stimulated by the work of van Laar and Scheer (1967) and Huijser and van Laar (1975), we undertook a study of crystals from different sources, each of which could be cleaved five to ten times in the same vacuum. This work confirmed that unpinned surfaces could be formed on cleaving n-type samples, in agreement with van Laar and coworkers, and established that the pinning previously seen was due to extrinsic surface states (Spicer *et al* 1976). By an extrinsic surface state, we mean a state due to an imperfection in the surface region (for example, a structural defect or an impurity; conversely, an intrinsic surface state is characteristic of the perfect surface). After studying the cleaved faces, we examined the effect of exposures to oxygen on a large number of cleaves and developed a large body of experimental data that will be reviewed in this paper (Pianetta *et al* 1978a,b).

The 'evaporation' substrate in figure 2 allows us to make one important reference measurement in this work: the determination of the position of the Fermi level at the surface, E_{Fs}, for the clean, cleaved sample or for the crystal after various amounts of oxygen or various metals were placed on the surface (Gregory and Spicer 1975, Chye *et al* 1977a, Chye *et al* 1980). Figure 3(*a*) gives a schematic indication of the valence structure of a metal, Au, that is used as the standard and of a semiconductor, GaAs, under study (Skeath *et al* 1979). The difficulty of locating the Fermi level E_F in a semiconductor is that since it probably lies in the band gap, there is little probability of

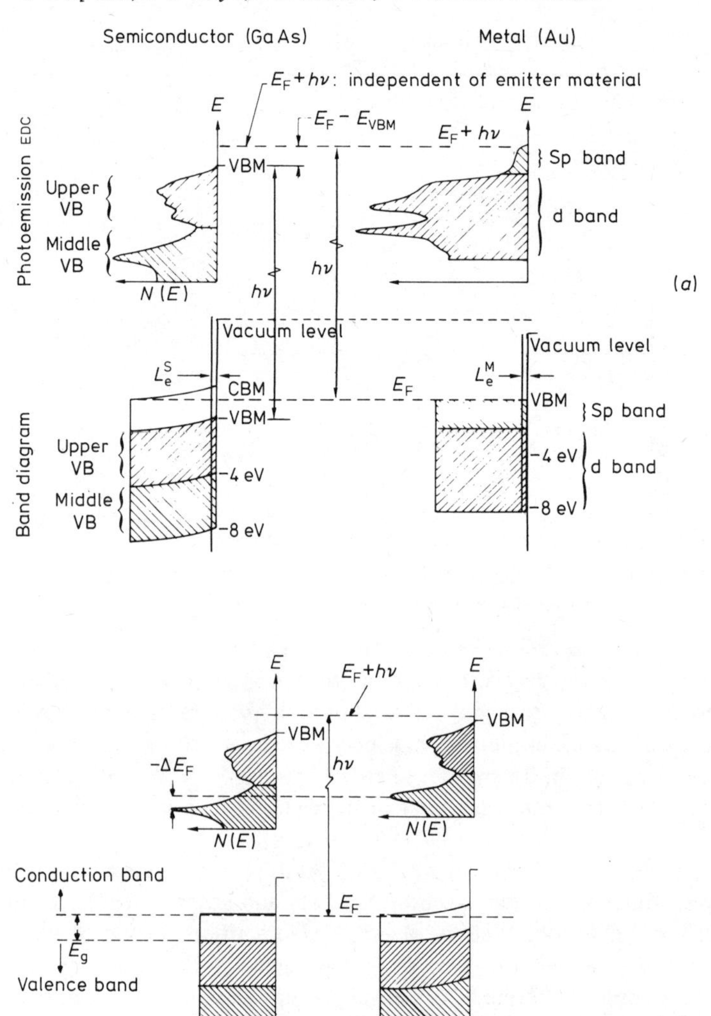

Figure 3. (*a*) Determination of the relative energies of the semiconductor valence band and the Fermi energy. A highly schematic band diagram of the metal reference is shown in the lower right-hand corner and above the band diagram is the corresponding photoemission EDC obtained with $h\nu = 21$ eV. A similar set of diagrams is given for the semiconductor, but with E_F at the conduction-band minimum (CBM) in the bulk and pinned at approximately midgap at the surface. The relative energies of the valence-band maximum (VBM) and E_F are obtained directly from the photoemission measurements. Note that the vacuum level at the sample surface may change without affecting peak positions in the EDC's; this is because all energies are measured relative to E_F. (*b*) The change in EDC's due to changes in the position of the Fermi level. The left-hand panel shows the situation for the flat bands in n-type material. The right-hand panel shows the changes due to movement of the Fermi level towards midgap; note that $E_F + h\nu$ changes by a constant amount but the valence band structure moves up in energy by an amount equal to the change in band bending.

exciting electrons from it. This difficulty is overcome by measuring the EDC at a given $h\nu$ for the standard provided by evaporating a Au film onto the 'evaporation' substrate of figure 2, a process that may be repeated if necessary. Since Au has electrons at the Fermi level, photoemission from this calibrates the energy analyser by determining the energy of the Fermi level relative to the 'effective' workfunction of the analyser. An EDC excited from the semiconductor with photons of the same energy is also shown in figure 3(*a*). Having calibrated the analyser, the absolute energy of the Fermi level relative to the EDC is known for all values of $h\nu$.

Figure 3(*b*) shows the changes that take place in the EDC from a given semiconductor due to changes in E_{Fs}. It can be seen from this figure that a change in the surface E_{Fs} of the semiconductor results in a shift in the energy of the valence (and core) levels of the semiconductor. By measuring the shifts in prominent structures in the semiconductor, changes in E_{Fs} can be followed.

4. Shifts of the Fermi level in n- and p-type GaAs, InP, and GaSb as oxygen or metals are added to the surface

As mentioned earlier, a critical parameter in the studies reported here is the position of the Fermi level E_{Fs} at the surface. Most important are the changes produced in E_{Fs} by adding metals or oxygen to the surface. For this reason, considerable attention was given in the last section (§3) to the method of measuring E_{Fs} directly using photoemission techniques. This method has been used throughout this work. In other recent work (Brillson 1978, 1979), techniques have been applied in which changes in E_{Fs} are not measured directly but are inferred from other measurements. It now seems clear that this has led to serious errors. The most obvious error was in assuming that a dipole of atomic or molecular dimensions could contribute to the height of the Schottky barrier. Electrons will tunnel through such a narrow barrier so that it will not contribute to the Schottky barrier height. It was also assumed that, by flooding a Schottky barrier with sufficient light, the band bending would be removed and flat-band conditions would be produced. This is an extremely dangerous assumption with little support from theory or previous experience (see the comments and discussion included as part of the *Proceedings of the 6th Annual Conference of Compound Semiconductor Interfaces*).

Figures 4–6 indicated changes in E_{Fs} for GaSb caused by addition of two quite different metals, Au (Chye *et al* 1978b) and Cs (Chye *et al* 1978), and the non-metal oxygen (Chye *et al* 1980, Spicer *et al* 1979a). Figure 4 shows changes in E_{Fs} caused by Cs or Au overlayers on n-type GaSb; in figure 5, we show the effect of oxygen on n- and p-type GaSb and figure 6 gives a comparison of the results for oxygen and Cs (going up to higher Cs exposures than those shown in figure 4). Data are not shown for the metals on p-GaSb since, as with oxygen, little change is produced.

The most striking result shown in figures 4–6 is the similarity of the results, independent of the adatom. This is a very surprising result and, as established by figure 7 (Spicer *et al* 1979a), which gives the final pinning position for various adatoms on GaAs, GaSb and InP, it is common to all of the III–V materials studied here. The pinning positions in figure 7 are consistent with those found for practical Schottky barriers to within about ±0·1 eV. Detailed curves for changes in E_{Fs} with deposition of adatoms for GaAs and InP can be found elsewhere (Chye *et al* 1978b, Pianetta *et al* 1978b, Skeath

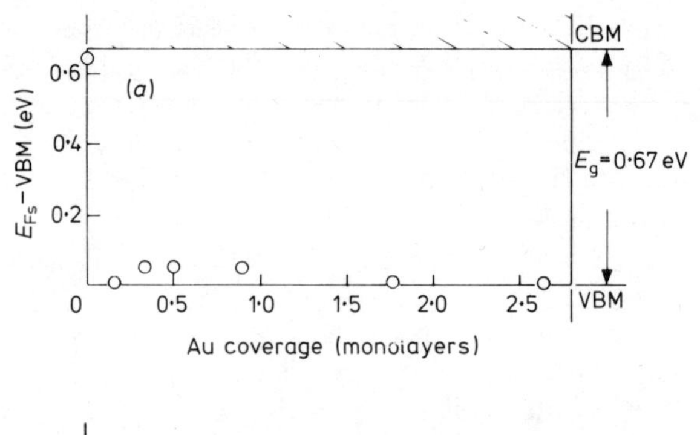

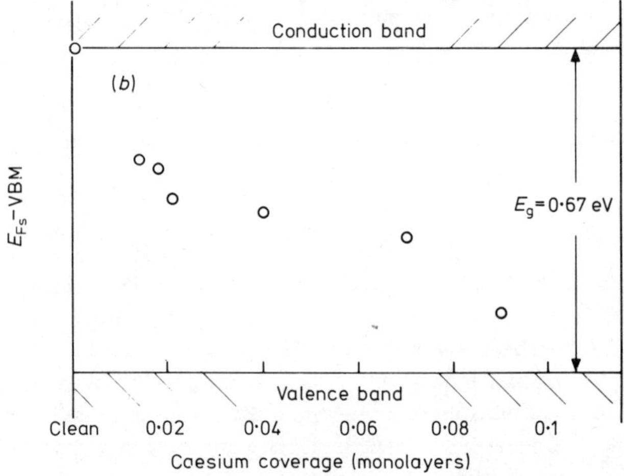

Figure 4. The change in E_{Fs} as a function of metal deposited for n-GaSb(110). Note change in scale from between Au (*a*) and Cs (*b*). The pinning is completed with only about 15% of a monolayer.

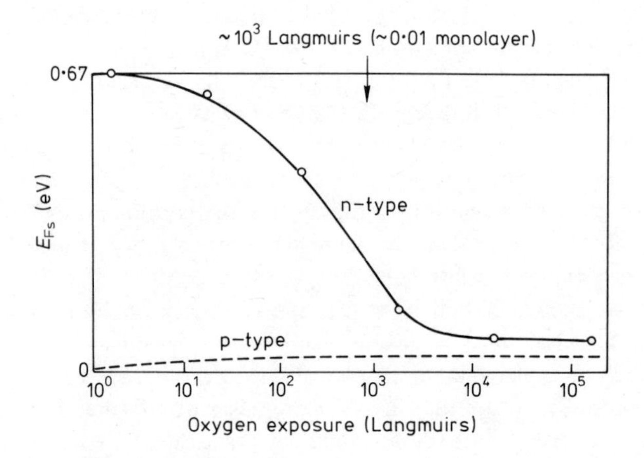

Figure 5. Change in E_{Fs} for GaSb with oxygen exposure. The pinning is completed with a few percent of a monolayer of oxygen.

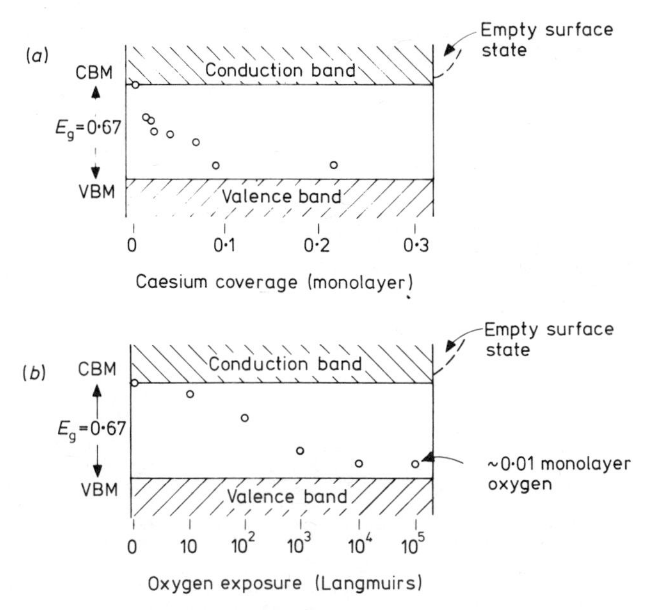

Figure 6. E_{Fs} plotted against metal (Cs, *a*) or oxygen (*b*) coverage for GaSb. Note the similarity in the final pinning position.

et al 1979). It is very difficult to explain these data in terms of levels produced directly by the various adatoms. It would be expected that the energy of such levels should depend, to an important extent, on the outer orbitals of the adatoms which are extremely different for Al, Cs, In, Au and oxygen. Thus we are driven to the conclusion that the levels that pin E_{Fs} are not induced directly by the adatom but are induced indirectly, for example by a disruption of the III–V lattice near the surface by addition of the adatom. Such a disruption could lead to lattice defect levels which could be responsible for the pinning (Lindau *et al* 1978, Spicer *et al* 1979a, b, c). It is now well established that the free (110) surface (as most surfaces of covalent semiconductors) is rearranged (Mark 1976, 1977, 1978, Spicer *et al* 1977a). Not only does this tend to sweep the surface states out of the band-gap region, it also leaves the surface in a highly strained condition (Spicer *et al* 1977a, b; see figure 8). This, together with the large heat of adsorption of the adatoms on the semiconductor surface, provides a mechanism for defect formation (Lindau *et al* 1978, Spicer *et al* 1979a). We will return later to a more detailed discussion of this phenomenon.

First, let us look at another fascinating detail of figures 4–6, the relatively small coverage of the adatom that is necessary to complete the pinning of E_{Fs}. For oxygen, it is only a few percent of a monolayer (Pianetta *et al* 1978a, b, Chye 1980); for the metals, it is 10 to 30% of a monolayer (Chye *et al* 1978b, Gregory and Spicer 1975, Lindau *et al* 1978). Thus, in all cases, it appears to be essentially an atomic process, i.e. the atoms are acting in an isolated 'atomic' fashion and not in terms of bands produced by a continuous overlayer. This is particularly important for the metals since in recent years there has been a very strong theoretical effort to calculate the metallic pinning position (and thus Schottky barrier height) using models based on an ideal, planar interface between slabs of semiconductor and metals of multimonolayer thickness. The present work shows that

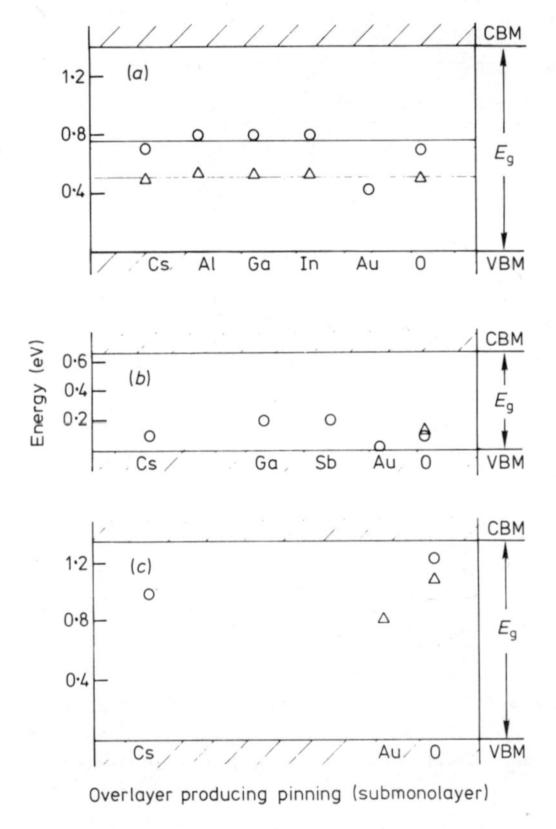

Figure 7. A summary of the final surface Fermi-level pinning positions obtained by experiments similar to that detailed in figure 3 for a wide variety of metals and oxygen on n-type (○) and p-type (△) GaAs(110) (*a*), GaSb(110) (*b*) and InP(110) (*c*). The absolute energy positions are located to no better than ±0·1 eV although relative positions may be better. Note the striking differences in the surface Fermi-level positions between the three semiconductors.

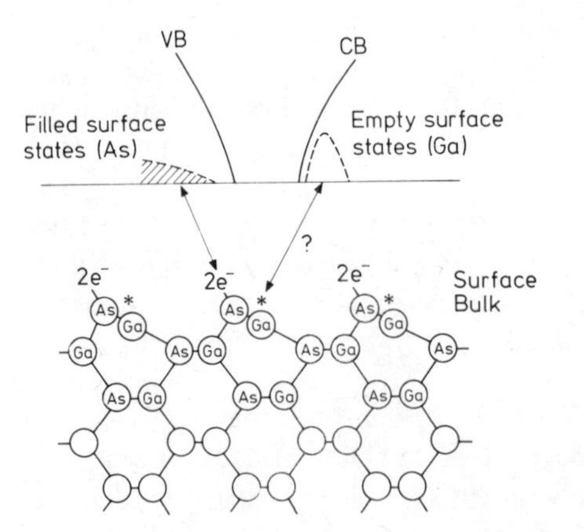

Figure 8. A schematic drawing of the lattice and electronic structure of the GaAs(110) face after rearrangement (relaxation). The As surface atoms have taken up a p^3 bonding arrangement (with two electrons in a $4s^2$ filled state), while the Gas has gone from the bulk sp^3 to a sp^2 arrangement. Since the p^3 bond angles are more acute than the bulk sp^3 bonds, the As moves outward; conversely, the Ga moves inward. The movements are large (large fractions of an angstrom). This rearrangement moves the filled and empty surface states out of the band gap. However, since the surface lattice is no longer lattice-matched to the bulk, there is strong strain at the surface.

the pinning takes place before a true continuous, conducting, metal overlayer is formed and thus that interactions between individual metallic atoms and the semiconductor should be considered, rather than interactions between an ideal conducting sheet or slab and the semiconductor.

In the course of this work, it was thought important to investigate the question of whether or not the metals formed clumps or single, dispersed atoms at low coverages.

Because the heat of adsorption of Cs on GaAs at low ($\lesssim 0{\cdot}2$ monolayer) coverages was known to be much higher (60 kcal mol^{-1}) than that for Cs on Cs (20 kcal mol^{-1}; Derrien and D'Avitaya 1977) and because the dipole–dipole repulsion between Cs atoms is well established, it is clear that clumping is not a problem with Cs.

For Au, such data are not available and so we measured the 'spin–orbit' splitting in the Au as a function of coverage. The atomic splitting is 1·5 eV, whereas that in solid Au is 2·3 eV. The increase in going from atomic to solid Au is attributable to interactions between the Au atoms (Chye *et al* 1977b) as they are brought together to form a solid with band structure. Figure 9 (Chye *et al* 1978b) shows the change in the measured spin–orbit splitting in going from 15% to many monolayers. It can be seen that at 15% coverage, the splitting is very close to the atomic value, whereas it approaches the value for solid Au at high coverages. (Presently we will show that semiconductor material is included in the Au; this prevents the value for the solid from being reached exactly.) The results at low coverage show that the Au is 'atomic' and highly dispersed on the semiconductor surface at the point of pinning. Perhaps more importantly, it gives evidence for a very large heat of adsorption of Au on GaAs and suggests that the heat of adsorption (the heat given up on adsorption) is as large as and probably larger than for Au on itself (87 kcal mol^{-1}). The heat of adsorption (also called heat of condensation) is then much larger than the heat of formation of compounds of Ga or As with Au; thus the heat of adsorption is truly a surface phenomenon that cannot be explained in terms of bulk thermodynamics. This is a very important result.

Knowledge of the heat of adsorption is critical for arguments that will be made later in this paper. However, the most important finding reported in this section should be re-emphasised. This is the result that pinning of the Fermi level at the surface is due to states induced indirectly by the adatoms and not to states associated directly with the adatoms (Lindau *et al* 1978, Spicer *et al* 1979a, b, c). A similar conclusion has been reached by Montgomery *et al* (1979). In the next section, we will examine experimental data that

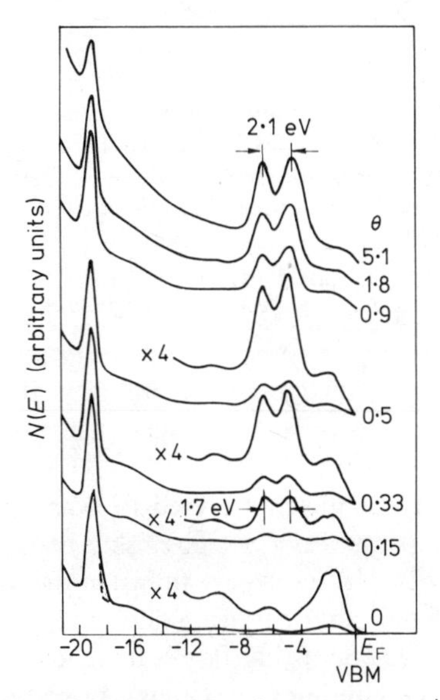

Figure 9. EDC's for GaSb as a function of Au coverage θ ($h\nu = 30$ eV). Note the increase in the Au spin–orbit splitting with coverage.

give direct evidence for the adsorption having a strong effect on the semiconductor, which can provide an indirect mechanism for the formation of defect levels in the semiconductor.

5. Evidence for the disruption of semiconductor surface region by adatoms

Figure 10 shows EDC's for GaSb taken with $h\nu = 120$ eV as Au is deposited (Chye *et al* 1978a,b). For $h\nu = 120$ eV, the escape depth for the Ga 3d and Sb 4d levels in Au is a few atomic layers. If the Au formed an ideal, planar interface with the GaSb, the Ga and Sb d intensity should decrease very rapidly with the addition of Au. Most strikingly, we would expect to see no Sb or Ga with 100 monolayers of Au. In contrast, the 3d

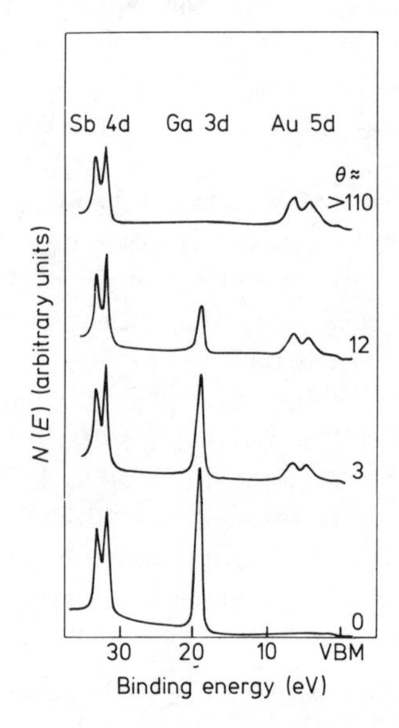

Figure 10. Photoemission spectra taken at a photon energy of 120 eV for GaSb with different Au coverages θ.

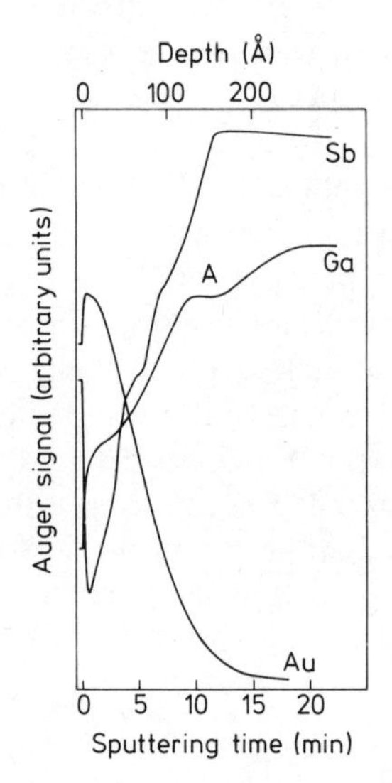

Figure 11. Compositional depth profile of Au-covered GaSb obtained using AES in combination with argon-ion etching. Note the build-up of Sb at the surface.

intensity falls off more slowly than expected at low coverages (although the Ga 3d intensity falls off noticeably more rapidly than the Sb 4d). At ~100 monolayers, there is no measurable Ga 3d intensity but the Sb 4d intensity is only slightly decreased. The only explanation for this behaviour is inclusion of semiconductor material in the Au, with the segregation of Sb to the surface and, perhaps, its removal in larger quantities than the Ga.

To check the photoemission results, sputter Auger techniques were used to examine the depth profile of the Au–GaSb Schottky barrier (Chye *et al* 1978b) using newly refined techniques (Helms *et al* 1978, Dupris *et al* 1979). The results are shown in figure 11. The surface enhancement of Sb is seen clearly at the outer surface. (This enhancement can be explained in terms of lowering the surface energy by surface segregation of the material with the lower cohesive energy, Sb, as first outlined by Gibbs.) It is also apparent that there is considerable intermixing between the GaSb and the Au with no ideal sharp metal–semiconductor interface.

Similar results (Chye *et al* 1978b) were found for GaAs and InP with the only difference that there was no strong surface segregation for these materials. Photoemission results for GaAs are shown in figure 12.

Figure 13 indicates a possible model for the creation of defects and the removal of semiconductor material into the metal (Spicer *et al* 1979a). This is based on the local energy released by adsorption of the metal onto the semiconductor surface. This energy is of the order of or larger than the semiconductor surface bond energy. Remembering also the highly strained nature of the free surface (see figure 8) and the local rearrangement that may be forced by the adsorbed metal atom, it does not seem impossible that occasionally (about 1% is probably necessary to explain the pinning) an atom is ejected from the semiconductor onto the surface by the arrival of a metal atom. Chemical interactions between the metal and either or both of the semiconductor atoms could also contribute to this process. The similarity of the pinning position for Al and In on GaAs, despite the strong replacement reaction for Al but not In (Skeath *et al* 1979), indicates that this mechanism is not important there. Williams *et al* (1978) give evidence for its importance for InP.

Let us now examine oxygen adsorption and oxide formation. By its very nature, oxide formation must remove semiconductor atoms into the oxide. Because of the difference between the chemistry of column III and V atoms, disruption of the III–V semiconductor lattice could be expected. In addition, our studies have demonstrated changes in the surface electronic structure that appear to be induced by disruption of the periodic lattice at the low coverages at which the Fermi level becomes pinned (Pianetta *et al* 1978b, Spicer *et al* 1977a,b). These are seen by examining the semiconductor valence structure just at the surface. Figure 14 shows the valence electronic structure of GaAs after exposure to progressively increasing amounts of oxygen. The key observation is the phase-like transformation in the valence states which occurred

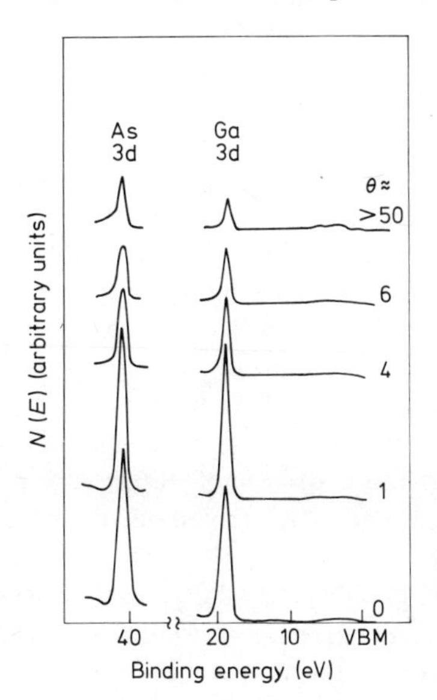

Figure 12. Photoemission spectra taken at a photon energy of 165 eV for GaAs with different Au coverages θ.

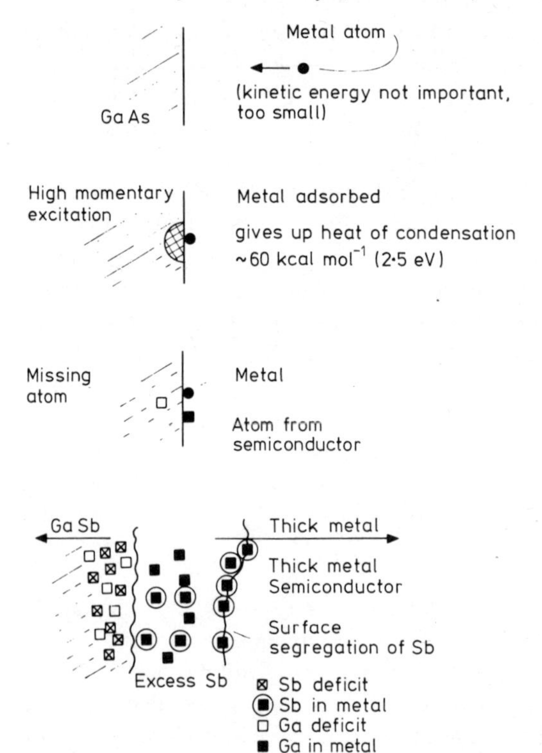

Figure 13. Schematic diagram of suggested defect mechanism due to deposition of metal atoms on clean III–V surfaces. This process (i.e. a defect must be formed) need occur only about once for every hundred metal atoms striking the surface in order to explain the Fermi-level pinning.

here for an oxygen exposure of 10^6 Langmuirs (about 3% of a monolayer, the approximate coverage at which E_{Fs} pinning is stabilised). Sharp structure for states within 4 eV of the valence-band maximum (VBM) is seen for exposures up to 10^6 Langmuirs. At and above this value, this structure disappears abruptly, suggesting a degree of disordering of the atoms in the surface region. Such a process would be ideal for the formation of defects in the surface region of the semiconductor. The exposure at which the 'order–disorder' transformation occurs varies somewhat (10^6–10^8 Langmuirs) (Pianetta 1976), probably because of the perfection of the cleaved surface (the variation in oxygen coverage is much less). However, the transformation always occurs as sharply as in figure 14.

LEED studies (Kahn *et al* 1980) are underway to investigate this phenomenon further. For the samples studied, exposures of 10^6 and 10^7 Langmuirs produced a slight weakening (and an increase in diffuse scattering) of the LEED pattern. This was not accompanied by a change in the shape of the intensity profiles which would substantiate any change in the surface atomic relaxation (see figure 8). However, an exposure of 10^8 Langmuirs produced a complete change in the LEED pattern. The diffuse background was greatly increased and the shape of the intensity profiles changed. In fact the I–V curves appeared consistent with an unrearranged bulk structure underlying a disordered surface layer 1–3 Å thick. More work is needed to correlate the LEED and UPS results, but the discarding of the outer layer of the GaAs surface appears clear.

Thus, both for metals and semiconductors, experiments indicate that a process takes place which might lead to formation of defect levels in the semiconductor surface (or interface) region. Detailed energy-level schemes will be given in the next section.

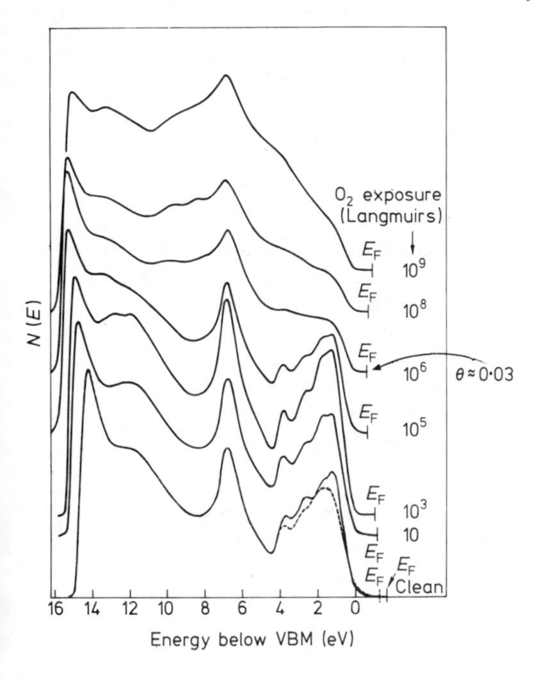

Figure 14. Energy distribution curves (EDC's) from the valence states of GaAs (sample LD1C) within a few atomic layers of the surface as a function of oxygen exposure. Up to 10^6 Langmuirs, the structure is relatively sharp and unchanged by the oxygen. Near 10^6 Langmuirs there is an abrupt loss of structure for $E > -5$ eV. This is associated with a partial disordering of the surface in which lattice defects may be formed.

6. Suggested energy-level scheme and defect assignment

From the data of figure 7, it is apparent that at least two levels at or near the surface are necessary to fit the experimental results for GaAs and InP, whereas one is probably sufficient for GaSb. Where two levels are needed, one must be a donor and the other an acceptor. Figure 15 shows the assigned levels and suggests the defect responsible for each (Spicer *et al* 1978). The level positions were obtained from the final pinning position due to either metal or oxygen adatoms. The data presently available show some relatively small variation with adatoms as well as treatment. Thus, a variation of ±0·1 eV is probably reasonable for the levels although, in some cases, they may be better established. Figure 15 indicates the simplest situation; interactions between the metal atoms and the defects, coalescence of simple defects into more complex structures and other 'geometric' effects may produce changes in the energy levels. Surprisingly, the empirical data indicate that these changes are small. Given in figure 15 is a suggestion regarding the simplest source of the defect: a missing column III or V atom. This is not to say that the defect is a simple vacancy, it may be more complicated, but that it is the result of a missing atom of the indicated species. The choice in figure 15 was made on the basis of a wide range of available data; only the principle of these will be mentioned below. The Sb deficit was chosen because the energy level corresponds to a bulk acceptor level known to be due to missing Sb (van der Muelen 1964, 1967, Effer and Etter 1964). When GaAs is cleaved, only a single defect level (an acceptor at 0·7 eV) is usually observed (Spicer *et al* 1976, Pianetta *et al* 1978b). This is assigned to a missing As atom on the quantitative argument that As is most likely to be lost in the cleaving process. The 1·2 eV level in InP is assigned to a missing In atom because of the reversal seen in the pinning position of E_{Fs} for n-type material as the oxygen exposure is increased (Chye *et al* 1980). This assignment, as well as

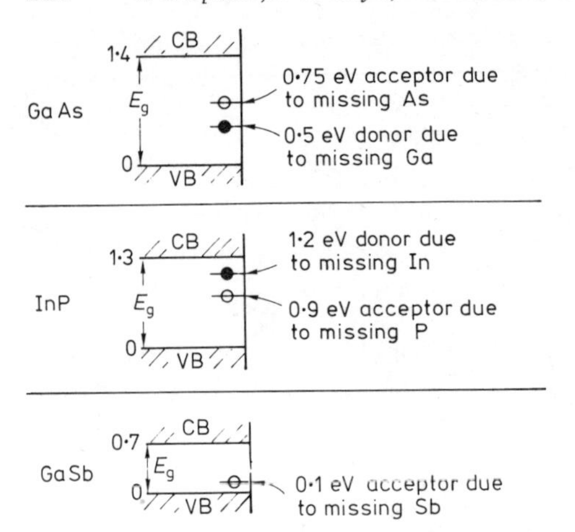

Figure 15. A model of extrinsic states produced near or at the surface by perturbing the surface through addition of metals or oxygen to the surface. Each level is tentatively associated with a deficit of an anion or cation; however, it is unlikely that the defect is a vacancy. More likely, it is an anti-site or more complicated defect.

that of 0·9 eV, to a P defect, appear consistent with the work of Farrow *et al* (1978); however, R H Williams (1979 private communication) has evidence to suggest that the level near 1·2 eV is due to a missing P rather than In. Further work is necessary to resolve this question.

7. Connection with states observed on real MOS device structures and conclusions

A critical question is whether contact can be made between this work and that on the structure of MOS devices. Correlations in this respect have been surprisingly good, as indicated by table 1 (Spicer *et al* 1979b). For more detailed correlations, the reader is referred to the literature and, in particular, to the paper by Wieder (1980) presented at this meeting and papers referred to therein.

It is not original to suggest that the differences between the III–V compounds and Si arise from the two-component nature of the III–V material; however, this cannot be overemphasised. In contrast to the III–V compounds, the cleavage surface of n-type Si is pinned by intrinsic surface states which are removed by oxygen adatoms and correspondingly E_{Fs} goes to the bulk position (Spicer *et al* 1979a, Wagner and Spicer 1974). Thus, the phenomena of oxygen adsorption and growth are very different and are reflected in the density of interface states.

Figure 16 shows models of the density of interface states for Si, GaAs, GaSb and InP. The density of states at the Si–SiO_2 interface is usually found to be roughly parabolic with the minimum near midgap. This distribution is usually attributed to strain due to the imperfect lattice match between the Si and the SiO_2. Such a background of states would also be expected in the III–V compounds and has been sketched into figure 16; however, superimposed on this are peaks associated with the defect-induced levels of figure 15. The magnitudes of these levels have been scaled to correspond to the available experimental data on GaAs and InP. None are available for GaSb.

Elsewhere (Chye *et al* 1980), detailed studies of the adsorption of oxygen on InP, GaAs and GaSb under different exposure conditions have been reported. However, one striking result is worth mentioning here. The resistance to the formation of bulk oxides (over a monolayer thick) with the presence of phase-separated column III and V oxides goes in the order GaSb (least resistance), GaAs and InP. GaSb forms bulk oxides when

Table 1. Fermi energy pinning position with respect to the conduction-band minimum after indicated surface treatment (the listed values are collected from the literature: see Spicer *et al* 1979a).

Material	Thick 'device' oxide	Submonolayer of of chemisorbed oxygen	Submonolayer to several monolayers of		
			Cs	Al, Ga, In	Au
GaAs (n-type)	0·83 0·8 0·5	0·65–0·8	0·6–0·8 0·7	0·6–0·7	0·9–1·0
InP (n-type)	0·14 0·087 0·075 0·1 0·1	0·05–0·2	0·15–0·45		

The similarity between the Fermi level pinning position for thick oxides formed by various techniques and that obtained by a fraction of a monolayer of oxygen should be noted. It should also be noted that the pinning position for Schottky barriers (the example here is caesium) is close to that for oxygen.

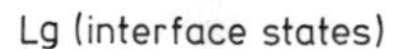

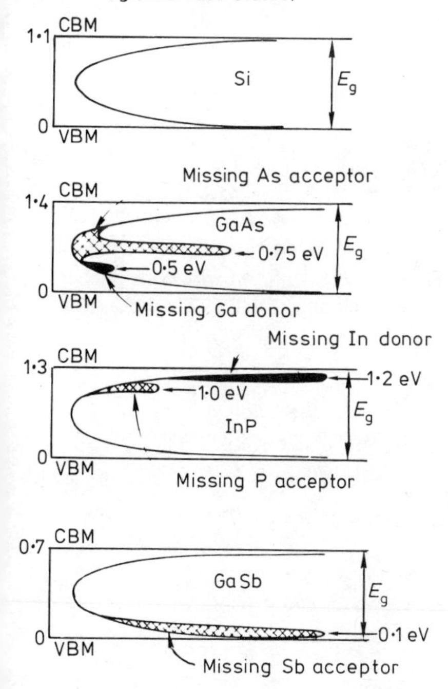

Figure 16. Model of interface states for GaAs, InP and GaSb MIS structures based on the present work.

exposed to molecular O_2 (in the ground state); GaAs only takes up about a monolayer of unexcited oxygen and shows no sign of formation of segregated bulk oxides (even when exposed to an atmosphere of O_2 for 15 min); however, when exposed to excited oxygen (probably containing very small amounts of atomic oxygen), thick oxides are formed and the segregation of oxides, as well as As loss, is observed (Pianetta *et al* 1978a). Even with excited oxygen, InP appears to accept only about a monolayer of oxygen and provides little evidence for the formation of phase-separated oxides (Chye *et al* 1980). This may be connected with the finding of Wieder *et al* (1980) that the density of interface states in InP MOS structures can be about an order of magnitude less than with GaAs. The fact that the dominant level in InP MOS structures lies about 0·1 eV below the conduction-band minimum is also fortuitous since, at room temperature, it will not provide nearly as long-lived a trap as the levels near midgap in GaAs.

The work outlined here was designed as a very fundamental study with the hope that it would give insight into Schottky barrier formation and the III–V oxide interface states. Much to our surprise, it appears that a one-to-one relationship can be drawn between our results and those from practical devices. As a result, insight can be gained into the material problems associated with those devices and how they can be overcome. Ultimately, we hope this work will help lead to an optimisation of the related devices by application of the basic knowledge developed in these and related studies.

References

Berglund C N and Spicer W E 1964 *Phys. Rev.* **136** A1030

Brillson L J 1978 *J. Vac. Sci. Technol.* **15** 1378

—— 1979 *Phys. Rev. Lett.* **42** 397

Cardona M and Ley L (eds) 1978 *Photoemission in Solids. General Principles* vol 1 (Berlin: Springer-Verlag)

Chye P W 1978 *PhD Dissertation* Stanford University, Stanford, California

Chye P W, Babalola I A, Sukegawa T and Spicer W E 1975 *Phys. Rev. Lett.* **35** 1602

Chye P W, Lindau I, Pianetta P, Garner C M and Spicer W E 1977b *Phys. Lett.* **63A** 387

—— 1978a *Phys. Rev.* **B17** 2682

Chye P W, Lindau I, Pianetta P, Garner C M, Su C Y and Spicer W E 1978b *Phys. Rev.* **B18** 5545

Chye P W, Su C Y, Lindau I, Garner C M, Pianetta P and Spicer W E 1980 submitted for publication

Chye P W, Sukegawa T, Babalola L A, Sunami H, Gregory P E and Spicer W E 1977a *Phys. Rev.* **B15** 2118

Derrien J and D'Avitaya 1977 *Surface Sci.* **65** 668

Dupris R D, Dapkus P D, Garner C M, Su C Y and Spicer W E 1979 *Appl. Phys. Lett.* **34** 335

Effer D and Etter P J 1964 *J. Phys. Chem. Solids* **25** 451

Farrow R F C, Cullis A G, Grant A J and Patterson J E 1978 *J. Crystal Growth* **45** 292

Feuerbacher B, Fitton B and Willis R F (eds) 1978 *Photoemission and the Electronic Properties of Surfaces* (New York: John Wiley and Sons)

Garner C M 1978 *PhD Dissertation* Stanford University, Stanford, California

Gregory P E and Spicer W E 1975 *Phys. Rev.* **B12** 2370

Gregory P E, Spicer W E, Ciraci S and Harrison W A 1974 *Appl. Phys. Lett.* **25** 511

Helms C R, Spicer W E and Johnson N M 1978 *Solid St. Commun.* **25** 673

Huijser A J and van Laar J 1975 *Surface Sci.* **52** 202

Kahn A, Kanani D, Mark P, Chye P W, Su C Y, Lindau I and Spicer W E 1980 submitted for publication

Lindau I, Chye P W, Garner C M, Pianetta P, Su C Y and Spicer W E 1978 *J. Vac. Sci. Technol.* **15** 1332

Lindau I, Pianetta P and Spicer W E 1976 *Phys. Lett.* **54A** 225

Lindau I and Spicer W E 1974 *J. Electron Spectrosc.* **3** 409

—— 1979 *J. Electron Spectrosc.* **15** 295

Mark P (ed) 1976 *Proc. 3rd Conf. Physics of Compound Semiconductor Interfaces. J. Vac. Sci. Technol.* **14** No. 4

—— 1977 *Proc. 4th Conf. Physics of Compound Semiconductor Interfaces. J. Vac. Sci. Technol.* **14** No. 4
—— 1978 *Proc. 5th Conf. Physics of Compound Semiconductor Interfaces. J. Vac. Sci. Technol.* **15** No. 4
Montgomery V, McKinley A and Williams R H 1979 *Surface Sci.* in press
Pianetta P 1976 *PhD Dissertation* Stanford University, Stanford, California
Pianetta P, Lindau I, Garner C M, Gregory P E and Spicer W E 1978a *Surface Sci.* **72** 298
Pianetta P, Lindau I, Garner C M and Spicer W E 1978b *Phys. Rev.* **B18** 2792
Skeath P R, Su C Y, Chye P W, Lindau I and Spicer W E 1979 *J. Vac. Sci. Technol.* **16** No. 4
Spicer W E 1958 *Phys. Rev.* **112** 114
—— 1976 in *Optical Properties of Solids: New Developments* ed B O Seraphin (Amsterdam: North-Holland) p631
Spicer W E, Chye P W, Garner C M, Lindau I and Pianetta P 1979c *Surface Sci.* in press
Spicer W E, Chye P W, Skeath P R, Su C Y and Lindau I 1979a *J. Vac. Sci. Technol.* **16** No. 4
Spicer W E, Lindau I, Gregory P E, Garner C M Pianetta P and Chye P W 1976 *J. Vac. Sci. Technol.* **13** 780
Spicer W E, Lindau I, Miller J N, Ling D T, Pianetta P, Chye P W and Garner C M 1977a *Phys. Scripta* **16** 388
Spicer W E, Lindau I, Pianetta P, Chye P W and Garner C M 1979b *Thin Solid Films* **56** 1
Spicer W E, Pianetta P, Lindau I and Chye P W 1977b *J. Vac. Sci. Technol.* **14** 85
Spicer W E And Simon R E 1962 *Phys. Rev. Lett.* **9** 385
van der Meulen 1964 *Solid St. Electron.* **7** 767
—— 1967 *J. Phys. Chem. Solids* **28** 25
van Laar J and Scheer J J 1967 *Surface Sci.* **8** 342
Wieder H 1980 *Insulating Films on Semiconductors 1979* (Inst. Phys. Conf. Ser. 50) pp234–50
Williams R H, Montgomery V and Varma R R 1978 *J. Phys. C: Solid St. Phys.* **11** L735

Surfaces and dielectric-semiconductor interfaces of some binary and quaternary alloy III-V compounds

H H Wieder

Electronic Material Sciences Division, Naval Ocean Systems Center, San Diego, California 92152, USA

Abstract. The surface properties and the semiconductor–dielectric interfacial properties of n- and p-doped InP and the quaternary alloy $In_xGa_{1-x}As_yP_{1-y}$ are described in terms of measurements made on metal–insulator–semiconductor (MIS) capacitors and depletion-mode and inversion-mode MIS field-effect transistors. It is shown that the surface Fermi-level pinning positions derived from such measurements are consistent with those determined from photoemission and Raman scattering measured on virgin InP surfaces oxidised under controlled conditions.

1. Introduction

The scientific and technological evolution of the intermetallic semiconducting III–V compounds has led, during the past few years, to a better although still incomplete understanding of the physics and chemistry of their surfaces. It has also affected the emerging technology of materials, processes and devices intended to provide a microwave analogue of the lower-frequency silicon discrete and integrated circuit metal–oxide–semiconductor (MOS) field-effect transistor (FET) technology. Of particular interest are the surfaces of the binary, ternary and quaternary III–V alloys with high electron mobility (whose fundamental bandgaps $E_g \geqslant 1$ eV) and the homo- and heterojunction interfaces between them or between any one of them and metal or dielectric overlayers.

Among the many important attributes of the Si-based MOS technology are the fortuitous dielectric and interfacial properties of steam-grown silicon oxide (SiO_x) on Si. To first order such dielectric layers may be considered as ideal; with a metal 'gate' deposited on the dielectric and an 'Ohmic' contact applied to the opposite surface of the Si wafer, the MOS structure may be represented in terms of the oxide capacitance in series with the surface space-charge capacitance produced by the redistribution of charges on and within the semiconductor. Surface and interface states, charge localisation and charge transport in the oxide and across the Si–SiO_x interface, workfunction differences and spatial inhomogeneities in the Si or SiO_x have required elaboration and refinement of this simple model; many specific features of the augmented theoretical model have been verified experimentally.

For the III–V compounds no such model is available as yet, because no fully compatible dielectric or oxide interface similar to Si–SiO_x has been found.

Native oxides are neither stoichiometric nor spatially homogeneous and are much too conductive to qualify as dielectric layers adequate for MOS structures. Preliminary results

0305-2346/80/0050-0234$03.00

can be obtained with electrochemical procedures such as those first described by Hartnagel (1976), Sugano and Mori (1974) and Hasegawa *et al* (1975). These authors demonstrated that dielectric layers grown in aqueous electrolytes by the anodisation of (100)- or (111)-oriented GaAs surfaces are suitable for a variety of simple measurements using metal–insulator–semiconductor (MIS) structures; although such anodically grown insulators are composites of more than one oxide, they may contain traces of their elemental constituents and are neither homogeneous nor isotropic.

Investigations have been performed in our laboratory on GaAs MIS structures in which the dielectric layers were made by anodisation; other GaAs MIS structures were made by the low-temperature pyrolysis of silane to form SiO_x and some were made by the pyrolysis of silane in the presence of ammonia to form Si_3N_4 dielectric layers. We have also investigated MIS structures with insulating layers deposited by means of low-energy neutralised ion-beam sputtering of SiO_2 targets and reactive sputtering of Si in nitrogen and argon to form Si_3N_4 dielectric layers on (100)- and (111)-oriented GaAs surfaces.

From the frequency dispersion measurements of capacitance against voltage (C–V), quasi-static measurements and conductance against voltage (G–V) data it appears that, at least to first order, the properties of such MIS structures are not dependent on the crystallographic orientation of their GaAs substrates. They are also not dependent on the type of insulating layer or the manner or condition of its preparation or subsequent annealing. Measurements were made in order to determine the density of surface states N_{ss} as a function of the surface potential ψ_s and the results obtained by Meiners (1979a,b) on representative n- and p-doped GaAs MIS capacitors are shown in figure 1. Similar measurements (made only on anodised dielectric MIS structures) were first reported by Hasegawa and Sawada (1977).

The position of the pinned surface Fermi level E_F^* can be calculated from the surface potential ψ_{so} ($V_g = 0$) shown in figure 1 and the bulk potential U_B; the free-carrier concentration required to determine U_B can be obtained (Many *et al* 1965) from deep-depletion C–V measurements. From these data E_F^* was calculated to be between 0·8 and 0·9 eV below the conduction band minimum (CBM) of n-type GaAs and between 0·7 and 0·8 eV below the CBM of p-type GaAs. These values of E_F^* are consistent, to within ±0·2 eV, with the Fermi-level pinning obtained by Spicer *et al* (1979) for clean (110)-oriented GaAs cleaved in ultra-high vacuum using x-ray photoemission (XPS) and ultra-

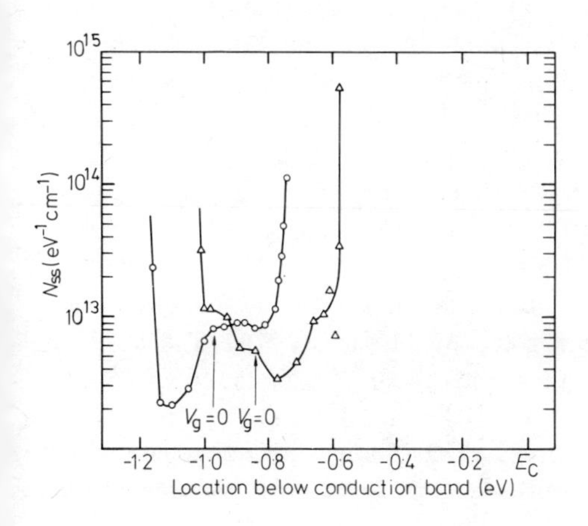

Figure 1. Dependence of the density of surface states on surface potential (at room temperature) in n-doped (○) and p-doped (△) GaAs calculated from C–V measurements (after Meiners 1979a, b).

violet photoemission (UPS) spectroscopic measurements to determine the shift in E_F^* from its bulk value after the chemisorption of a fractional monolayer of oxygen.

Figure 1 shows that variations in the surface potential of either n-type or p-type GaAs MIS capacitors are restricted to ~0·45 eV within the lower half of the bandgap. Neither surface accumulation nor inversion can be achieved with applied electric fields of the order of 10^6 V cm^{-1} because of the steep increase in N_{ss} as ψ_s is displaced from ψ_{so} towards either of the band edges.

The surface properties of InP and GaAs are qualitatively similar. However, there are important quantitative differences between them and also between their dielectric–semiconductor interfaces. Anodisation of InP in a 0·1 M KOH solution was first described by Wilmsen (1975). He used Auger electron spectroscopy to investigate the surface chemistry of anodic and thermally generated oxide layers on InP. Subsequently Wilmsen and Kee (1978) refined and augmented these measurements by the use of ESCA techniques. They found the anodic oxides to be primarily In_2O_3 with relatively little P_2O_5. The In–O and P–O bonding penetrate InP to approximately the same depth, in contrast with the anodic oxides of GaAs in which the Ga–O bond extends deeper into the GaAs than the As–O bonding. Thermal oxides appear to be similar in most respects to the anodic oxides except that elemental P was found to be localised at the interface.

Lile and Collins (1976) used salicylic acid, sodium salicylate in ethyl alcohol, to anodise InP and investigated the properties of corresponding MIS capacitors by means of *C–V* measurements. They found that for n-type InP MIS structures at room temperature and $V_g = 0$, the surface is depleted, $\psi_{so} \simeq 0{\cdot}47$ V and $N_{ss} \leqslant 3{\cdot}6 \times 10^{11}$ eV^{-1} cm^{-2}, a value which remains essentially constant from ~0·13 eV to ~0·9 eV from the conduction band into the band gap. Auger analysis shows that such layers contain substantial amounts of C and elemental P near the interface. None of the MIS capacitors exhibited the increase in capacitance above their minima to be expected in the surface inversion regime of quasi-static or low-frequency *C–V* measurements. Surface inversion has been observed experimentally by Roberts *et al* (1977) on MIS capacitors using Langmuir films made by deposition of organic layers of the Cd salts of stearic and arachnidic acids on InP substrates with electron concentrations of $10^{15} < n < 10^{16}$ cm^{-3}. With *C–V* and *G–V* measurements at 30 Hz, they determined the dependence of the density of surface states to be $N_{ss} \cong 3 \times 10^{11}$ eV cm^{-2} over a large portion of band gap; the total variation in the surface potential was between 0·8 and 0·9 V. Lorenzo *et al* (1979) investigated anodic oxidation of InP in a citric acid–ethylene glycol electrolyte and used these data, in conjunction with differential Hall measurements, to determine the carrier concentration profiles of ion-implanted InP. The dependence of the anodic oxidation process of InP on the nature of the electrolyte and the crystallographic orientation of the substrate, together with the anodisation of (111)-oriented $In_xGa_{1-x}As_yP_{1-y}$ in various electrolytes, have been described by Williams *et al* (1978); however, no data are available as yet on the electrical properties of the corresponding MIS structures.

The electrical properties of InP–SiO_x MIS structures made by the low-temperature pyrolysis of silane on n-type InP were investigated by Messick (1976). The insulating layers were found to be hard and glassy with a resistivity $> 10^{15}$ Ω cm at room temperature and breakdown field strengths between 5×10^6 and 10^7 V cm^{-1}. However, such layers exhibit instabilities of their *C–V* parameters that are dependent on gate voltage and may be attributed to charge injection and trapping of electrons in the oxide as well as ionic conduction in the SiO_x. Under optimum conditions, the average density of surface states, determined from *C–V* measurements, is $N_{ss} \simeq 2 \times 10^{11}\ eV^{-1}\ cm^{-2}$. Fritzsche (1978) found that the addition of HCl to silane during low-temperature chemical vapour deposition of SiO_x on n-type InP is an appropriate method for making MIS structures with a low density of surface states in the vicinity of the conduction band edge. The interface properties were investigated by means of *C–V* and *G–V* measurements under fast sweep and DC bias conditions. Fritzsche found evidence of deep donor-type states that were neutral in accumulation and positively charged in depletion and which limit the deviation of the steady state surface potential to ~0·4 V, which corresponds to a surface Fermi level ~0·5 eV below the conduction band. These states appear to be independent of the vapour deposition process (with or without HCl) and of any annealing temperature or parameters of the gaseous environment. The density of surface states, $N_{ss} \simeq 5 \times 10^{11}\ eV^{-1}\ cm^{-2}$, is a weak function of the surface potential.

C–V measurements in the frequency range 10^2 Hz–1 MHz at gate-voltage sweep rates of ~0·1 V s^{-1}, augmented by high-frequency admittance data, were used by Meiners (1979a,b) to determine the surface and interface properties of pyrolytically deposited SiO_x–InP MIS structures. Additional *C–V* measurements at 1 MHz with gate-voltage sweep rates of ~400 V s^{-1} were made in order to evaluate the MIS capacitors in deep depletion. The free-carrier concentration calculated from the data was found to be in good agreement with values determined from Hall measurements and the corresponding high-frequency capacitance minimum is consistent with the asymptotic value of the minimum capacitance calculated theoretically. Analysis of the quasi-static *C–V* curve using the method of Berglund (1966) yields a value for the total change in surface potential of ~1·0 V. This data has been used, in conjunction with results obtained from the high-frequency *C–V* measurements, to derive the surface-potential-dependent densities of surface states shown in figure 2. These are in good qualitative and fair quantitative agreement with the N_{ss} against ψ_s results obtained by Pande and Roberts (1979) on anodised dielectric-layer InP MIS capacitors.

Although the microwave power gain of a depletion-mode GaAs metal–insulator–semiconductor field-effect transistor (MISFET) is comparable to that of a metal–semiconductor–field-effect transistor (MESFET) of the same geometrical configuration (Lile and Collins 1976, Tokuda *et al* 1977), the frequency response of the MISFET is limited by a high density of slow surface states irrespective of the type or preparation of dielectric layers investigated to date. Furthermore, the high density of surface states of a GaAs

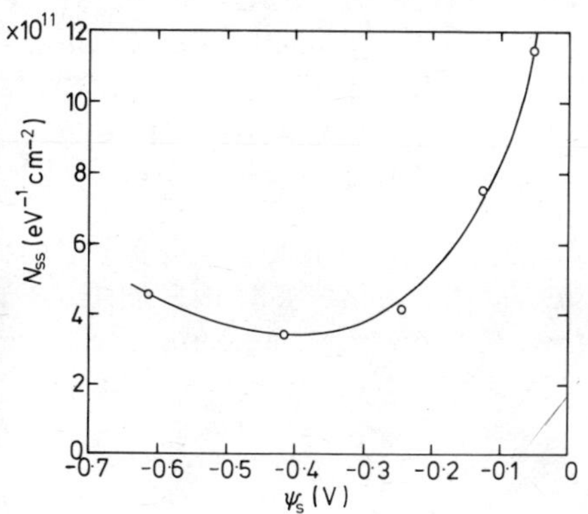

Figure 2. Dependence of the density of surface states on the surface potential (at room temperature) in n-type InP (after Meiners *et al* 1979).

MISFET limits its transconductance (Lile 1978) and holds little promise for the eventual use of these structures in microwave digital integrated circuits.

2. InP MIS and MISFET structures

The first depletion-mode InP MISFET with power gain in the microwave region as well as at lower frequencies was made in our laboratory and has been described by Messick *et al* (1978). The prototype mesa structure shown in figure 3 was made by photolithographic and lift-off procedures using a 0·35 μm thick, n-type InP layer grown by liquid-phase epitaxy (LPE) on a (100)-oriented Fe-doped semi-insulating InP substrate. Source and drain contacts were made by alloying a Au–Ge/Ni eutectic to the epilayer and a dielectric layer of SiO_x 0·2 μm thick was deposited upon it by the low-temperature pyrolysis of silane. The drain–source separation was $2l \simeq 8$ μm, width $w \simeq 260$ μm and the length of the vacuum-deposited aluminium gate electrode was $l \simeq 4$ μm.

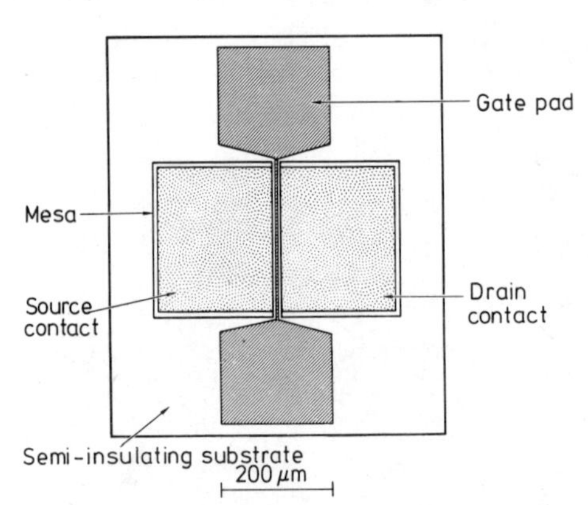

Figure 3. Mesa-type InP MISFET made of an n-type epitaxial layer on a Fe-doped semi-insulating (100)-oriented InP substrate.

Figure 4 shows the transistor characteristics of such a depletion-type MISFET. A typical device has a current gain of 8 dB, a noise figure of ~10 dB at 1 GHz and a current-gain cut-off frequency of 6 GHz. The length-to-width, $2(l/w)$, aspect ratio of this transistor makes it suitable (Wieder 1971) for determining the electron mobility profile of the conduction channel by means of field-effect controlled differential magnetoresistance measurements. The carrier concentration profile of the InP epilayer is considered to be independent of position to within 1 Debye length of the semi-insulating interface (Lile and Collins 1979). The channel current i in figure 4(*b*) is a linear function of the source–drain voltage v; the channel conductance, which is considered to be uniform over its entire length, decreases from its initial value ($V_g = 0$) with increasing surface depletion.

A magnetic field B perpendicular to the transistor surface generates a Hall field in a direction transverse to both the current density in the channel and the magnetic field vector. The Hall field is electrostatically short circuited by the closely spaced source and drain electrodes. The resistance measured between the source and drain electrodes increases with B^2, in a manner similar to a Corbino disc, in which the electrodes provide a conductive short circuit of the Hall field. If the 'physical magnetoresistance' is negligible, then the geometrical magnetoresistance is

$$\Delta R/R_0 = \xi \mu^2 B^2 \qquad (1)$$

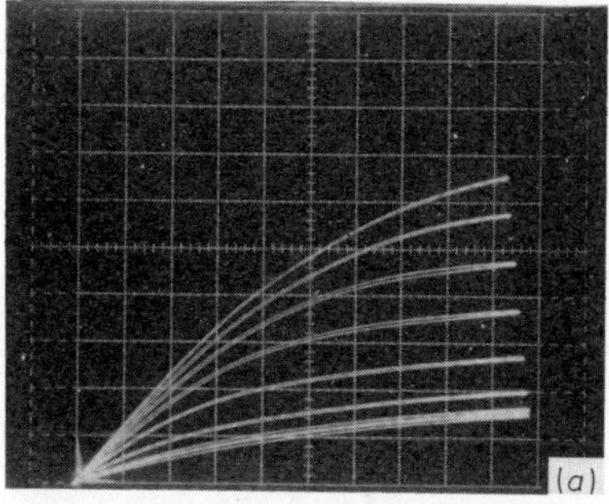

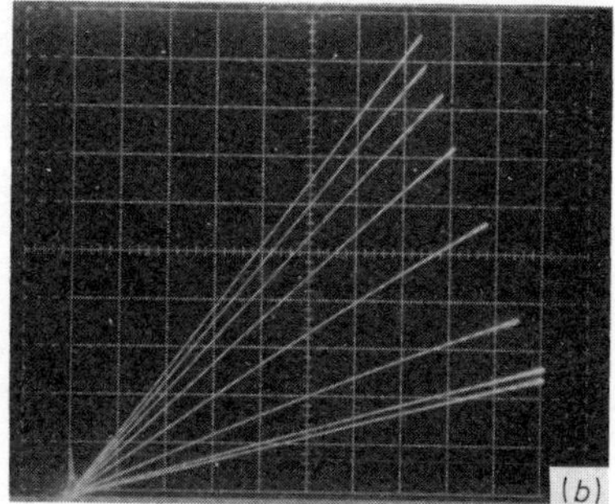

Figure 4. Characteristic transistor curves for a depletion-mode InP MISFET. (*a*) The source–drain potential made large enough to include the channel pinch-off region (*b*) the expanded 'triode' region of the same transistor. Scales: (*a*) 10 mA per vertical division, 1 V per horizontal division, 2 V per step, β or g_m = 5 ms per division; (*b*) 1 mA per vertical division, 100 mV per horizontal division, 2 V per step, β or g_m = 500 μs per division.

where μ is the Hall mobility and ξ is a parameter that depends on the electron scattering processes. A variational calculation of ξ for polar scattering (Howarth and Sondheimer 1953), the dominant scattering mechanism for InP at 300 K, indicates that, to first order, $\xi \simeq 1$. Experimentally this parameter can be determined from combined Hall and magnetoresistance measurements as the ratio of the respective mobilities.

The mobility profile of the InP MISFET from figure 4 was determined from measurements of $\Delta R/R_0$ made from $B \simeq 0$ to $B = 1{\cdot}0$ T in 0·05 T steps for each of the values of V_g shown in figure 4. The electron mobility, calculated from the slope of $(\Delta R/R_0)$ against B^2, assuming $\xi = 1$, is $\mu = 3050\ \mathrm{cm^2\ V^{-1}\ s^{-1}}$; it is not a function of V_g. Thus, within the error of measurement of ~10%, μ is not a function of the channel thickness. The Hall mobility measured on the epilayer before construction of the MISFET was $\mu_H = 2900\ \mathrm{cm^2\ V^{-1}\ s^{-1}}$ and thus $\xi = 1{\cdot}111$. This may be compared with $\xi = 1{\cdot}03 \pm 0{\cdot}07$ determined from magnetoresistance and Hall measurements made by Jervis and Johnson (1970) on GaAs epilayers.

An electron mobility independent of depletion depth was found by Jay *et al* (1978) using Schottky-barrier gate-controlled galvanomagnetic measurements made on van der Pauw-type (1958) clover-leaf structures on GaAs epilayers. Poth (1978) has compared the electron mobilities determined from Hall effect and Schottky-barrier gate-controlled Corbino disc magnetoresistance measurements made on GaAs epilayers. The magnetoresistance mobility is in good agreement with the Hall mobility and in substantially better agreement with theoretical expectations than the drift mobility derived from *C–V* measurements (Pucel and Krumm 1976).

If the channel conductivity $\sigma = ne\mu$ (where the electron density $n = 3 \times 10^{16}\ \mathrm{cm^{-3}}$ and $\mu \simeq 3 \times 10^3\ \mathrm{cm^2\ V^{-1}\ s^{-1}}$ is constant and independent of V_g, then the MISFET conductance in figure 4(*b*) is primarily a function of the thickness of the channel depletion layer. In terms of figure 5, let d be the epilayer thickness, δ_0 the depletion layer thickness and $i_0(v)$ the current in the channel for a specific source–drain potential v (in the range shown in figure 4*b*) for $V_g = 0$. Then the channel conductance is

$$G_0(v) = \sigma w(d - \delta_0)/2l. \tag{2}$$

For a fixed source–drain voltage and arbitrary gate voltages, the depletion layer is presumed to have the same length l and width w (under the gate electrode) but a thickness

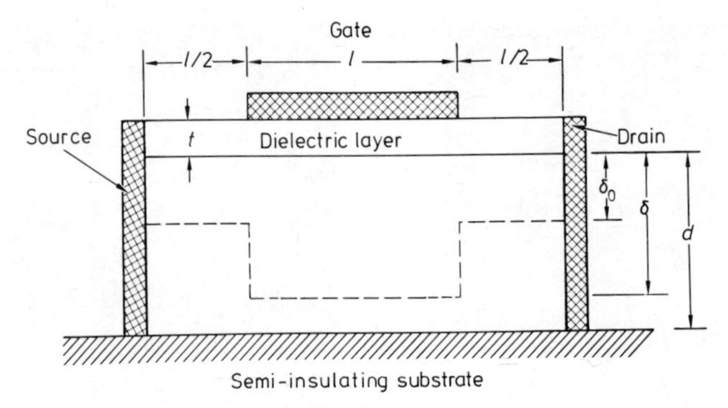

Figure 5. Model for the triode region (figure 4*b*) of a depletion-mode MISFET. Length of channel = $2l$; gate length = l; depletion depth for zero gate voltage = δ_0; depletion depth for an arbitrary gate voltage = δ; epilayer thickness = d; thickness of dielectric layer = t.

δ that is dependent on gate voltage. The conductance is

$$G(v) = \sigma w/l \, \{(d-\delta)(d-\delta_0)/[2d-(\delta+\delta_0)]\}. \tag{3}$$

The ratio of the currents $\alpha = i(V_g)/i(V_g = 0)$ is, therefore

$$\alpha = 2(d-\delta)/[2d-(\delta+\delta_0)]. \tag{4}$$

Assuming the applicability of the depletion approximation (Grove 1967)

$$\delta_0 = (2\epsilon_s \epsilon_0 \psi_{so}/en)^{1/2} \tag{5}$$

where the dielectric constant of the semiconductor $\epsilon_s = 12{\cdot}5$, the permittivity of free space $\epsilon_0 = 8{\cdot}854 \times 10^{-14}$ F cm^{-1} and the surface potential for $V_g = 0$ is $\psi_{so} \cong 0{\cdot}05$ V in terms of figure 2. The surface potential for an arbitrary gate voltage is

$$\psi_s = \psi_{so}\left[\frac{2d}{\delta_0}\left(\frac{1-\alpha}{2-\alpha}\right) + \frac{\alpha}{2-\alpha}\right]^2. \tag{6}$$

With $\delta_0 = 6{\cdot}65 \times 10^{-6}$ cm and $d = 3{\cdot}5 \times 10^{-5}$ cm, ψ_s can be determined as a function of α from figure 4(*b*). Assuming that $v \ll V_g$ and neglecting the workfunction difference and electron affinities of the MIS structure

$$V_g = (t/\epsilon_D \epsilon_0)(\Delta Q_s + \Delta Q_{ss}) + \Delta\psi_s \tag{7}$$

where t is the thickness and ϵ_D the dielectric constant of the SiO_x layer, $\Delta Q_s = en(\delta - \delta_0)$ is the difference between the charge densities per unit area of the space-charge region for an arbitrary value of ψ_s and that for ψ_{so}, ΔQ_{ss} is the difference between the charge density per unit area of surface states $Q_{ss}(\psi_s)$ and $Q_{ss}(\psi_{so})$ and $\Delta\psi_s = \psi_s - \psi_{so}$. The density of surfaces states N_{ss} per unit surface area per electron volt can be determined from equations (6) and (7) and the relation

$$N_{ss} = (1/e)(\Delta Q_{ss}/\Delta\psi_s) \tag{8}$$

is obtained. The dependence of N_{ss} on ψ_s shown in figure 6 was calculated for the two different values of ψ_{so}. The N_{ss} against ψ_s curve corresponds to deep depletion because the MISFET parameters used for these calculations are those of figure 4(*b*) measured at

120 Hz. A portion of the curve is in good agreement (considering the gross approximations used) with the branch of the N_{ss} against ψ_s data in figure 2 for the range of values of ψ_s over which deep-depletion and equilibrium C–V measurements are likely to coincide. Thus InP MISFET structures have values of N_{ss} nearly one order of magnitude smaller than the corresponding values for GaAs MISFET structures and a negligible low-frequency dispersion of their transconductances, compared with the large dispersion observed for GaAs (Lile and Collins 1979).

3. The surface Fermi level of n-type InP

The surface Fermi level E_F^* of n-type InP can be determined by means of a diagram such as that shown in figure 7. The intrinsic Fermi level E_{Fi} can be calculated (Smith 1968) from

$$E_{Fi} = -(E_g/2) + 4kT/(3e) \ln (m_p^*/m_n^*) \tag{9}$$

where E_g (300 K) = 1·34 eV, k is Boltzmann's constant, the hole effective mass $m_p^* = 0{\cdot}4 m_0$ and the electron effective mass $m_n^* = 0{\cdot}07\, m_0$; consequently E_{Fi} (300 K) = 0·65 eV. The

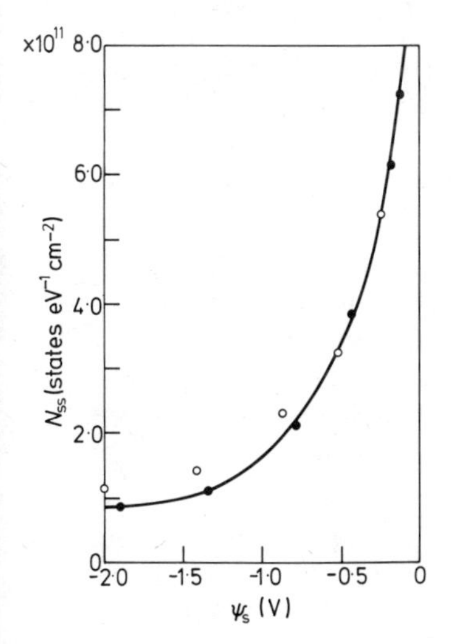

Figure 6. Dependence of the density of surface states on the surface potential using the model shown in figure 5 and transistor parameter values from figure 4(b). • ψ_{so} = 0·05 V; ○ ψ_{so} = 0·10 V.

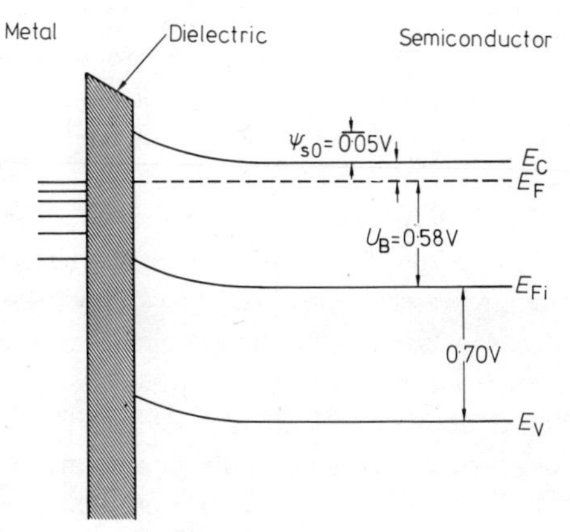

Figure 7. Equilibrium band diagram for n-type InP at room temperature. The pinned surface Fermi level E_F^* is 0·12 eV. E_g (300 K) = 1·34 eV; n_i (300 K) = 9·54 × 10⁶ cm⁻³; n (300 K) = 3 × 10¹⁶ cm⁻³.

bulk potential U_B is

$$U_B = (kT/e) \ln (n/n_i) \tag{10}$$

where the intrinsic carrier concentration n_i (300 K) = $9{\cdot}546 \times 10^6$ cm^{-3}; for an electron concentration n (300 K) = 3×10^{16} cm^{-3}, $U_B = 0{\cdot}58$ eV. The Fermi level is $E_F = 0{\cdot}65 - 0{\cdot}58 = 0{\cdot}07$ eV. If the surface potential for $V_g = 0$ is $\psi_{so} = 0{\cdot}05$ V, in accordance with figure 2, then $E_F^* = 0{\cdot}12$ eV. This is in good agreement with the pinning position $E_F^* = 0{\cdot}14$ eV found by Spicer *et al* (1979) using XPS on (110)-oriented cleaved InP surfaces after their exposure to a fractional monolayer of O_2. It is also in fair agreement with the results obtained by Williams and McGovern (1975), $\psi_{so} = 0{\cdot}075$ V, by means of photovoltage measurements on (100)-oriented InP surfaces etched in bromine–methanol. Pinczuk *et al* (1979) used Raman scattering from the coupled-plasmon LO phonon modes on etched ⟨111⟩ B-oriented InP surfaces and found $\psi_{so} \simeq 0{\cdot}2$ V and thus $E_F^* \simeq 0{\cdot}27$ eV. To first order therefore, the surface Fermi level pinning of n-type InP is $0{\cdot}1 < E_F^* < 0{\cdot}3$ eV and is a weak function of crystallographic orientation.

4. The surface Fermi level of p-type InP

In contrast to the low surface barrier of n-InP, acceptor-doped InP has a large surface barrier. This is evident in the Raman scattering measurements made by Pinczuk *et al* (1979) on ⟨111⟩B-oriented etched and oxidised p-type InP surfaces which yield $\psi_{so} \simeq 0{\cdot}80$ V and $E_F^* = 0{\cdot}24$ eV and the XPS measurements made by Spicer *et al* (1979) on p-type (110)-oriented surfaces which gave the position of the pinned surface Fermi level with respect to the conduction band minimum as $E_F^* = 0{\cdot}34$ V.

We have made charge-carrier transport measurements on the inversion layers present on (100)-oriented acceptor-doped $p = 5{\cdot}07 \times 10^{13}$ cm^{-3} InP MIS structures in order to determine E_F^*. Field-effect controlled van der Pauw-type measurements were made (Meiners and Wieder 1979), similar to those described by Tansley (1975), using specimens processed in accordance with the configuration shown schematically in figure 8. Four orthogonal equispaced blocking contacts were made to the substrate by the vacuum deposition of Sn : Au in a 1 : 4 (by weight) ratio, with subsequent alloying at 325 °C for 1 h. These n–p junctions exhibit rectifying characteristics with negligible current flow (<1 μA) up to their breakdown potential, $20\text{ V} < V_b < 50$ V. After the contact annealing process a layer of SiO_x 0·1 μm thick was deposited on the InP surface by sputtering of quartz; the symmetrical cross-shaped aluminium gate electrode, shown in figure 8, was aligned symmetrically with respect to the junction contacts.

Haeusler and Lippmann (1968) have shown that for a symmetrical cross-shaped Hall generator, the magnetoresistance of geometrical origin introduces an error of less than 0·3% in the expected linear relation between the Hall voltage v_H and the transverse magnetic field B in the range $B = 0$–1 T. De Mey (1973, 1974) has made a theoretical analysis of the symmetrical cross by conformal representation using the Schwartz–Christofell transformation. He found that the error in linearity does not depend on the Hall mobility, the magnetic field or the extent of the contacts made to the cross arms. Furthermore, the broad contacts allow a higher current to be used than do point contacts, without introducing thermal and thermoelectric errors in the Hall measurements. The symmetrical cross is, in fact, a variant of the clover-leaf specimen configuration

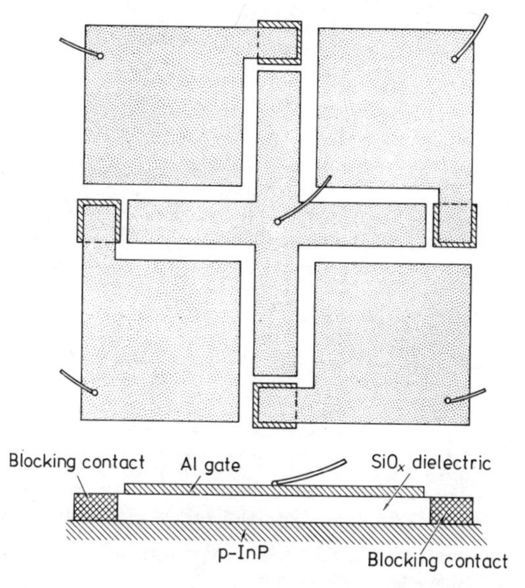

Figure 8. Schematic diagram of a symmetrical-cross MIS structure for galvanomagnetic measurements on surface inversion in p-InP. n-type blocking contacts are made to the substrate and contact pads are attached; van der Pauw-type measurements are made as a function of transverse magnetic field and gate voltage.

used for van der Pauw-type (van der Pauw 1958) resistivity and Hall effect measurements. The applicability of van der Pauw's relations to specimens that are inhomogeneous in their thickness has been analysed theoretically by Pauwels (1971). To first order, the perturbation introduced by small-scale inhomogeneities can be neglected.

The sheet resistivities and sheet Hall coefficients of the inversion layer were measured as a function of gate voltage and these measurements were then used to calculate the electron densities per unit surface area $\Delta N_s(V_g)$ and electron mobilities $\mu_s(V_g)$. For $V_g = 0$, $\Delta N_s = 1{\cdot}45 \times 10^{11}\ \mathrm{cm^{-2}}$ and $\mu_s \simeq 400\ \mathrm{cm^2\ V^{-1}\ s^{-1}}$. Neglecting the non-parabolicity of the conduction band and assuming a Maxwell–Boltzmann distribution for the charge carriers, the surface potentials that correspond to measured values of ΔN_s can be calculated from

$$\Delta N_s = \left(\frac{\epsilon_s \epsilon_0}{2kT}\right)^{1/2} \int_{\psi_s}^{0} \{n_B[1-\exp(e\psi/kT)]\,d\psi\}\,\{P_B[\exp(-e\psi/kT)+e\psi/kT-1] + n_B[\exp(e\psi/kT)-e\psi/kT-1]\}^{-1/2} \tag{11}$$

where P_B is the bulk acceptor concentration and $n_B = n_i^2/P_B$. Equation (11) cannot be solved analytically. Numerical integration by Simpson rule and a piecewise polynomial interpolation (Conte and de Boor 1972) has been used to solve this formula and yields the value $\psi_{so} = 0{\cdot}993$ V.

E_F^* can be determined from the diagram shown in figure 9. The bulk potential $U_B = (kT/e)\ln(P_B/n_i)$ is $U_B\,(300\ \mathrm{K}) = 0{\cdot}40$ V and the intrinsic Fermi level position with respect to the conduction band is $E_{Fi} = 0{\cdot}65$ eV. The position of the intrinsic Fermi level above the valence band is $E_g - E_{Fi} = 0{\cdot}69$ eV and the position of the Fermi level E_F above the valence band is 0·69–0·40 = 0·29 eV; therefore

$$E_F^* = E_g - \psi_{so} - 0{\cdot}29. \tag{12}$$

If the value $\psi_{so} = 0{\cdot}99$ V, calculated from the gate-controlled Hall measurements, is used in equation (12), then $E_F^* \simeq 0{\cdot}06$ eV. However, if the usual approximation made for

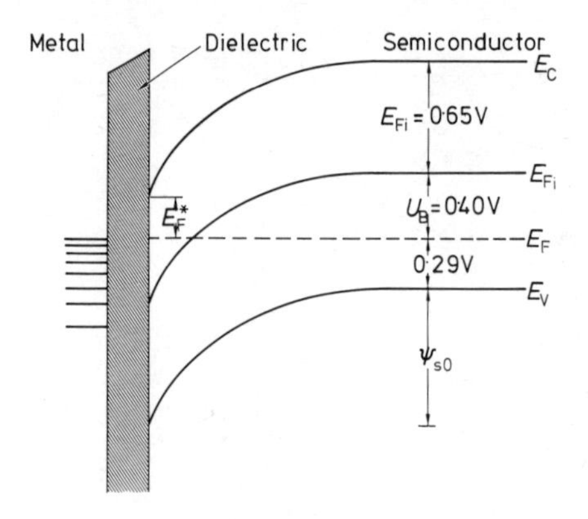

Figure 9. Equilibrium band diagram of p-type InP at room temperature. For a surface potential $\psi_{so} = 0{\cdot}8$ V, the pinned surface Fermi level is $E_F^* = 0{\cdot}25$ eV. n_i (300 K) = $9{\cdot}54 \times 10^6$ cm^{-3}; $m_n^* = 0{\cdot}07\, m_0$; $m_p^* = 0{\cdot}4\, m_0$; E_g (300 K) = 1·34 eV; P_B (300 K) = $5{\cdot}067 \times 10^{13}$ cm^{-3}.

inversion layers (Grove 1967), $\psi_{so} \simeq 2U_B = 0{\cdot}8$ V, is used, then $E_F^* = 0{\cdot}25$ eV. This value is in better agreement with that determined by Spicer *et al* (1979) for (110)-oriented p-doped InP, $E_F^* = 0{\cdot}3$ eV. *C–V* measurements made by Meiners (1979a, b) gave $U_B = 0{\cdot}58$ V and $\psi_{so} = 0{\cdot}73$ V; therefore $E_F^* = 0{\cdot}5$ eV. Pinczuk *et al* (1979) using Raman scattering on (111)-oriented oxidised free surfaces of p = 10^{18} cm^{-3} InP found $\psi_{so} \simeq 0{\cdot}8$ V; this leads to $U_B = 0{\cdot}66$ V and $E_F^* = 0{\cdot}5$ eV. The variations in E_F^* are essentially the same as those found by Williams *et al* (1978) for Schottky barriers; they may depend more on the details of surface preparation and surface treatment than crystallographic orientation or specific material parameters.

5. Inversion-mode transistors

Inversion-mode n-channel InP MISFET structures were first reported by Lile *et al* (1978a, b). Symmetrical source and drain n–p junction contacts separated by ~4 μm were made on acceptor-doped InP with p ≃ 10^{16} cm^{-3} and a hole mobility μ_p (300 K) ≃ 95 cm^2 V^{-1} s^{-1}. A dielectric layer of pyrolytically deposited SiO_x ~ 0·1 μm thick and a vacuum-deposited Al gate electrode ~ 3·5 μm long and ~200 μm wide, processed photolithographically, were made in the geometrical configuration described by Meiners *et al* (1979). Figure 10 shows the transistor characteristics of such a device. The source–drain current for zero gate bias is negligible; it increases with increasing positive gate voltage and source–drain voltage. The transconductance per gate width, 10 mS mm^{-1}, is within a factor of three of the value obtained for a depletion-type MISFET made on n-type

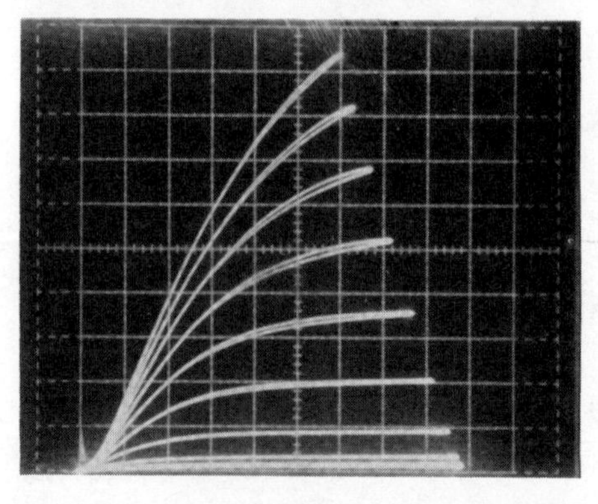

Figure 10. Characteristic transistor curves for an inversion-mode n-channel InP MISFET. Scales = 500 μA per vertical division; 500 mV per horizontal division; 1 V per step; β or g_m = 500 μs per division.

InP. The transconductance of the MISFET shown in figure 10, together with the gate capacitance per unit area of $\sim 3 \times 10^{-8}$ F cm^{-2}, suggests an inversion layer mobility of ~ 400 cm^2 V^{-1} s^{-1}, a value that is consistent with the galvanomagnetic measurements made on the symmetrical cross structure.

The reason for the low mobility of the electrons in the inversion layer compared to their mobility in the depletion-mode MISFET has not yet been determined. It might, however, arise from scattering from surface roughness and from centres localised at the dielectric–semiconductor interface (Ferry 1979). A surface-to-bulk mobility ratio $\mu_s/\mu_B \simeq 0{\cdot}15$ was found by Dinger (1977) for n-channel inversion layers present on p-type Ge MIS structures at ~ 80 K. This ratio is a function of the density of surface states per unit energy and it increases with increasing band bending and decreasing temperature. In contrast to this, the inversion-layer electron mobility of Si MOS structures $\mu_s/\mu_B \simeq 0{\cdot}4$ for $N_{ss} \simeq 10^{11}$ cm^{-2} eV^{-1} at 300 K.

The transconductance of the n-channel inversion-mode MISFET is frequency-independent, like that of the depletion-mode InP MISFET, but in contrast with the low transconductance at low frequencies of a GaAs MISFET. For a source–drain voltage $V_{sd} = 5{\cdot}5$ V and $V_g = 4{\cdot}5$, the power gain is 7 dB at 1 GHz and the power gain cut-off frequency is 2·5 GHz.

We have not succeeded in making a p-channel inversion-mode MISFET structure on n-type InP. This is consistent with our C–V measurements, made on n-type InP MIS structures with pyrolytically deposited or sputtered SiO_x layers, which cannot be driven into inversion probably because of the large hole recombination rate at the dielectric–semiconductor interface.

We have made inversion-mode MISFET structures on the Zn-doped quaternary alloy $In_{0{\cdot}84}Ga_{0{\cdot}16}As_{0{\cdot}34}P_{0{\cdot}66}$ with a band gap E_g (300 K) = 1·14 eV, p = 2·34 × 10^{16} cm^{-3} and $\mu_p = 71$ cm^2 V^{-1} s^{-1} grown by LPE on Fe-doped semi-insulating InP. The n–p junction contacts, the geometrical configuration and photolithographic processing were of the same type as those used for the InP inversion-mode MISFET. The transconductance is $\sim 25\%$ smaller than that of a comparable InP MISFET; however, neither the material quality nor the junction metallurgy have yet been fully developed. It is to be expected that a judicious compromise can be made between the electron mobility, the surface potential and the band gap of ternary and quaternary allow MISFET structures.

6. Discussion

Among the most important results that have emerged from fundamental and applied research on the surfaces of InP and its interfaces with metals and dielectrics is the striking similarity between Schottky surface barriers and the pinning position of the surface Fermi levels of oxidised free surfaces and MIS structures. This similarity suggests a physical mechanism that is not a function of the specific properties of the adsorbed atomic or molecular species or the crystallographic orientation of the InP surfaces. Available evidence indicates that InP, like GaAs, has no intrinsic surface states in the band gap; filled P surface states are present in the valence band and empty In states are present in the conduction band. It seems likely that pinning of the surface Fermi level is caused by extrinsic surface states localised at surface imperfections and stoichiometric defects. Oxygen is initially adsorbed at such residual lattice disorder sites (Mark and Creighton

1975). Additional disorder is produced because of the release of exothermic adsorption energy and oxygen reacts with the elemental constituents to form stable In_2O_3, primarily volatile oxides of P and a smaller quantity of stable P_2O_5 (Mark and Creighton 1979).

The formation of Schottky barriers on semiconductor surfaces is dominated (Brillson 1978) by the chemical reaction and interdiffusion of the metal and atomic constituents of the semiconductor. Williams *et al* (1978) have shown that the metal–(n-type)–InP barrier height is a function of the heat of reaction of the metal electrode with phosphorus. Compounds formed between Ni, Fe or Al with P have negative heats of reaction and form Ohmic contacts to InP while Cu, Ag, and Au, which have positive heats of reaction, produce surface barriers of ~0·5 eV on InP. Al and Cu, whose heats of reaction with P are close to that of InP, have barrier heights that are a strong function of the surface temperature. McCaldin *et al* (1976) have shown that the sum of the Au electron and hole barriers is equal to the energy gap of III–V compounds and that the Au Schottky barriers for holes depend only on the electronegativity of the anion. Furthermore, the surface Fermi level of oxygen chemisorbed on the (110) plane of cleaved n-type InP, $0{\cdot}05 < E_F^* < 0{\cdot}2$ eV, is comparable to the barrier obtained by the deposition of one or several monolayers of Cs, $0{\cdot}15 < E_F^* < 0{\cdot}45$ eV (Chye *et al* 1976). It has been suggested (Spicer *et al* 1979) that this similarity is the result of the interaction between the adatoms and InP. Lattice defects in the form of dissociated vacancies, complexes and/or anti-site defects (Van Vechten 1975) are generated and the surface states responsible for E_F^* are related to the specific properties of these defects and are not dependent on the characteristics of the adatoms.

The displacement of the surface Fermi levels from their equilibrium bulk positions as a function of oxygen exposure, shown in figure 11, were determined experimentally by Spicer *et al* (1979) by XPS measurements on cleaved (110)-oriented n- and p-doped InP surfaces. For exposures $<10^5$ Langmuirs the displacement of E_F^* towards the valence band of the n-type material suggests that electron traps are generated by chemisorption of oxygen onto surface P atoms, in analogy with the theoretical calculations of Barton *et al* (1979) and the transient capacitance spectroscopic measurements of Majerfeld *et al* (1978) and White *et al* (1978) of electron traps located 0·20, 0·24 and 0·34 eV below the conduction band minimum. Increasing the oxygen exposure above 10^5 Langmuirs

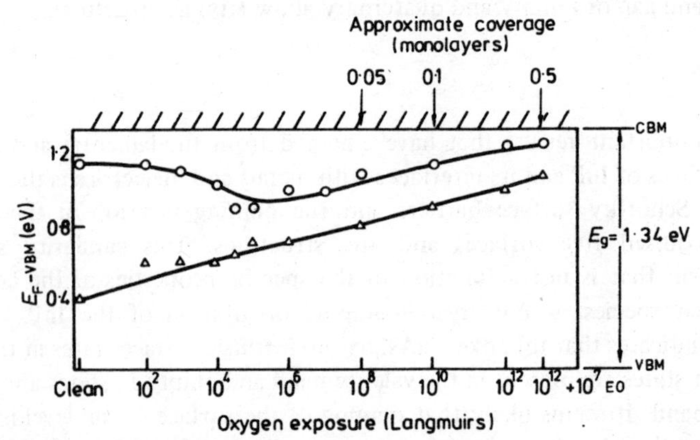

Figure 11. Displacement of the bulk Fermi level of (110)-oriented cleaved n-doped (○) and p-doped (△) InP after oxidation (from Spicer *et al* 1979).

generates electrons from donor centres; this makes E_F^* rise towards its equilibrium position near the conduction band. The minimum of the curve shown in figure 11 is consistent with the position of the deep-interface donor states at ~0·5 eV deduced by Fritzsche (1978) from *C–V* measurements on MIS structures. The monotonic rise of the Fermi level toward the conduction band of the p-type material in figure 11 with oxygen exposure suggests the dominant role of hole traps.

The native oxide that grows spontaneously on InP surfaces is about 1 to 1·5 nm thick, as determined from ellipsometric measurements. It is assumed that the oxide–semiconductor interface is not perturbed by the deposition of an SiO_x layer onto the native oxide. Langan *et al* (1978) and Langan and Viswanathan (1979) have demonstrated the importance of maintaining the integrity of the native oxide of InSb MIS structures made by low-temperature pyrolysis of silane in the presence of oxygen. Inferior-quality MIS devices are produced if the native oxide (In_2O_3) is reduced by Si due to the heterogeneous nucleation of SiO_x. Differential sputtering and XPS spectra indicate the presence of elemental In, In_2O_3 and a different In oxidation state at the interface. Superior-quality MIS devices are produced if the silane disproportionation reaction is homogeneous, with deposition of SiO_2 and preservation of the integrity of the native oxide; this is evident in the XPS spectra of such layers, the low densities of surface states, $<10^{10}\ eV^{-1}\ cm^{-2}$, and the absence of hysteresis in *C–V* measurements made on corresponding MIS structures.

The SiO_x pyrolysis process used in our laboratory for InP MIS and MISFET is, as yet, of lower quality. ESCA and AES analyses made by C W Wilmsen (1979 private communication) indicate the presence of In in the SiO_x layers produced by the partial dissociation of the native In_2O_3 oxide. The perturbation of this oxide and the inhomogeneous distribution of In in the SiO_x layers may be responsible for the hysteresis and anomalous frequency dispersion of the *C–V* curves in accumulation (Meiners *et al* 1979) and their relatively large DC conductance.

A similar frequency dispersion of the MIS capacitance has been observed experimentally by Kamimura and Sakai (1979) on InP MIS structures with $Al(OC_3H_7)_3$. Conductance and capacitance measurements made on InP–SiO_x MIS structures were used by Zeisse (1979) to fit the experimentally observed frequency dispersion of the capacitance in the accumulation region to a Maxwell (1892)–Wagner (1913) model of an inhomogeneous dielectric. Similar interfacial polarisation phenomena have been noted by Snow and Deal (1966) in the analysis of silicon MIS structures with a composite P_2O_5–SiO_2 dielectric layer in contact with an SiO_2 layer.

7. An overview

Based on results obtained thus far on (100)- and (111)-oriented GaAs and InP MIS capacitors made of arbitrary dielectric layers, it may be stated that:

(i) The density of surface states of InP near the midgap is at least one order of magnitude smaller than that of GaAs.
(ii) The surface Fermi level of InP is pinned in the vicinity of its conduction band while that of GaAs is pinned at or below the midgap.
(iii) In view of the similarity of Fermi-level pinning of spontaneously oxidised InP surfaces, MIS structures and Schottky barriers, it is presumed that the surface states

responsible for pinning do not depend on the specific attributes of the adatoms but on lattice defects present on the virgin surfaces and additional defects produced by the chemical reaction between the adatoms and InP.

(iv) The total deviation of the surface potential in GaAs is limited to ~0·4 V in the lower half of the band gap whereas the surface potential of InP can be displaced by at least 1 V within the band gap.

(v) Neither accumulation nor flat band can be reached on GaAs surfaces with electric fields of the order of 10^6 V cm^{-1}; however, flat band and accumulation are achieved with relative ease on InP surfaces.

(vi) A continuum of fast surface states is present in GaAs. Frequencies higher than 0·1 GHz are required for *C–V* measurements if charge exchange with surface states is to be prevented; for InP frequencies of the order of 1 MHz are sufficient to prevent surface-state interactions.

(vii) Depletion-mode GaAs and InP MISFET structures can be made with microwave power gain comparable to those of MESFET structures of some geometrical configurations. However, the low-frequency transconductance of a GaAs MISFET has a strong frequency dispersion whereas for InP, negligible dispersion occurs.

(viii) n-channel inversion-mode transistors have been made on epitaxial layers of p-type InP quaternary alloy $In_xGa_{1-x}As_yP_{1-y}$. No inversion-mode p-channel transistors on n-type InP have been reported as yet and no reliable evidence has been presented, as yet, on inversion-mode transistors made of GaAs.

(ix) Depletion- and inversion-mode InP MISFET structures may be used as test structures for probing the electronic structure of the dielectric–semiconductor interface and for determining the mobility profile of the source–drain conduction channel.

Although the dielectric layers and the dielectric–semiconductor interfaces require improvement, the results obtained so far, together with the higher saturated electron velocity, shorter transit time and higher thermal conductivity of InP compared with GaAs clearly indicate the potential and inherent advantages of a microwave MISFET integrated circuit technology based on InP.

Acknowledgments

I am indebted to my associates L Meiners and D Lile for much help and support. I wish to record in particular my gratitude to Professor Spicer and his students and to Professors McGill, Goddard, D Smith and their students for many stimulating and thought-provoking discussions.

References

Barton J J, Goddard W A III and McGill T C 1979 *Proc. 6th Conf. Physics of Compound Semiconductors: J. Vac. Sci. Technol.* **16** in press

Berglund C N 1966 *IEEE Trans. Electron. Devices* **ED-13** 701

Brillson L 1978 *Phys. Rev. Lett.* **40** 160

Chye P W, Lindau I, Pianetta P, Garner C M, Su C Y and Spicer W E 1976 *Phys. Rev.* **B13** 4439

Conte S D and De Boor C 1972 *Elementary Numerical Analysis* (New York: McGraw Hill) pp 232, 290

De Mey 1973 *Archiv Elektron. Ubertrag.* **27** 309
—— 1974 *Archiv Elektron. Ubertrag.* **28** 335
Dinger R 1977 *Thin Solid Films* **43** 311
Ferry D K 1979 *Thin Solid Films* **56** 243
Fritzsche D 1978 *Electron. Lett.* **14** 51
Grove A S 1967 *Physics and Technology of Semiconductor Devices* (New York: John Wiley and Sons) p267
Haeusler J and Lippmann H J 1968 *Solid St. Electron.* **11** 173
Hartnagel H L 1976 *J. Vac. Sci. Technol.* **13** 860
Hasegawa H, Forward K and Hartnagel H L 1975 *Electron. Lett.* **11** 53
Hasegawa H and Sawada T 1977 *Proc. 7th Int. Vacuum Congr., Vienna* (PO Box 300/A-1082, Vienna) p549
Howarth D J and Sondheimer E H 1953 *Proc. Phys. Soc.* **A219** 53
Jay P R, Crossley I and Cardwell M J 1978 *Electron. Lett.* **14** 190
Jervis T R and Johnson E F 1970 *Solid St. Electron.* **13** 181
Kamimura K and Sakai Y 1979 *Thin Solid Films* **56** 215
Langan J D and Viswanathan C R 1979 *Proc. 6th Conf. Physics of Compound Semiconductors: J. Vac. Sci. Technol.* **16** in press
Langan J D, Viswanathan C R, Merilainen C A and Santarosa J F 1978 *Proc. Int. Electron Devices Mtg, Washington DC* to be published
Lile D L 1978 *Solid St. Electron.* **21** 1199
Lile D L and Collins D A 1976 *Appl. Phys. Lett.* **28** 554
—— 1979 *Thin Solid Films* **56** 225
Lile D L, Collins D A, Meiners L G and Messick L 1978a *Electron. Lett.* **14** 657
Lile D L, Collins D A, Messick L and Clawson A R 1978b *Appl. Phys. Lett.* **32** 247
Lorenzo J P, Davies D E and Ryan T G 1979 *J. Electrochem. Soc.* **126** 118
Majerfeld A, Wada O and Choudhury A N M M 1978 *Appl. Phys. Lett.* **33** 957
Many A, Goldstein Y and Grover N B 1965 *Semiconductor Surfaces* (Amsterdam: North Holland) p136
McCaldin J O, McGill T C and Mead C A 1976 *Phys. Rev. Lett.* **36** 56
Mark P and Creighton W F 1975 *Appl. Phys. Lett.* **27** 400
—— 1979 *Thin Solid Films* **56** 19
Maxwell J C 1892 *Electricity and Magnetism* Vol. 1 (Oxford: Clarendon) p452
Meiners L G 1979a *Dielectric–Semiconductor Interfaces of GaAs and InP: Colorado State Univ. Rep. No.* SF19
—— 1979b *Thin Solid Films* **56** 201
Meiners L G, Lile D L and Collins D A 1979 *Proc. 6th Conf. Physics of Compound Semiconductors: J. Vac. Sci. Technol.* **16** in press
Meiners L G and Wieder H H 1979 to be published
Messick L 1976 *J. Appl. Phys.* **47** 4949
Messick L, Lile D L and Clawson A R 1978 *Appl. Phys. Lett.* **32** 494
Pande K P and Roberts G G 1979 *Proc. 6th Conf. Physics of Compound Semiconductors: J. Vac. Sci. Technol.* **16** in press
Pauwels H J 1971 *Solid St. Electron.* **14** 1327
Pinczuk A, Ballman A A, Nahory R E, Pollack M A and Worlock J M 1979 *Proc. 6th Conf. Physics of Compound Semiconductors: J. Vac. Sci. Technol.* **16** in press
Poth H 1978 *Solid St. Electron.* **21** 801
Pucel R A and Krumm C F 1976 *Electron. Lett.* **12** 240
Roberts G G, Pande K P and Barlow W A 1977 *Electron. Lett.* **13** 581
Smith R A 1968 *Semiconductors* (Cambridge: Cambridge UP) p78
Snow E H and Deal B E 1966 *J. Electrochem. Soc.* **113** 263
Spicer W E, Chye P W, Skeath P R, Su C Y and Lindau I 1979 *Proc. 6th Conf. Physics of Compound Semiconductors: J. Vac. Sci. Technol.* **16** in press
Spicer W E, Lindau I, Pianetta P, Chye P W and Garner C M 1979 *Thin Solid Films* **56** 1
Sugano T and Mori J 1974 *J. Electrochem. Soc.* **121** 113

Tansley T L 1975 *J. Phys. E: Sci. Instrum.* **8** 52
Tokuda H, Adachi Y and Ikoma T 1977 *Electron. Lett.* **13** 761
van der Pauw L J 1958 *Philips Res. Rep.* **13** 1
Van Vechten J A 1975 *J. Electrochem. Soc.* **122** 419–23
Wagner K W 1913 *Ann. Phys., Lpz.* **40** 817
White A M, Grant A J and Day B 1978 *Electron. Lett.* **14** 409
Wieder H H 1971 *Hall Generators and Magnetoresistors* (London: Pion) p25
—— 1978 *J. Vac. Sci. Technol.* **15** 1498
Williams J O, Wright P J and Elmorsi M A 1978 *J. Mater. Sci.* **13** 2292
Williams R H and McGovern I T 1975 *Surface Sci.* **51** 14
Williams R H, Montgomery V and Varma R R 1978 *J. Phys. C: Solid St. Phys.* **11** L735
Wilmsen C W 1975 *Crit. Rev. Solid St. Sci.* **5** 313
Wilmsen C W and Kee R W 1978 *J. Vac. Sci. Technol.* **15** 1513
Zeisse C R 1979 *Proc. 6th Conf. Physics of Compound Semiconductors: J. Vac. Sci. Technol.* **16** in press

Chemical vapour deposited SiO_2-InP interface†

C W Wilmsen‡, J F Wager‡ and J Stannard§

‡ Colorado State University, Fort Collins, Colorado 80523, USA

§ Naval Research Laboratory, Washington DC 20375, USA

Abstract. Chemical vapour deposition (CVD) of SiO_2 appears to be an attractive method of forming an insulator for InP MIS devices. There are, however, several possible mechanisms that can degrade the SiO_2–InP interface during fabrication. The results of an investigation of the chemistry of formation of the CVD SiO_2–InP interface are presented. ESCA depth profiles of samples fabricated at 340 and 450 °C show that thermal oxidation of the InP occurs in the CVD reactor, that In diffuses through the SiO_2 layer but P does not and that there is evidence for the decomposition of the SiO_2 by InP. These processes are expected to degrade the dielectric properties of the SiO_2 and surface conduction in the InP.

1. Introduction

InP is a possible alternative to GaAs for high-speed semiconductor devices (Wieder 1978). The initial MOS work reported for InP (Wilmsen 1975) indicated that the surface potential could be gate-controlled for anodic oxide and anodic oxide–sputtered SiO_2 double-layer insulators. The anodic oxide by itself does not have good DC insulating characteristics, but sputtering SiO_2 directly on to the InP surface does not allow variation of the surface potential. Recently there have been a number of reports (Messick 1976, 1979, Fritzsche 1978) on the successful use of CVD SiO_2 on InP for MOS capacitors and MISFET structures. MOS capacitors fabricated on p-type InP with a CVD SiO_2 layer have exhibited inversion behaviour and inversion layer conduction (Meiners 1978, Stannard and Henry 1979).

In order to fabricate good-quality CVD SiO_2 layers on InP, it is necessary to optimise several properties of this MOS system. These properties include the DC insulating characteristics of the SiO_2, the AC frequency dispersion attributable to the quality of the SiO_2 and the SiO_2–InP interface and the electronic properties of the SiO_2–InP interface which are determined by the interfacial width, chemistry, etc. The device properties may be controlled by optimising the processing steps during fabrication, such as the substrate temperature, the background gases present in the reaction chamber and the preparation of the InP surface.

For example, in choosing an optimum CVD deposition temperature, several considerations are important. First, in order to obtain the highest-quality SiO_2 dielectric layer, it is necessary to deposit at a high temperature. Conversely, however, deposition should be at a relatively low temperature since the InP surface is known to be thermally unstable and will evaporate at low temperatures. Also, since the diffusion of In through the SiO_2 could be a problem, a low deposition temperature would be appropriate to reduce this diffusion to a minimum. These factors indicate that there must be a compromise in the

† Supported in part by ONR.

deposition temperature. Indeed, it has been found that deposition of SiO_2 above 350 °C leads to a poor MOS device, although deposition at or below this temperature yields good results.

This paper models the formation of the CVD SiO_2–InP interface and presents ESCA profiles which verify the major concepts of the model.

2. Sample fabrication

The SiO_2 was deposited on the InP in a standard CVD reactor used for Si wafer processing. The InP wafers were first polished with a bromine–methanol solution, rinsed in methanol and stored in dry N_2. For the deposition, the cleaned samples were placed in the carousel of the CVD reactor and the entire system purged with N_2 gas. A N_2 curtain continued to flow over the InP while the material was heated to the deposition temperature and the conditions for proper CVD deposition of SiO_2 were established at another position on the carousel. These two steps required approximately 5 min, after which time the carousel rotated the InP wafer to the SiO_2 deposition position. The deposition rate was approximately 30 Å min^{-1}. For this investigation special samples with only about 200 Å of SiO_2 were deposited at temperatures of 340 and 450 °C. Deposition at 340 °C corresponds to the best SiO_2–InP MOS capacitors made by this process so far. Deposition at 450 °C was chosen to accentuate the effects predicted by the model. The thin SiO_2 layer (200 Å) provides better Auger/ESCA profiling conditions than the usual layer of oxide 1000–2000 Å thick.

3. Interface formation

Based on the above cleaning/deposition sequence (which is similar to that used by other workers), the initial surface will be covered with a 10–20 Å thick layer of air-grown oxide and be contaminated with hydrocarbons and possibly trace elements from the cleaning/rinsing solutions and the air. Heating the InP could cause evaporation of its surface or the initial oxide but it could also cause the growth of thermal oxide in the residual O_2 of the reactor. Upon deposition of the SiO_2 at raised temperatures, the InP could react with the SiO_2 or even diffuse through the depositing layer.

3.1 Thermal decomposition and oxidation of the surface

The work of Bayliss and Kirk (1976), Farrow (1974, 1975) and Lum and Clawson (1979) has shown that, when heated in vacuum, the surface of InP decomposes and loses phosphorus which causes the surface to pit. On the other hand, when heated in an oxygen atmosphere, the surface oxidises readily if the temperature is sufficiently high (Wager and Wilmsen 1979). It is significant that the rates of both of these phenomena rise rapidly near 340–350 °C as illustrated in figure 1. Silicon wafers are also known to pit because of evaporation when heated in a non-oxidising atmosphere. When the Si surface is oxidised, the SiO_2 seals the surface and prevents thermal etching (Burger and Donovan 1967). Similarly, an oxide should help to seal the InP surface. However, the thermal oxide of InP is far more complex than that of Si, as can be seen from the ESCA profiles of 38 and 177 Å thick layers of thermal oxides of InP grown at 400 °C (figure 2). The 177 Å thick layer of oxide clearly shows the collection of excess phosphorus at the

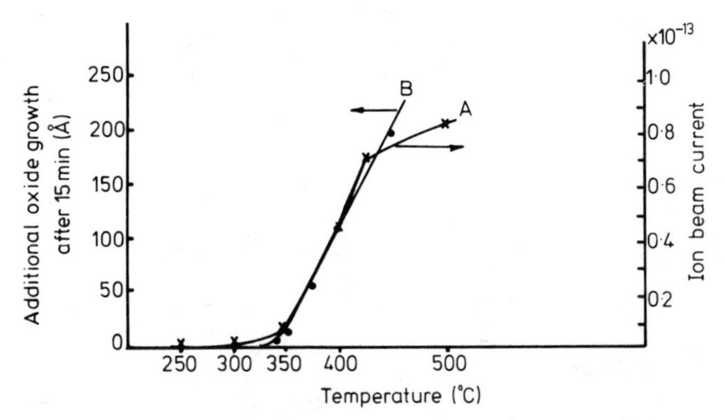

Figure 1. Thermal evaporation of P from the InP surface (Bayliss and Kirk 1976; curve A, right-hand ordinate) and the thickness of a thermal oxide layer after 15 min (Wager and Wilmsen 1979; curve B, left-hand ordinate) plotted against temperature.

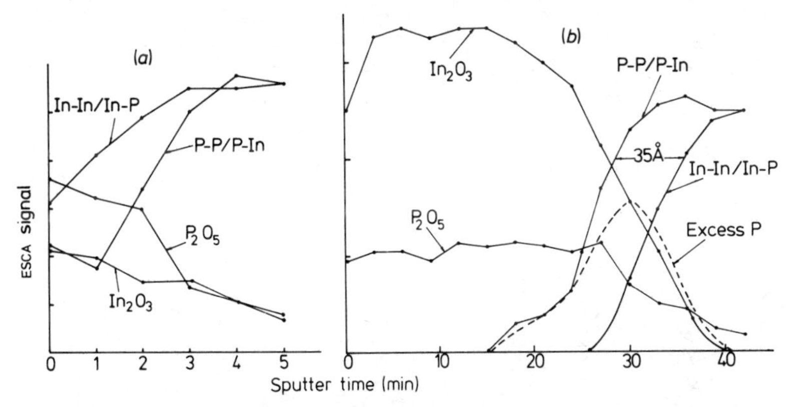

Figure 2. ESCA depth profiles for oxide layers (*a*) 38 Å thick and (*b*) 177 Å thick grown on InP at 400 °C in dry O_2.

interface and an oxide layer that has an In_2O_3/P_2O_5 ratio of about 3. The 38 Å thick oxide layer on the other hand appears to have excess In and an In_2O_3/P_2O_5 ratio of about 0·6. This suggests that thermal oxidation proceeds by island growth which means that thin layers will not seal the surface and will also create a non-uniform interface. In both cases, the thermal oxide would appear to create an undesirable interface for MOS devices.

With regard to the initial 10–20 Å of oxide, we have observed that placing the wafer in the oxidising furnace for a very short time (less than a minute) may cause a reduction in the measured thickness of oxide, which indicates that the initial oxide is not stable. Trace contaminants on the surface are thought to remain at the interface during the deposition of SiO_2. The net effect of contamination on the electronic properties of the interface is not known.

Two other factors can affect the CVD SiO_2–InP MOS device: reactions between the SiO_2 and the InP and the interdiffusion of the elements. From the growth of InP crystals in quartz crucibles (Mullin *et al* 1974, Popov 1976, Baumann *et al* 1976), it is known that elemental silicon enters the InP and resides on In sites. This reaction does not appear to have been investigated in detail as it has for the GaAs–SiO_2 reaction. Cochran and Foster

(1962a,b) reported that the following reaction for GaAs proceeds with a value of $\Delta H \simeq 170\,\mathrm{kcal\,mol^{-1}}$:

$$4\mathrm{Ga}_{(\mathrm{from\ GaAsP})} + \mathrm{SiO_2} \leftrightharpoons \mathrm{Si}_{(\mathrm{into\ GaAs})} + 2\mathrm{Ga_2O}.$$

A similar reaction is expected to occur between InP and SiO_2. This reaction does not involve any In–Si or In–SiO_2 compound formation but instead leaves the In or In oxide free to diffuse through the SiO_2. Indeed Grove *et al* (1964) reported that In diffuses very rapidly through thermally grown SiO_2. It has recently been reported by Jain *et al* (1978) that Ga diffuses through SiO_2 as Ga_2O; thus it may be expected that In will be found in the CVD SiO_2 in an oxidised state.

The chemistry of phosphorus–SiO_2 is considerably different in that P will substitute for Si in silicate glasses, P_2O_5 does not phase-separate from SiO_2 and P diffuses very slowly through SiO_2 (Doremus 1973). Extrapolation of the diffusion data of Sah *et al* (1959), as shown in figure 3, indicates that at 350 °C, P will diffuse only 2 Å into thermally grown SiO_2 and at 450 °C only 13 Å. Although this extrapolation is not strictly valid, it does point out that P diffuses slowly in SiO_2.

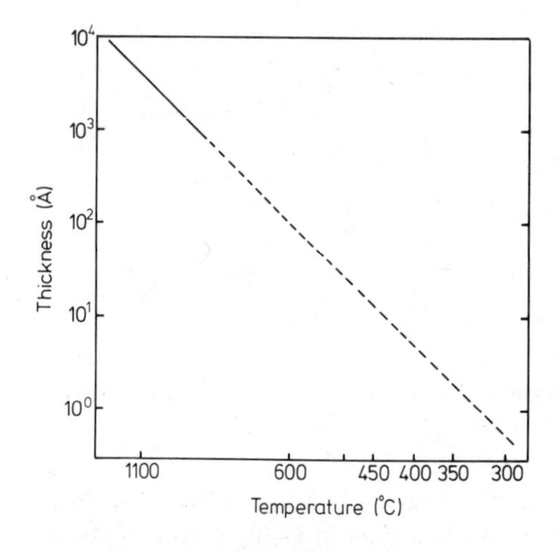

Figure 3. The thickness at which thermally grown SiO_2 completely masks the diffusion of P (from Sah *et al* 1959).

4. Experimental results

The ESCA and Auger profiles of samples fabricated as described in §2 were determined to examine the extent of the diffusion, surface oxidation, etc described in the previous section. The SiO_2 deposited at 340 °C appeared uniform in colour across the wafers. Ellipsometric measurements verified that the SiO_2 layer was of a uniform thickness of 220 Å. The layer deposited at 450 °C, however, had a non-uniform colour and the thickness varied between 150 and 400 Å. Auger analysis provided much less information about the SiO_2 on InP than did the ESCA measurements and thus only the latter results are presented here. The ESCA sputter profiles result from recording the phosphorus $2P_{3/2}$ and the indium $3d_{3/2}$ lines at intervals through the SiO_2–InP sample.

An ESCA profile for a sample fabricated at 450 °C is shown in figure 4. The profile is spread out because of the non-uniformity of the SiO_2 and the large (about 2 mm) area of the x-ray beam. Nonetheless, certain points are clear; there is a thick In_2O_3/P_2O_5 layer

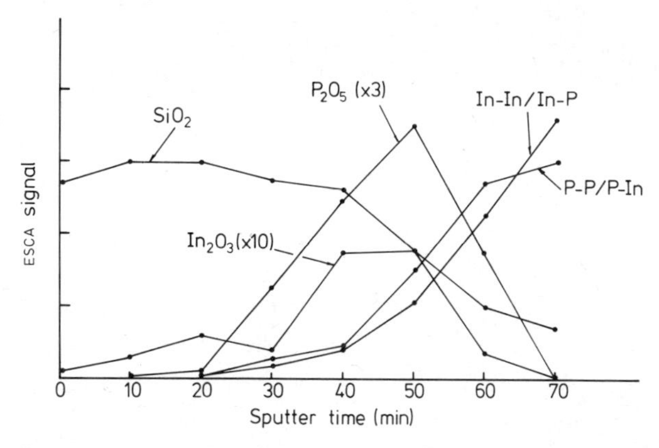

Figure 4. ESCA depth profile of a CVD SiO_2 layer ~ 200 Å thick deposited on InP at 450 °C.

at the interface and there is In_2O_3 distributed throughout the SiO_2 layer but there is no detectable P in the SiO_2. This establishes that a thermal oxide can grow readily in the CVD reactor and that In can diffuse through the SiO_2 layer (apparently as an oxide) while the P does not diffuse significantly.

The ESCA profile (figure 5) of the sample fabricated at 340 °C reveals a similar result with differences that are consistent with the discussion given in §3. The profile indicates that In is distributed throughout the SiO_2 layer (including the surface) at a concentration of about 1%, the In concentration increases sharply approximately 50 Å before the first detectable P signal and there is no detectable P in the bulk of the SiO_2. Not shown explicitly in figure 5 are the bonding states of In and P. No P_2O_5 was observed in the SiO_2 or at the interface; however, the major portion of the In in the SiO_2 layer and at the interface is bonded to oxygen, as illustrated in figure 6. These data further emphasise the fact that In diffuses readily through SiO_2 whereas P does not. The lack of a P_2O_5 layer at the interface indicates that there was no significant growth of thermal oxide. This is in keeping with the thermal oxidation rates of InP at 340 °C (figure 1). The presence of a significant concentration of an In oxide (thought to be In_2O_3) at the

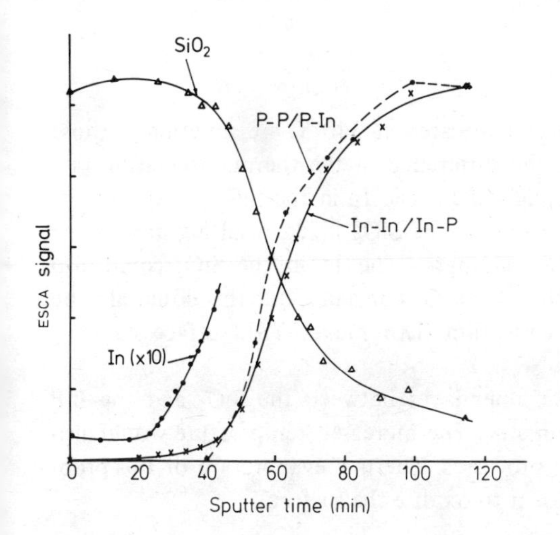

Figure 5. ESCA depth profile of a CVD SiO_2 layer 220 Å thick deposited on InP at 340 °C.

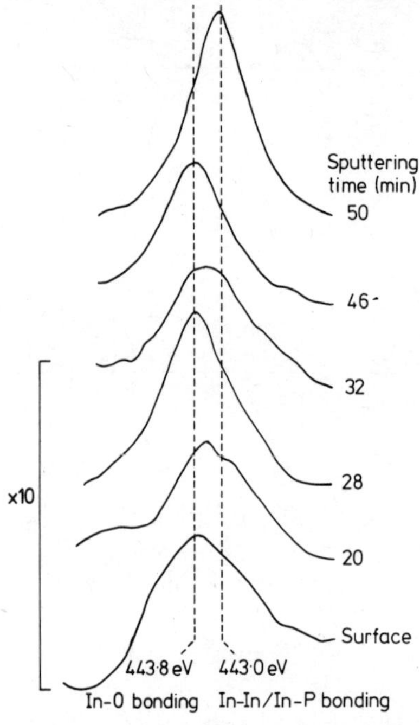

Figure 6. In $3d_{5/2}$ ESCA lines after sputtering at various depths into the deposited SiO_2.

interface is suggestive of decomposition of the SiO_2 into In oxide and Si. No free Si was detected, however. Presumably the Si enters the InP and the In oxide diffuses into the SiO_2. This process would leave a phosphorus residue which must be trapped at the interface since it does not readily diffuse into either the SiO_2 or the InP. The ESCA lines cannot resolve the difference between P–P and P–In bonding and therefore we must rely on observing the difference between the magnitudes of the In and P peaks. This is not reliable for small excess concentrations, particularly after sputtering through the SiO_2 layer. There are too many uncertainties to know whether the apparent excess of P indicated at the interface in figure 5 is in fact real or is just an inaccuracy of the measurement. Further work is required to resolve this point.

5. Discussion and conclusions

The reasons why CVD SiO_2–InP MOS devices fabricated at 340 °C are superior to those fabricated at 450 °C appear to arise from the difference in the thermal oxidation rate of InP and the rate of diffusion of In through SiO_2. The In in the SiO_2 is expected to cause a degradation of the dielectric properties of the SiO_2 layer, resulting in a higher DC leakage current and a lower breakdown strength. The In in the SiO_2 could also account for the frequency dispersion of the dielectric constant, but this could also be caused by the incorporation of water into the film (Doremus 1973), surface states or interfacial contamination.

The thermal oxide creates a conductive inner layer between the SiO_2 and the InP and causes deviations from surface stoichiometry. The increased temperature would also accelerate the diffusion and decomposition processes. Thermal evaporation of P is probably not a problem unless there is no O_2 present to oxidise the surface.

It is likely that Si enters the InP and changes the surface concentration of dopant. Since Si is always an n-dopant in InP, the surface of a p-type MOS capacitor could be brought closer to inversion, as occurs for the ion-implanted V_T control of Si MOSFET structures. The added doping could also increase the surface impurity scattering, which in turn would reduce the surface electron mobility. Since the inversion layer is only tens of angstroms thick, only weak diffusion of Si is required to affect the mobility of the inversion layer.

The fact that In diffuses rapidly through the SiO_2 and that Si can dope the surface of InP suggests that SiO_2 is not the ideal choice for an MOS dielectric on InP unless the deposition temperature can be reduced below 340 °C. However, further work on interface formation is required before the SiO_2–InP interface can be fully understood. The rate and kinetics of the decomposition of SiO_2 in the presence of InP needs particular attention.

References

Baumann G G, Benz K W and Pilkuln M H 1976 *J. Electrochem. Soc.* **123** 1232–5
Bayliss C R and Kirk D L 1976 *J. Phys. D: Appl. Phys.* **9** 233–44
Burger R M and Donovan R P 1967 *Fundamentals of Silicon Integrated Device Technology* vol 1 (Englewood Cliffs: Prentice Hall)
Cochran C N and Foster L M 1962a *J. Electrochem. Soc.* **109** 144–8
—— 1962b *J. Electrochem. Soc.* **109** 149–54
Doremus R H 1973 *Glass Science* (New York: Wiley)
Farrow R F C 1974 *J. Phys. D: Appl. Phys.* **7** 2436–48
—— 1975 *J. Phys. D: Appl. Phys.* **8** L87–9
Fritzsche D 1978 *Electron. Lett.* **14** 51–2
Grove A S, Leistiko O and Sah C T 1964 *J. Phys. Chem. Solids* **25** 985–92
Jain G C, Prasad A and Chakravary D C 1978 *Phys. Stat. Solidi* **A46** K151–5
Lum W Y and Clawson A R 1979 *J. Appl. Phys.* **50** 5296–301
Meiners L G 1978 *Electron. Lett.* **14** 657–9
Messick L 1976 *J. Appl. Phys.* **47** 4949–50
—— 1979 *Solid St. Electron.* **22** 71–6
Mullin J B, Royle A, Straushan B W, Tuftan P J and Williams E W 1974 *J. Cryst. Growth* **13/14** 640
Popov A S 1976 *Phys. Stat. Solidi* **A37** K53–6
Sah C T, Sello H and Tremere D A 1959 *J. Phys. Chem. Solids* **11** 288–98
Stannard J and Henry R L 1979 *Appl. Phys. Lett.* **35** 86
Wager J F and Wilmsen C W 1979 *J. Appl. Phys.* to be published
Wieder H H 1978 *J. Vac. Sci. Technol.* **15** 1498–1506
Wilmsen C W 1975 *Crit. Rev. Solid St. Sci.* **5** 313–7

Interface studies on InP MIS inversion FET's with SiO_2 gate insulation

D Fritzsche

Forschungsinstitut der Deutschen Bundespost beim FTZ, D 6100 Darmstadt, Federal Republic of Germany

Abstract. Results on n-channel inversion-type InP MISFET devices with doped CVD SiO_2 are reported. Inversion mobilities as high as 1000 $cm^2 V^{-1} s^{-1}$ have been achieved, exceeding the values for silicon by 20%. Device characteristics at 300 and 77 K indicate a sufficiently low density of interface states. Electrical drift similar to the behaviour of thin-film FET's is explained by the tunnelling of thermally excited electrons into the insulator and appears to be a serious problem for this device.

1. Introduction

Over recent years the success of the MISFET concept for integrated circuits has attracted increasing interest in the hope that it might be adaptable to GaAs and InP. By exploiting the speed advantages of these materials over Si, it is hoped that integrated MISFET circuits for a gigabit data rate will become feasible. Possible applications are seen in high-speed optical-fibre communication systems.

So far, results with GaAs have been disappointing; this is attributed to a high density of interface states. It seems likely that InP will be more useful, as discussed by Lile and Collins (1979). Measurements on n-InP MIS capacitors with pyrolytically deposited SiO_2 indicate a minimum density of interface states of about $5 \times 10^{11} cm^{-2} eV^{-1}$. Within a limited energy range, this density of states was drastically reduced by Fritzsche (1978) using HCl in the deposition process. The essential feature of the InP/doped SiO_2 interface achieved is the low density of interface states near the conduction-band minimum as

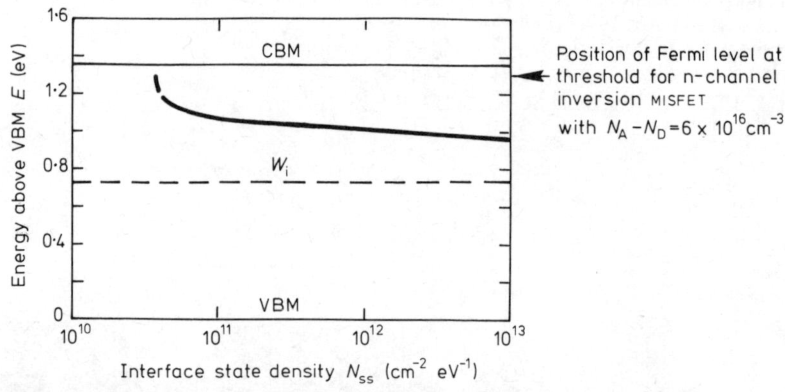

Figure 1. Distribution of interface states for InP/doped SiO_2 according to C–V and G–V data on n-InP MIS capacitors.

0305-2346/80/0050-0258$01.00

outlined in figure 1. If the energy range within which this low interface state density exists extends to the conduction-band minimum, then inversion-mode MISFET operation would be almost unaffected by interface states. The Fermi level at the surface of p-InP could then be shifted by the field of a gate electrode, within a suitable energy range, into strong inversion. This means that a n-channel inversion-mode MISFET could be switched on. The first demonstration of an inversion-mode InP MISFET using undoped SiO_2 and alloyed source and drain contacts was described recently by Lile *et al* (1978). Apart from a rather low electron mobility in inversion of about 400 $cm^2\,V^{-1}\,s^{-1}$, no anomalous behaviour was reported.

In the present paper results on p-InP MISFET devices with doped SiO_2 and n^+-contacts are given. Device operation at 300 and 77 K, electron mobilities in inversion and observed current drifts are discussed to characterise the interface.

2. Device fabrication

All data given in this paper refer to MISFET's fabricated on [100]-oriented p-InP wafers with a carrier density $N_A - N_D = 6{\cdot}2 \times 10^{16}$ cm^{-3} from Metals Research (Great Britain). The wafers were covered with a thin n^+-InP epilayer (1–2 μm, $N_D \simeq 10^{18}$ cm^{-3}) by vapour-phase or liquid-phase epitaxy. Subsequent mesa etching with a solution of 1% bromine in glycol–methanol (1 : 2) using a SiO_2 mask was used to produce the source and drain contacts. The resulting structure of this mesa-type MISFET is depicted in figure 2.

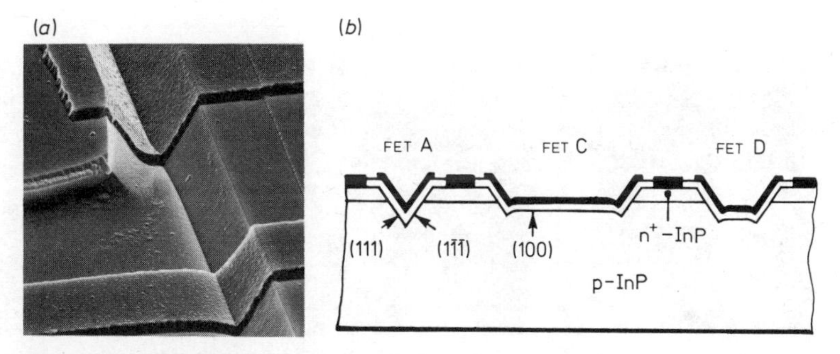

Figure 2. Structure of the MISFET's investigated. (*a*) Electron scanning micrograph of v-groove and (100) bottom; (*b*) schematic cross section showing three of the four FET's within the test structure.

According to the observed angles, the mesa slopes in the channel region are believed to be (111) and $(1\bar{1}\bar{1})$ surfaces with slow etching rates. The groove bottom was at least 1·5 μm below the n^+–p junction within the p-InP. Just before deposition of the gate oxide, the wafers were cleaned with organic solvents and then treated by an etching sequence consisting of 5 min H_2O_2 at 60 °C, 2 min H_3PO_4 at 40 °C and 15 s buffered HF. Each step was followed by a short H_2O rinse and a N_2 dry blow. The SiO_2 was deposited in a cold wall reactor (ASM, Netherlands) with the deposition parameters 340 °C, N_2 40 l min^{-1}, SiH_4 12 $cm^3\,min^{-1}$, O_2 50 $cm^3\,min^{-1}$. For deposition of doped SiO_2, the wafers were treated in the reactor 5 min before deposition with 0·2% HCl and 1 min before deposition the HCl rate was halved and 0·05% PH_3 added. The first 30 nm were deposited in the presence of HCl and PH_3. The final thickness of the SiO_2 was 150 nm.

Addition of PH_3 was mainly intended to retard the degradation of InP, as discussed by Clawson *et al* (1979). It was originally thought that HCl attacks only the native oxide on InP; but Gaind and Kasprzak (1979) found also with Si that HCl influences the SiO_2 deposition process at 1000°C, lead to a reduced density of interface states and oxide charge. Gate metallisation was achieved with Al. Source and drain contacts consisted of a multilayer system of In, Sn, Au, Ni, Au and were structured by a lift-off technique. A double layer of In and Au formed the back contact. The contacts were annealed in N_2 at 280 °C for 30 min. For operation at 77 K, the back contact, although Ohmic at 300 K, had to be improved by soldering with InZnAu.

The test structure contained a pair of identical FET's with a common drain (FET A, B: channel width $W = 238$ μm, channel length $L = 12\cdot5$ μm) and a pair of FET's with different channel lengths (FET C, D: $W = 383$ μm, $L = 93$ and $16\cdot3$ μm respectively). The insulator capacitance C_i was 24·8 nF cm^{-2}.

3. Measurements and analysis

Measurements were made on wafers on a light-shielded probe station. A general impression of the signal behaviour is given in figure 3. With a short pulse the response appears to be quite normal with distortions resulting only from capacitance effects (figure 3*a*). With the pulse duration greater than 1 μs, a substantial decrease in the drain current becomes obvious (figure 3*b*). This drift is absent at 77 K where the FET displays stable operation (figure 3*c*). Compared to the value at 300 K, the drain current at 77 K is

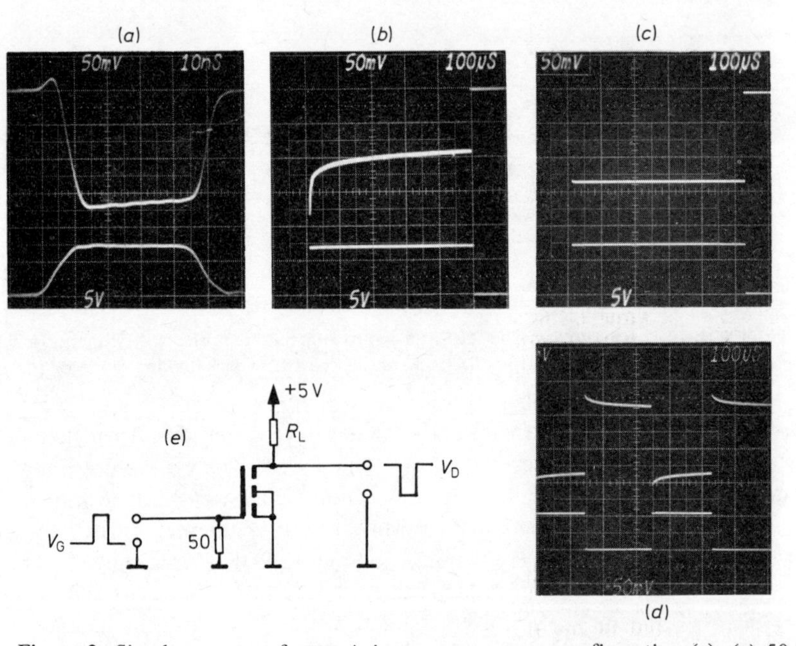

Figure 3. Signal response of FET A in common source configuration (*e*). (*a*) 50 ns 7 V gate pulse (5 V/division) at 300 K, $R_L = 50\ \Omega$; (*b*) 500 μs 7 V gate pulse, otherwise as (*a*); (*c*) same as (*b*) but at 77 K; (*d*) small-signal CW operation at 300 K, 2·5 kHz 50 mV gate signal (50 mV/division) at 6 V gate bias, $R_L = 5\ k\Omega$. Upper traces: drain voltage response, 50 mV/division; lower traces: gate voltage.

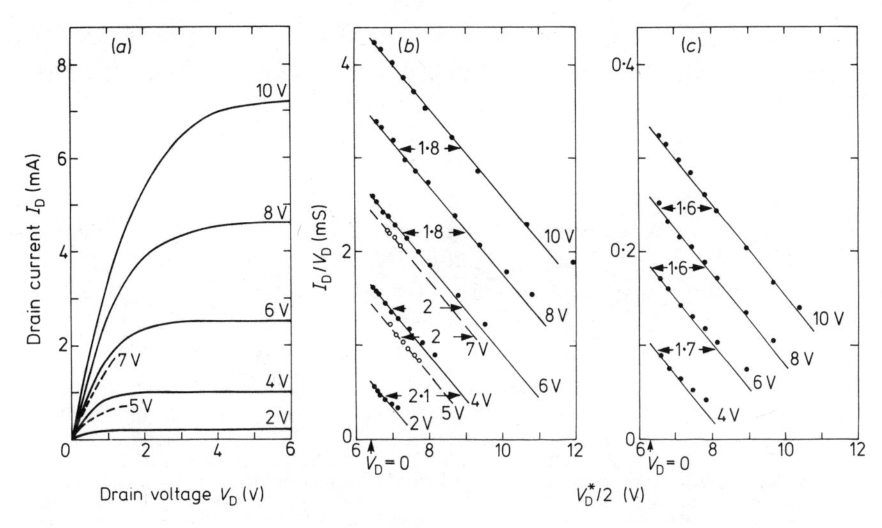

Figure 4. Drain current–voltage characteristics of MISFET's with pulsed gate voltage as parameter. (*a*) Usual $I_D - V_D$ presentation for FET A at 300 K (——) and 77 K (-----); (*b*) and (*c*) I_D/V_D versus $V_D^*/2$ as given in text for FET's A (●—● 300 K, ○--○ 77 K) and C respectively.

reduced by 25% because of an increase of 1 V in the threshold voltage. For small-signal operation at 300 K, the strong current decrease seen with large-signal operation is substantially suppressed (figure 3*d*). As a consequence of this behaviour, the current transport properties at 300 K had to be investigated under pulse conditions to avoid the effects of this drift.

3.1. Device characteristics in pulse operation

The drain current–voltage ($I_D - V_D$) characteristics at 300 K were measured 100 ns after the leading edge of a 300 ns gate pulse with a repetition rate of 1–200 Hz. Some measurements were made at 77 K with pulses of 100 ns–100 ms duration where no drift was observed.

Typical results for $I_D - V_D$ characteristics for FET A are given in figure 4(*a*). The curves display the usual behaviour of a MISFET with a good saturation beyond pinch-off. The threshold voltage is approximately 1 V, increasing to 2 V at 77 K. Leakage currents at zero gate voltage are below 0·2 μA.

The behaviour of these transistors was compared with the basic MISFET theory based on Shockley's gradual-channel approximation model. This was done by plotting the I_D/V_D ratios against a voltage $V_D^*/2$ which is defined below. Starting from the basic $I_D - V_D$ equation for the unsaturated mode, this ratio is given by

$$I_D/V_D = (W/L)\mu^* C_i [V_G - V_{FB} - 2\psi_B - (V_D^*/2)]$$

where

$$V_D^* = V_D + \tfrac{4}{3}\sqrt{2}(K/V_D)[(V_D + 2\psi_B)^{3/2} - (2\psi_B)^{3/2}]$$

$$K = (1/C_i)[\epsilon_s \epsilon_0 q(N_A - N_D)]^{1/2}.$$

With the effective mobility μ^*, the flat-band voltage V_{FB}, the potential difference ψ_B between the bulk Fermi level and the intrinsic Fermi level W_i, this gives a linear relationship between I_D/V_D and V_D^*. For the calculations, the following specific values were used: $\epsilon_s = 12{\cdot}5$ at 300 K and 12·05 at 77 K; $2\psi_B = 1{\cdot}15$ V at 300 K and 1·38 V at 77 K; other values are given in §2. The results are shown in figure 4(*b*) and (*c*) for FET's A and C. The diagrams display sets of straight lines where the parallel shift $\Delta V_D^*/2$ roughly equals the change in the gate voltage ΔV_G as required by theory. This demonstrates that these InP MISFETs are adequately described by the basic MISFET model.

From the slope in figure 4(*b*) and (*c*) the effective electron mobility in the inversion layer was calculated using the equation given above. For FET A this yielded a rather large mobility $\mu^* = 1000\ \mathrm{cm^2\,V^{-1}\,s^{-1}} \pm 20\%$ at 300 K and also at 77 K. This compares rather favourably with the value $\mu^* = 600$–$800\ \mathrm{cm^2\,V^{-1}\,s^{-1}}$ for Si. For FET C a smaller mobility $\mu^* = 500\ \mathrm{cm^2\,V^{-1}\,s^{-1}}$ was found. This difference was observed on all [100]-oriented wafers and may possibly be attributable to different structures of the surface. For FET A the channel surface consists of only (111) and $(\bar{1}\bar{1}\bar{1})$ facets whereas for FET C the largest part of the channel surface is a (100) surface (see figure 2). This explanation is supported by the temperature independence of the mobility; indicating the limiting influence of a surface scattering process which may be related to surface roughness as discussed by Ferry (1979).

To investigate the density of interface states, a rather sensitive and decisive test is the increase in threshold voltage between 300 and 77 K. Assuming Boltzmann statistics, the Fermi level for threshold has to rise by approximately 50 meV. If during this rise no interface states are occupied, the resulting increase ΔV_{th} can be calculated, according to Sze (1969), as

$$\Delta V_{th} = \Delta[2\psi_B + K(4\psi_B)^{1/2}] = 0{\cdot}75\ \mathrm{V}.$$

This is a lower estimate because of the approximation arising from the Boltzmann statistics. Experimentally an increase $V_{th,\,exp} = 1$ V was found, which excludes the presence of a high density of interface states at the conduction-band edge.

3.2. *Drift effect*

In all of the MISFET's investigated, a current drift was found. This holds also for FET's with undoped SiO_2 as gate insulator. The general feature of this drift effect is revealed in a lg–lg presentation of the channel conductivity measured over the linear range of the transistor plotted against time, as shown in figure 5. By inspection of these curves it becomes obvious that after a certain time $t_0 = 1$–$10\ \mu$s, the conductivity drift follows a power law over several decades of time. This can be written in the form

$$\ln[(\sigma_0 - \Delta\sigma)/\sigma_0] = s \ln(t/t_0) \qquad t > t_0$$

or, for $\Delta\sigma \ll \sigma$

$$-(\Delta\sigma/\sigma_0) = s \ln(t/t_0).$$

The slope s of these drift traces remains essentially constant when the voltage of the gate pulse starting from zero voltage is increased (figure 5*a*). Slightly different slopes are found for the different FET's A, C and D all within the same test structure. The smallest drift observed so far has been approximately 15% per time decade.

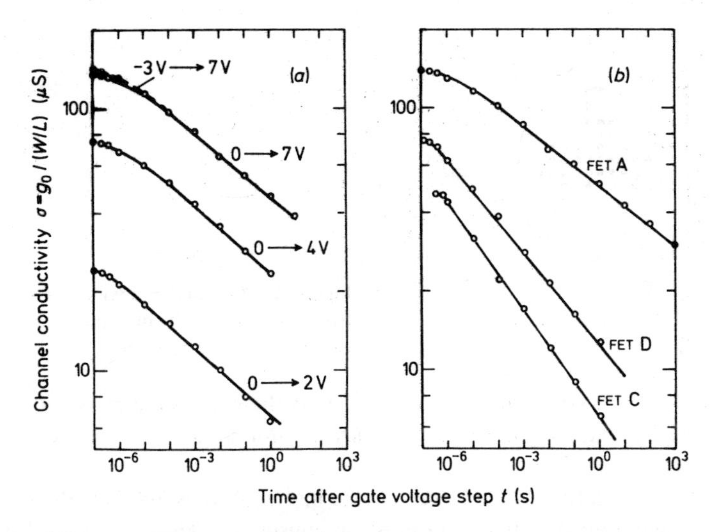

Figure 5. Drift characteristics of the channel conductivity ($g_0 = I_D/V_D$ at $V_D = 100$ mV). (*a*) Gate voltage step as parameter for FET A; (*b*) different FET's with gate voltage step $0 \rightarrow 7$ V.

After the drift, the MISFET's recovered at zero gate voltage within a time comparable to the pulse duration. Only after a pulse duration exceeding some seconds was a negative bias stress necessary for fast restoration of the original state. In attempting to determine the process(es) involved in the drift, the interface states can be excluded on basis of their time dependence. Using recent data for the capture cross section and thermal velocity, values for the time constants for charge trapping by interface states of 10^{-10}–10^{-6} s are obtained, i.e. much faster than observed.

For thin-film FET's, where a similar drift is encountered, Koelmans and de Graaff (1967) discussed what might occur in terms of a simple model with monoenergetic traps being charged as they shift below the Fermi level. This model yields the number of electrons Δn trapped in the time interval between t_0 and t:

$$\Delta n = (N_{tr}/a) \ln (t/t_0)$$

where N_{tr} is the trap density and a the tunnelling constant. With $\Delta n \sim \Delta\sigma$, this relationship gives the logarithmic time dependence that approximately describes the InP MISFET's. However, this model cannot explain the decrease in the drift at low temperatures.

To account for this temperature dependence, it is assumed that only the electrons in the distribution tail (i.e. thermally excited electrons) can tunnel into oxide traps, as outlined in figure 6. In this model the trap density increases sharply within a few kT/q above the InP conduction-band minimum. By this means an essential fraction of the inversion-layer electrons can tunnel into the insulator at 300 K. With the increase in the filled trap region $x(t)$, the tunnelling probability decreases, leading to the observed time response. At low temperatures the electron density in the distribution tail is greatly reduced and so is the loss of electrons due to tunnelling. The observed differences in drift properties (figure 5*b*) may be related to variations in the trap distribution that are

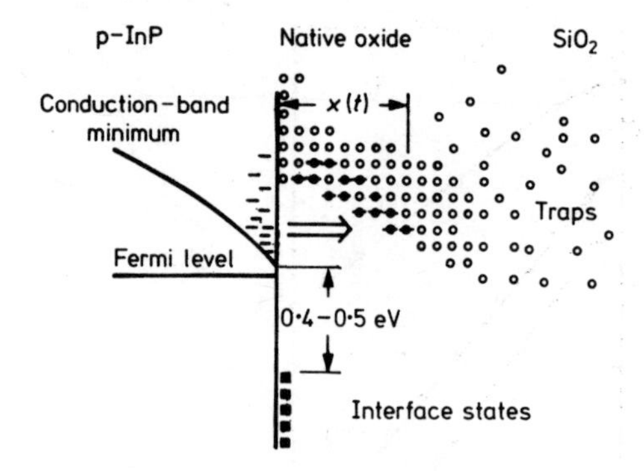

Figure 6. Proposed model for the energy band structure in the drift phase after application of a positive gate bias.

dependent on the surface structure. This suggests that these traps are mainly defects of the unavoidable native oxide on InP and that they are not situated within the deposited SiO_2.

To complete the description of the interface, the interface traps 0·4–0·5 eV below the conduction-band minimum (see figure 1) must be considered. These traps are found to have no influence on operation as a n-channel MISFET, but prevent the structure from being used as a p-channel device. They are believed to be interface states because of their fast response after charging by negative bias and then discharging by positive bias (figure 5*a*, gate voltage step $-3\ V \rightarrow 7\ V$). Over this energy range surface states of high density have also been reported by Spicer *et al* (1979) due to adsorption of oxygen on the cleaved p-InP surface.

4. Conclusions

It has been shown that n-channel inversion-mode InP MISFET's can be fabricated with inversion-layer mobilities up to $1000\ cm^2\ V^{-1}\ s^{-1}$; this exceeds the values for Si by 20%. Low densities of interface states near the conduction band, low threshold voltages and satisfactory reproducibility of the technology are positive aspects of this MISFET system. Electrical drift is, however, a serious problem and was shown to be due to tunnelling of thermally excited electrons into the native oxide.

Acknowledgments

The author is pleased to acknowledge the contributions of K W Benz and E Kuphal to the epitaxial technology and for advice on processing. Valuable discussions with G Weimann are also gratefully acknowledged. Thanks are also expressed to colleagues of the Research Group FI 43 for sample preparation and to K Balcar for numerous measurements.

References

Clawson A R, Lum W Y and McWilliams G E 1979 *J. Crystal Growth* **46** 300–3
Ferry D K 1979 *Thin Solid Films* **56** 243–52
Fritzsche D 1978 *Electron. Lett.* **14** 51–2
Gaind A K and Kasprzak L A 1979 *Solid St. Electron.* **22** 303–9

Koelmans H and de Graaff H C 1967 *Solid St. Electron.* **10** 997–1005
Lile D L and Collins D A 1979 *Thin Solid Films* **56** 225–34
Lile D L, Collins D A, Meiners L G and Messick L 1978 *Electron. Lett.* **14** 657–9
Spicer W E, Lindau I, Pianetta P, Chye P W and Garner C M 1979 *Thin Solid Films* **56** 1–18
Sze S M 1969 *Physics of Semiconductor Devices* (New York: Wiley-Interscience) ch 10, p519

A study of deposited dielectrics and the observation of n-channel MOSFET action in InP

A J Grant, D C Cameron, L D Irving, C E Greenhalgh and P R Norton
Royal Signals and Radar Establishment, St Andrews Road, Malvern, Worcestershire, UK

Abstract. A study has been made of the properties of the interface between InP and dielectrics of plasma-deposited silicon nitride and silicon dioxide. It was found that carrier inversion could be achieved on n-type substrates and a density of surface states of the order of 10^{12} states cm^{-2} eV^{-1} was calculated. Deposition of dielectrics caused the formation of donor states in the semiconductor. n-channel enhancement-mode MOSFET action was observed in a device grown on semi-insulating Fe-doped material using a silicon dioxide dielectric and a self-aligned gate technique.

1. Introduction

There has been increasing effort applied to the production of microwave FETs using indium phosphide as the active layer as it has a higher peak electron velocity and a lower density of interface states than GaAs. Recently Schottky-barrier MESFETs (Gleason *et al* 1978) and depletion MISFETs with dielectric layers of pyrolitic silicon dioxide (Messick *et al* 1978) have been produced that have shown power gain at microwave frequencies. For the production of MISFETs the dielectric–InP interface must have as low a density of surface states as possible in order to allow the energy bands to bend sufficiently to produce a surface inversion layer. A study has been made of the properties of dielectrics of silicon nitride and silicon dioxide produced by the plasma-enhanced chemical vapour deposition (PECVD) method and their interface with both bulk-grown n- and p-type material and n-type epitaxial layers. A prototype n-channel enhancement MOSFET has been fabricated using silicon dioxide insulator and a self-aligned gate structure on Fe-doped semi-insulating InP.

2. Dielectric growth and properties

The silicon nitride films were deposited by the plasma-enhanced reaction of silane and nitrogen at pressures of 0·05–0·2 Torr. The reaction chamber was a 10 cm diameter silica tube pumped by a 100 l min^{-1} rotary pump and the reaction was excited by a 27 MHz RF induction coil. The ratio of nitrogen to silane was within the range 200 : 1 to 50 : 1 and growth rates of the order of 500 Å h^{-1} were obtained.

The silicon dioxide films were grown by the plasma oxidation of tetraethylorthosilicate (TEOS) which was introduced into the reaction chamber by bubbling oxygen carrier gas through it at room temperature. The plasma in this system was excited by a capacitively coupled RF discharge. Deposition was carried out at a background pressure of approximately 1 Torr and growth rates of around 2000 Å h^{-1} were obtained. The nitride and oxide films had similar resistivities and breakdown characteristics, with

0305-2346/80/0050-0266$01.00

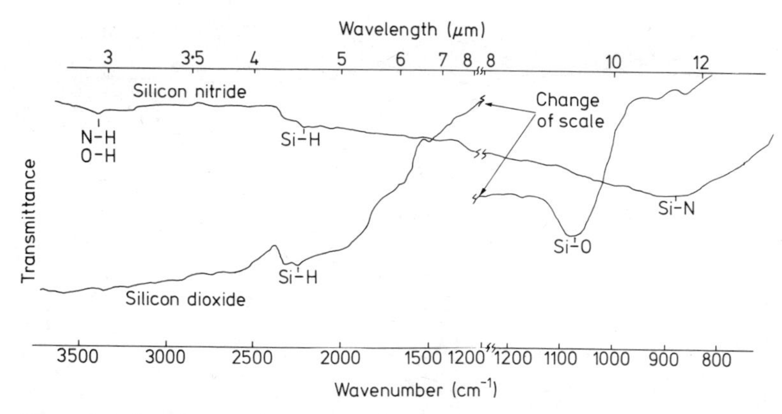

Figure 1. Infrared absorption spectra of plasma-deposited silicon nitride and silicon dioxide.

resistivity in the range 10^{12}–$10^{13}\ \Omega$ cm and breakdown occurring at fields of about 5×10^6 V cm^{-1}. The leakage current through the films depended on the applied voltage according to $\lg I \propto V^{1/2}$, indicating field-enhanced thermal activation of trapped carriers (Frenkel–Poole conduction). The dielectric constant of the nitride films varied from 4 to 8 and in some cases showed considerable dispersion with frequency, whereas the oxide dielectric constant remained fixed at approximately 3·9.

Infrared absorption measurements were made over the range 700–3500 cm^{-1} (2·8–14 μm) using a Pye Unicam SP1200 double-beam spectrophotometer. The nitride films showed a broad absorption peak at about 900 cm^{-1} (11 μm) as shown in figure 1, corresponding to the Si–N bond stretching mode. This is shifted slightly towards lower frequency compared with the absorption peak of crystalline Si_3N_4 (Hu 1966), and is also broadened, indicating a looser and more amorphous structure. There is no evidence of a peak due to Si–O bonding but there is a small peak at 3380 cm^{-1} due to N–H or O–H bonding and one at 2200 cm^{-1} which may be due to the Si–H bond. The absorption spectrum of the silicon dioxide films had a strong peak at 1070 cm^{-1} (9·35 μm) caused by the Si–O bond stretching mode. The position of the absorption peak in conventional thermally grown oxide on silicon is 1093 cm^{-1} (9·15 μm) (Pliskin and Lehman 1965). Thus the TEOS films have a less dense structure. A peak is evident at 870 cm^{-1}, indicating a slight deficiency of oxygen in the SiO_2 structure (Pliskin *et al* 1967) and a small peak occurs at 2250 cm^{-1}, probably due to Si–H bonding.

3. MIS diode studies

MIS structures were grown using the dielectrics described above on bulk-grown liquid-encapsulated Czochralski (LEC) indium phosphide, both n- and p-type, and on n-type epitaxial layers grown by VPE on n^+ substrates. The n-type slices had back contacts of NiGeAu, the p-type slices had NiZnAu contacts and the gate electrodes were 0·5 mm aluminium dots. The capacitance–voltage characteristics were measured by the AC capacitance technique while a slow bias voltage sweep was applied.

Figure 2 shows the C–V characteristics of an MIS diode with a dielectric of approximately 1000 Å of silicon nitride on an n-type epitaxial layer with a doping concentration of 5×10^{15} cm^{-3}. The dielectric constant shows no dispersion with frequency and the

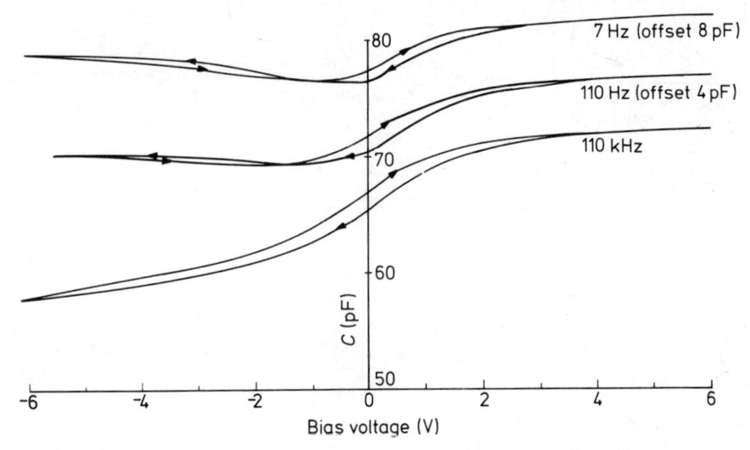

Figure 2. *C–V* characteristics of a silicon nitride–epitaxial n-InP diode. Sample CV506; dielectric 1000 Å Si_3N_4; bias sweep rate 100 mV s^{-1}.

positive bias capacitance shows the true accumulation value. At frequencies of 110 Hz and lower, inversion can be seen to occur in negative bias. The density of surface states was calculated by the method of Berglund (1966) and a minimum value of about 1×10^{12} states cm^{-2} eV^{-1} was obtained. The curves show hysteresis in a clockwise direction which is normally considered to be caused by charge injection into the insulator. The semiconductor space-charge capacitance was calculated for the measured doping concentration of the epitaxial layer and compared with the measured semiconductor capacitance; the result indicated that the actual doping of the epilayer must have been considerably greater. The variation in surface potential from +6 V to −6 V bias was found by integrating the low-frequency capacitance curve and a value of approximately 0·60 V was determined. Diodes fabricated on bulk undoped n-type (8×10^{15} cm^{-3}) material using silicon nitride gave the *C–V* characteristics shown in figure 3. The dielectric constant exhibited considerable dispersion with frequency, but inversion of the surface carriers was observed at 1·1 Hz and slight inversion was seen at 11 Hz when the diode was illuminated. Diodes grown on the bulk undoped material using TEOS silicon dioxide dielectric showed no dispersion of the dielectric constant and displayed the normal accumulation–depletion behaviour in positive and negative bias, but no inversion down to a frequency of 11 Hz. However, slight inversion was seen at 11 Hz under illumination. The semiconductor doping density obtained by measuring the capacitance–voltage characteristics in deep depletion using the high-frequency (110 kHz) curve showed that in both n-type bulk samples the doping was also greater than expected.

Diodes grown on p-type bulk indium phosphide with a doping concentration of 3×10^{16} cm^{-3} using silicon nitride dielectrics accumulated in negative bias and depleted in positive bias at frequencies of 110 Hz and above. Modelling studies showed that the values of capacitance obtained in depletion could only be explained by assuming a hole concentration in the depletion region that was reduced compared with the measured bulk value. At 11 Hz a large increase in capacitance occurred in positive bias which may be caused by strong carrier inversion.

The characteristics of both the nitride and oxide MIS diodes on n-type substrates show that in every case the carrier concentration is increased over that of the pre-

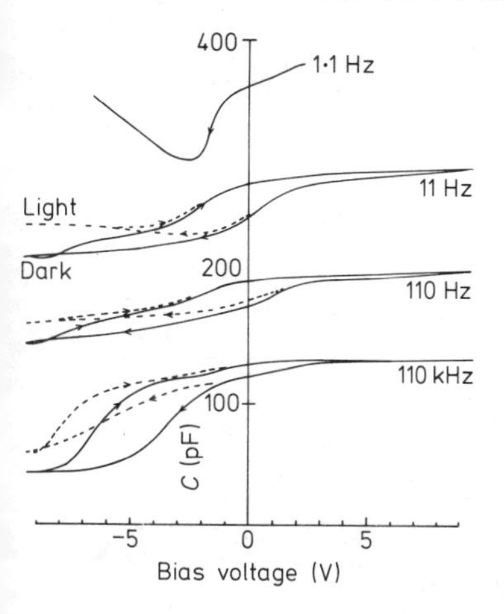

Figure 3. C–V characteristics of silicon nitride–bulk n-InP diode. Sample VIPC-201 PD35A; dielectric 1000 Å Si_3N_4; bias sweep rate 60 mV s^{-1}.

deposited indium phosphide; samples produced on p-type substrates show a decrease in carrier concentration. The results on p-type material are consistent with those reported by Pinczuk *et al* (1979), in which Raman scattering studies indicate that a large amount of downward band bending occurred at the surface of p-type InP. This increase in donor concentration may be attributable to phosphorus depletion caused by bombardment by ions during the dielectric deposition process.

4. MOSFET studies

A prototype n-channel enhancement-mode MOSFET has been made on a Fe-doped bulk semi-insulating substrate, using a TEOS silicon dioxide dielectric. The source and drain contacts were NiGeAu alloy pads and the 6 μm long gate was defined by a self-aligned technique. The characteristics of this device are shown in figure 4 where the gate voltage is stepped from 0 to +7 V. (The first two curves at 1 and 2 V can be seen faintly.) These characteristics took a few seconds to establish themselves; the drain current was initially at a low level. This device obviously has a large number of slow states at the dielectric–semiconductor interface. It is encouraging, however, to observe MOSFET action on a semi-insulating substrate where the large number of deep levels due to the Fe pin the Fermi level near midgap in the bulk.

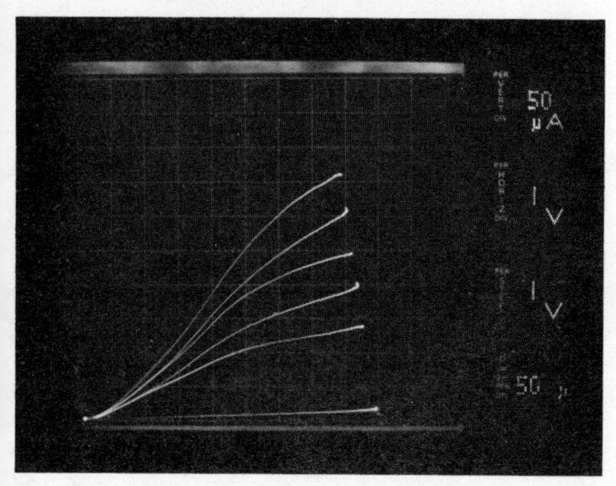

Figure 4. Output characteristics of InP–silicon dioxide MOSFET. I_{DS} = 500 μA/division; V_{DS} = 1 V/division; V_{GS} = 1 V/step.

5. Conclusion

Dielectric films can be deposited on indium phosphide by the plasma-enhanced reaction method and can give interfaces with densities of surface states of the order 10^{12} states $cm^{-2}\,eV^{-1}$. The deposition process, however, appears to cause the formation of donor-like states at the semiconductor surface leading to an increase in n-type doping and compensation of p-type doped semiconductors.

Enhancement-mode n-channel MOSFETS can be made on semi-insulating material, showing that the energy bands can be considerably bent even in the presence of a large number of deep-level traps due to Fe, which in bulk InP are known to pin E_F at about 0·6 eV below the conduction band.

References

Berglund C N 1966 *IEEE Trans. Electron Devices* **ED-13** 701
Gleason K R, Dietrich H B, Henry R L, Cohen E D and Bark M L 1978 *Appl. Phys. Lett.* **32** 578
Hu S M 1966 *J. Electrochem. Soc.* **113** 693
Messick L, Lile D L and Clawson A R 1978 *Appl. Phys. Lett.* **32** 494
Pinczuk A, Ballman A A, Nahory R E, Pollack M A and Worlock J M 1979 *Proc. PCSI6 Asilomar. J. Vac. Sci. Technol.* to be published
Pliskin W A, Kerr D R and Perri J A 1967 *Thin Films* **4** 257
Pliskin W A and Lehman H S 1965 *J. Electrochem. Soc.* **112** 1013

Composite Al_2O_3 and native oxide on GaAs and InP, incorporating enhanced group III oxides for surface passivation

S Hannah† and B Livingstone
Department of Electrical and Electronic Engineering, The University, Newcastle upon Tyne, NE1 7RU, UK

Abstract. Investigations into the MIS characteristics of anodic oxides on GaAs and InP have led to the development of a wet anodic method that enables good dielectric layers to be grown. The resulting MIS capacitance–voltage characteristics exhibit, for the first time, low-frequency-type behaviour, indicating that anodisation is feasible as a method of surface passivation of GaAs and InP. The results presented indicate that for the n-type semiconductors, the surface potential for GaAs may be varied between flat-band and strong inversion and for InP both strong inversion and accumulation are possible.

1. Introduction

GaAs and InP have become increasingly important as microwave device semiconductors because of their high electron mobility and high saturation velocity. This, together with the outstandingly successful Si MIS technology, has created the need for a good reliable surface passivation technique for these semiconductors. Until recently, all attempts to produce such a surface passivating layer have resulted in leaky dielectrics and/or surface-state-rich semiconductor surfaces. To date the best results for GaAs have been reported by Chang *et al* (1979) who formed a native dielectric layer in a plasma of O_2 in CF_4. For InP, the best dielectrics produced were described by Fritzsche (1978), who used chemical vapour deposition of SiO_2 onto InP surfaces.

In this paper, anodisation in an electrolyte as a technique for forming a dielectric is considered and three types of oxide layers have been investigated. Firstly, a native oxide formed by anodisation of the clean semiconductor surface was studied. Both the native oxides on GaAs and InP exhibit poor MIS characteristics; those of the GaAs are consistent with a large number of surface states and those on InP are primarily due to a leaky oxide. It is for these reasons that investigations of composite oxides on these semiconductors have been made. Two types of composites have been studied: the first involves anodisation of a deposited layer of Al on the clean semiconductor surface, the second, the chemical reduction of a thin native oxide before the Al deposition, with formation of the dielectric by the anodisation of the structure. It is the latter of these composites that exhibits promising characteristics.

† Technische Hochschule Darmstadt, Institut für Hochfrequenztechnik, 6100 Darmstadt, Merckstrasse 25, West Germany.

0305-2346/80/0050-0271$02.00

2. MIS diode preparation

A convenient method of assessing the quality of a surface passivating layer is to measure the capacitance–voltage characteristics of MIS diodes. The following section describes the procedures adopted to produce MIS diodes for each of the three types of passivating oxides.

All the oxides were grown under constant-current conditions at a temperature of approximately 5 °C using an electrolyte first reported by Hasegawa and Hartnagel (1976) and consisting of a mixture of 3% tartaric acid solution and propylene glycol in the ratio 1:2. The growth of the oxides was monitored by measuring the overpotential across the anodisation cell as a function of time.

As the initial step for all the oxides, the semiconductor substrates were cleaned in hot acetone, trichlorethylene and methanol in sequence, and then blown dry in a stream of nitrogen. Following this the surface on which the oxides were to be grown was etched; for GaAs the etchant was a mixture of 2% by weight of NaOH in water and 1·2% by volume H_2O_2 in a ratio of 1:1, whereas for InP the etchant consisted of a mixture $HNO_3 : HCl : H_2O$ in the ratio 1:2:10. A typical etch time was 2 min, which removed approximately 5000 Å of material; after etching the substrates were rinsed in de-ionised water and again blown dry in nitrogen.

Growth of the native oxide took place after the etching of the semiconductor surface. The current density for GaAs native oxide was 100 μA cm^{-2} and that for InP was 200 μA cm^{-2}. The growth constants and limitations of the oxide growth for GaAs and InP have been reported by Hasegawa and Hartnagel (1976) and by Colquhoun and Hartnagel (1977) respectively.

The first type of compositie oxide (MAOS structures) was formed initially by deposition of a layer of Al, by thermal evaporation at a pressure of 10^{-6} Torr to a typical thickness of 500 Å, onto the etched semiconductor surface. Following the Al deposition the composite oxides were formed by anodisation. The effect of anodising such a structure on GaAs has been reported by Bayraktaroglu *et al* (1977). The composite oxide thus formed depends on the current density used and the amount of GaAs consumed after the complete oxidation of the Al layer. Growth of such oxides on InP resisted all attempts; the partially anodised Al film was seen to flake off the InP surface. This behaviour is assumed to be a function of the sticking qualities of the Al onto the etched InP surface.

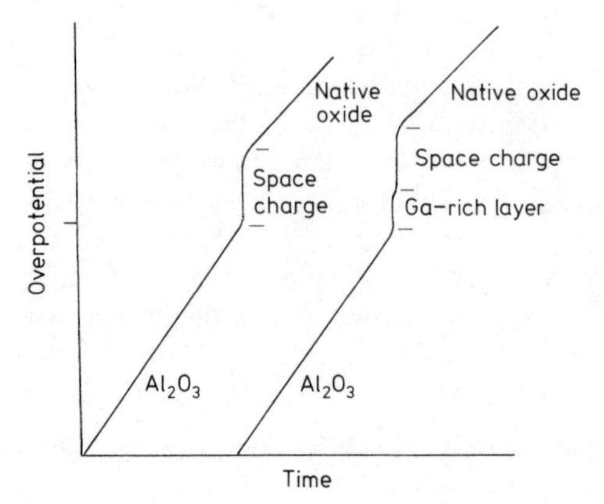

Figure 1. Schematic diagram for the overpotential time curves for MAOS composite and new composite oxide on GaAs.

The second type of composite oxide is more complex than the above and great care had to be taken to ensure that the semiconductor surface remained clean during the process.

2.1. Step 1

After the initial etching of the semiconductor surface, a thin layer of native oxide, 100 Å thick, was grown on the surfaces of both the GaAs and InP with current densities of 1 mA cm^{-2} and 200 μA cm^{-2} respectively, under strong light conditions. As these experiments were performed on n-type substrates, strong light was used to ensure a uniform oxide film.

The substrates were then removed from the electrolyte and cleaned in hot acetone, the samples were again blown dry and microscopically inspected for stains and particles.

2.2. Step 2

The oxides were then annealed in hydrogen at 1 atm for 10 min. The temperature for the GaAs substrate was 600 °C, which was sufficiently high to reduce the oxides to their metallic elements and also to boil off the arsenic leaving behind a Ga-rich layer in intimate contact with the GaAs. The temperature for the InP substrates was 400 °C, which produced the best device characteristics. The effect of a temperature of 600 °C on the InP was to produce severe pitting of the surface.

2.3. Step 3

A layer of Al was deposited by evaporation at 10^{-6} Torr to a typical thickness of 500 Å·

2.4. Step 4

The structures were then anodised in dull light in the same manner as for the MAOS structures, at a current density of 100 μA cm^{-2} for the GaAs and 200 μA cm^{-2} for the InP. The growth curve for these oxides contained a new feature when compared with that of a similar composite oxide of the first type. A schematic diagram for the two structures, for comparison, is shown in figure 1. The initial linear region for both cases is characterised by a straight line and is due to the formation of Al_2O_3. The new feature occurring with the second composite is believed to be due to the re-anodisation of, for GaAs, the Ga-rich layer and is characterised by a steep rise (approximately 4 V for an initial oxide layer of 100 Å) followed by an inflection. The second sharp rise corresponds to the sharp rise in the MAOS growth curve and is due to the formation of a space-charge layer in the GaAs surface (this space-charge voltage does not occur under strong light conditions). The final linear portion of the curve is due to the continued oxidation of the underlying substrate. This type of composite oxide was found to be readily and reproducibly grown on InP, unlike the attempts to grow MAOS type oxides and the overpotential–time characteristics show similar behaviour to those for GaAs.

2.5. Step 5

Ohmic contacts were then deposited onto the reverse surface of the substrates. For n-type GaAs, Au–Ge was used and for n-type InP, Au–Ge–Ni contacts were formed. The Ohmic contact metals had been omitted during the previous stages of this type of oxide growth to avoid any possible contamination and degradation during the initial annealing cycle. The penultimate step for all the structures was to anneal the oxides (and Ohmic contacts) in nitrogen for 15 min, at a temperature of 350 °C for GaAs (the optimum annealing cycle for native oxides as determined by Weiss *et al* 1977), and at 300 °C for InP (the optimum temperature for native oxides as determined by Colquhoun and Hartnagel 1976). Finally Al field-plates were deposited by thermal evaporation through a metal screen. A typical field-plate was 1500 Å thick and 10^{-3} cm^2 in area, with typically 50 diodes per substrate.

3. Capacitance–voltage measurements

C–V characteristics were measured with a two-phase Brookdeal Lock-in Amplifier model 9502 in conjunction with a variable-frequency signal generator and a ramp voltage generator. A reference signal of 250 mV RMS was supplied to the lock-in amplifier and was capacitively divided to supply a measuring signal of 25 mV RMS to the diode under test. The calibration of the amplifier was achieved using air capacitors of known value. The *C–V* curve was monitored on a *X–Y* recorder. Low-temperature measurements were obtained by cooling the test rig with liquid nitrogen; the temperature was estimated by a thermocouple mounted near the substrate under test. Theoretical *C–V* curves were computed using the manufacturers quoted substrate background doping and by measuring the oxide capacitance at low frequencies and high accumulating bias.

Figure 2 shows *C–V* characteristics for native oxides on GaAs, for an n-type substrate with background doping of 2×10^{16} cm^{-3} and a diode area of 2×10^{-3} cm^2. Figure 2(*b*) shows the room-temperature behaviour plotted against frequency and figure 2(*a*) the effect of lowering the temperature. It can be seen that the essential feature at high frequencies and low temperatures is that the *C–V* curve becomes flat, indicating that the surface potential is virtually pinned. A comparison of experimental *C–V* curves with the theoretical results indicates that for n-type GaAs with background doping of 2×10^{16} cm^{-3}, the surface potential is virtually pinned at -0.4 eV, which corresponds to depletion, and differs from the value obtained by Hasegawa and Hartnagel (1976) whose

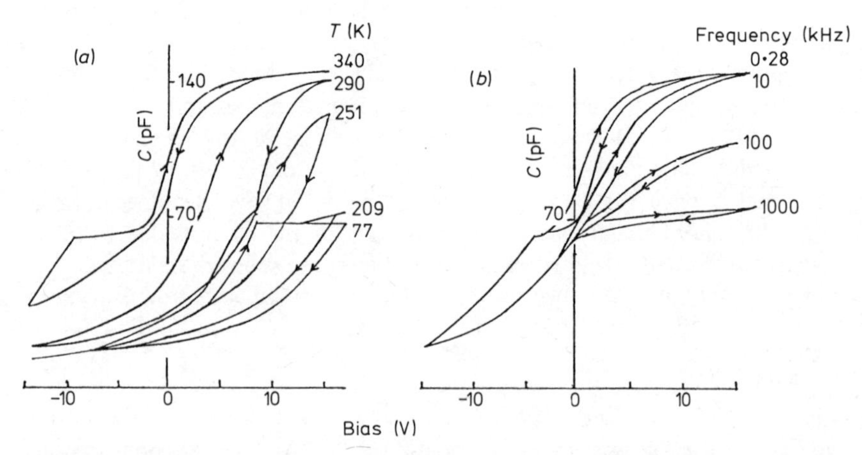

Figure 2. *C–V* characteristics of native oxide on GaAs. (a) Effect of temperature (data obtained at 1 kHz) and (b) effect of frequency (data obtained at 300 K). Sample B2; oxide layer 1000 Å thick; n-type GaAs; background doping 2×10^{16} cm^{-3}.

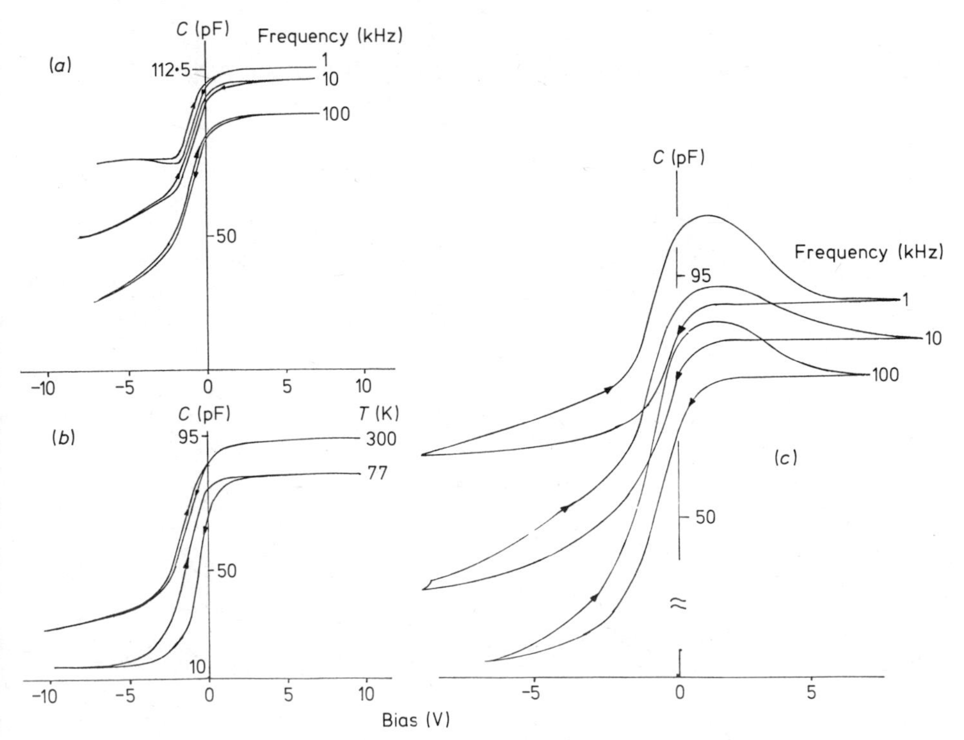

Figure 3. *C–V* characteristics of native oxides on InP. (*a*) Effect of frequency (data obtained at room temperature), (*b*) effect of temperature (data obtained at 100 kHz) and (*c*) effect of ageing (data obtained at room temperature). Sample e28, dull light, area = $0{\cdot}717 \times 10^{-3}\,cm^2$.

results indicate pinning in weak inversion. Results for p-type GaAs indicate that for background doping of $10^{17}\,cm^{-3}$, the surface potential is virtually pinned at 0·5 eV. *C–V* characteristics for native oxides on InP are shown in figure 3 for an n-type substrate with background doping of $10^{15}\,cm^{-2}$ and a diode area of $0{\cdot}717 \times 10^{-3}$ cm. Figure 3(*a*) shows the room-temperature behaviour plotted against frequency immediately after removal from the final annealing cycle. It can be seen that there is a large frequency dispersion in the oxide capacitance in accumulation, whereas in inversion the capacitance does not level off, indicating that there may be a leaky oxide. DC leakage measurements show that with an electric field of $2 \times 10^6\,V\,cm^{-1}$ across the oxide, the leakage current density was $\sim 10^{-3}\,A\,cm^{-2}$. The temperature variation of the *C–V* plots is shown in figure 3(*b*) and as can be seen, the curves become steeper when in inversion and the capacitance levels off almost exactly at the theoretical high-frequency capacitance. From a comparison of theoretical data and the experimental *C–V* curves it was found that there was no pinning of the surface potential. However, after leaving the test diode in air for a period of 1 week, the *C–V* characteristics changed to those shown in figure 3(*c*). These changes are produced as a result of the native oxide absorbing water from the surrounding atmosphere. A short annealing cycle of 150 °C for 30 min returns the *C–V* characteristics to those shown in figure 3(*a*).

Similar measurements on MAOS devices on GaAs indicate that the surface potential is again virtually pinned. Figure 4 shows the effect of temperature on the *C–V* characteristic for a device with an Al_2O_3 1400 Å layer, overlaying a layer of native oxide estimated to be between 50 and 100 Å thick. The value of the surface potential was found to be dependent on the amount of native oxide incorporated into the composite. For thin oxides, of the order of 50 Å thick, surface potentials were estimated to be -0.8 and 0·95 eV for n-type (10^{16} cm^{-3}) and p-type (10^{17} cm^{-3}) GaAs respectively. For oxide layers of thickness greater than 100 Å, the surface potential became pinned at -0.55 and 0·7 eV for n-type (10^{16} cm^{-3}) and p-type (10^{17} cm^{-3}) GaAs respectively.

The third type of oxide on GaAs displays *C–V* characteristics that are again dependent on the amount of GaAs consumed during the oxide growth. If the amount of GaAs consumed is limited to between 10 and 40 Å, as indicated in figure 5 (for a diode pro-

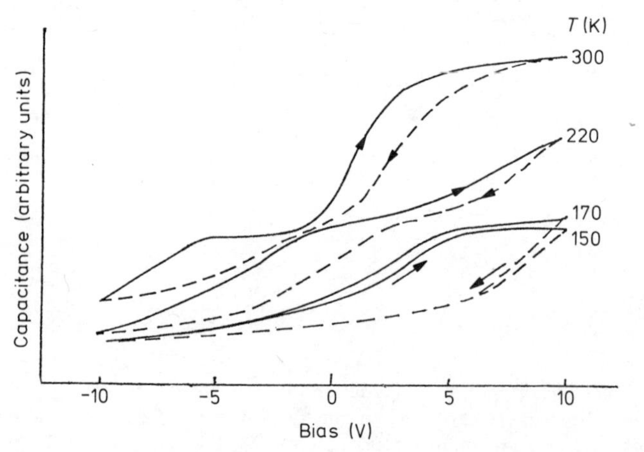

Figure 4. Effect of temperature on the *C–V* characteristics of a MAOS diode. Sample SJA5, frequency 1 kHz, dull light.

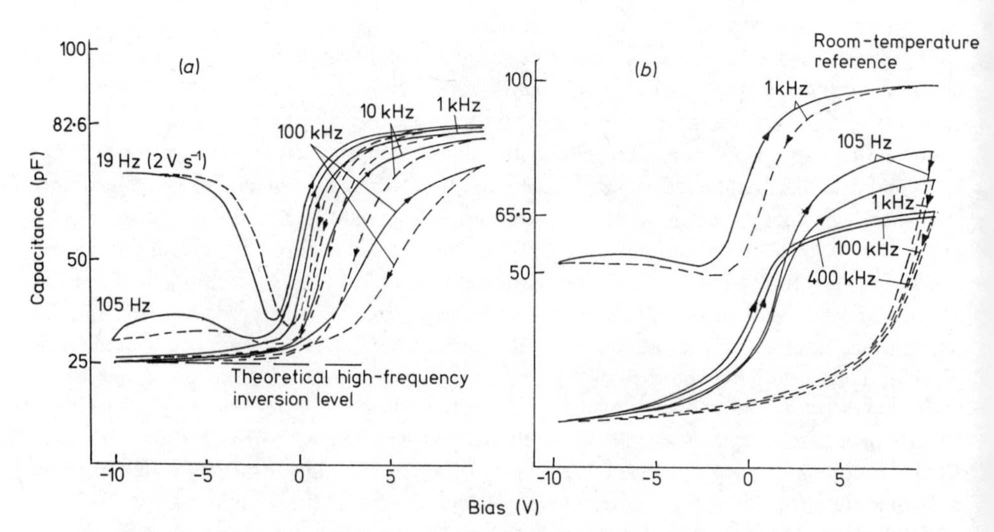

Figure 5. Effect of frequency at 300 K (*a*) and 77 K (*b*) on the *C–V* characteristics of new composite oxide on GaAs. (*a*) Sample 17/Ga, area 0.9×10^{-3} cm^2, dull light; (*b*) sample 12/Ga, area 1×10^{-3} cm^2, dull light.

duced on n-type GaAs, background doping 2×10^{16} cm^{-3}, total oxide thickness 750 Å, diode area 9×10^{-4} cm^2), the surface potential is no longer pinned. Figure 5(*a*) shows the characteristics as a function of frequency and it can be seen that for frequencies above 10 kHz, the inversion capacitance coincides with the theoretical high-frequency inversion capacitance; for low frequencies below 10 kHz, the inversion capacitance exhibits the expected low-frequency behaviour, characterised by a minimum in the curve. Deep depletion, which occurs for even relatively slow ramp speeds with the two previous types of oxides, is not observed with this oxide until a sweep rate of the order of 200 V s^{-1} is applied. Figure 5(*b*) shows the effect of frequency at low temperatures on a similar diode with a total oxide thickness of 700 Å and an area of 10^{-3} cm^2. The accumulation capacitance is seen to vary, but the frequency dispersion saturates at 400 kHz and at this point the estimated surface potential is 0·05 eV, which corresponds to almost flat-band conditions.

If the amount of GaAs consumed is greater than 50 Å, then the *C–V* characteristics again deteriorate and eventually become the same as for MAOS devices incorporating thick native oxide layers. Surface potential pinning is exhibited for n-type GaAs (background doping 10^{16} cm^{-3}) at a value of −0·5 eV.

Similar results for n-type InP are shown in figures 6(*a*) and (*b*), where the background doping of the substrate is again 10^{15} cm^{-3} and the area and oxide thickness of the test diode are $0{\cdot}813 \times 10^{-3}$ cm^2 and 750 Å. It can be seen from figure 6(*a*) that low-frequency-type behaviour starts at around 1 MHz and shows both strong inversion and accumulation. Deep depletion is not observed for ramp speeds up to 100 V s^{-1}.

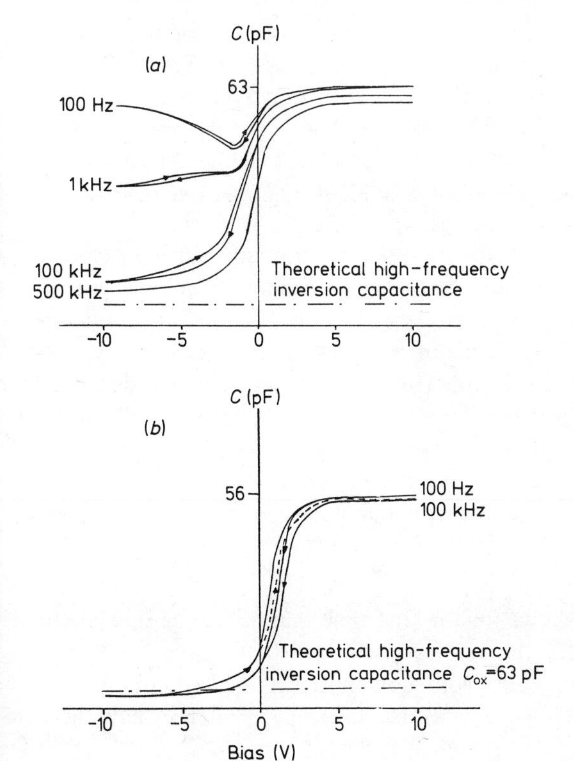

Figure 6. Effect of frequency at room temperature (*a*) and 77 K (*b*) on the *C–V* characteristics of composite oxides on InP. Sample e22, area $0{\cdot}813 \times 10^{-3}$ cm^2, dull light.

There is again some frequency dispersion of the oxide capacitance, as with the native oxide structures, although it is reduced and saturates at approximately 500 kHz. The effect of temperature on the *C–V* characteristics is shown in figure 6(*b*) and again both strong inversion and strong accumulation are observed, with the inversion capacitance approaching the theoretical high-frequency value. It is clear from these results that there is no pinning of the surface potentials. DC leakage measurements showed that with a field of 2×10^6 V cm^{-1} across the oxide, the leakage current was $\sim 10^{-8}$ A cm^{-2}.

Similar to GaAs, if the amount of InP consumed is greater than 50 Å, the *C–V* characteristics again deteriorate, i.e. greater frequency dispersion and lower minority-carrier generation occur. The breakdown field of these composite oxides on both GaAs and InP was found to be $\sim 4 \times 10^6$ V cm^{-1}.

4. Discussion

The surface potential pinning that occurs with both native oxide and MAOS structures on GaAs is associated with a large number of surface states at the oxide interface. With the MAOS structures the distribution of these traps is seen to be a function of the amount of GaAs consumed; recent work by Bayraktaroglu *et al* (1979), who anodised Al on p-type GaAs in a solution of ammonium pentaborate, has shown that if the growth is terminated so that virtually no native oxide is formed, then no surface pinning occurs. The nature and cause of the surface traps are not fully understood, but work by Breeze *et al* (1979) has shown that near a native oxide–GaAs interface, a region exists that contains an amount of non-oxidised or partially oxidized As.

It is believed that the improvements observed with the second type of composite oxide arise mainly from the reduction in the amount of As at the oxide interface, because of the formation of the Ga-rich layer by reduction in hydrogen. Experiments involving the annealing of a clean GaAs surface in hydrogen at 600 °C did not show the same improvements as those in which an oxide layer was grown on the surface before annealing. This is probably because the vapour pressure of As in the oxide is much greater than that of As in GaAs (Ishii and Jeppson 1977) at these temperatures.

The poor characteristics of InP native oxides have been associated with poor leakage characteristics of the oxide which do not allow the build-up of an inversion on accumulation charge. The attempts to grow MAOS-type oxides on the InP surface failed, probably because of poor sticking qualities. The new type of composite oxide, however, exhibits excellent capacitance–voltage characteristics which may be associated with the decreased conductivity of the insulator on the surface. However, similar oxides on an InP surface that was simply annealed in hydrogen, although they could be grown, did not show the same type of capacitance–voltage characteristics, indicating that the role of a reduced oxide in intimate contact with the InP is important in obtaining good MIS characteristics.

5. Conclusions

A method has been presented which shows for the first time that it may be possible to use wet anodisation as a technique for the surface passivation of GaAs and InP. The results have shown that for GaAs strong inversion and almost flat-band conditions have been achieved, whereas for InP both strong inversion and accumulation have been obtained.

To date, few attempts have been made to optimise the various steps involved in producing the devices and it is hoped that the MIS characteristics may still be improved and extended to p-type semiconductors.

Acknowledgments

The authors would like to thank the Science Research Council who supplied the finance, Dr C A Walley, Head of the Department of Electrical and Electronic Engineering, for the use of the facilities, Dr C R Whitehouse for his helpful encouragement and Dr A J Grant of RSRE who supplied the InP.

References

Bayraktaroglu B, Hannah S J and Hartnagel H L 1977 *Electron. Lett.* **13** 45

Bayraktaroglu B, Schuermay F L S and Grant J T 1979 *Proc. 6th PCSI, Monterey, California* to be published

Breeze P A, Hartnagel H L and Sherwood P M A 1979 *J. Electrochem. Soc.* to be published

Chang R P H, Coleman J J, Polak A J, Fellman L C and Chang C C 1979 *Appl. Phys. Lett.* **34** 237

Colquhoun A and Hartnagel H L 1977 *Surface Technol.* **5** 291–302

Fritzsche D 1978 *Electron. Lett.* **14** 51–82

Hasegawa H and Hartnagel H L 1976 *J. Electrochem. Soc.* **133** 713

Ishii T and Jeppson B *J. Electrochem. Soc.* **124** 1784

Weiss B, Kohn E, Bayraktoroglu B and Hartnagel H 1977 *Gallium Arsenide and Related Compounds (Edinburgh) 1976* (Inst. Phys. Conf. Ser. 33a) p168

GaAs surface passivation using Si_3N_4: interface characteristics

B Bayraktaroglu, W M Theis and F L Schuermeyer
Air Force Avionics Laboratory, WPAFB, Ohio 45433, USA

Abstract. Si_3N_4 layers with a low oxygen content and formed by plasma-enhanced deposition (PED) were used for the passivation of GaAs surfaces. The electrical properties of the Si_3N_4–GaAs interface were analysed by variable-frequency C–V and G–V measurements. Each C–V curve showed hysteresis effects arising from charges stored in the insulator. Partial removal of the hysteresis was possible with a short annealing cycle. The density of surface states was estimated to be in the region of 10^{11} cm^{-2} eV^{-1} with a broad peak close to 10^{12} cm^{-2} eV^{-1}, 0·8–0·9 eV from the valence-band edge.

1. Introduction

Deposited dielectric films such as CVD Si_3N_4 (Foster and Swartz 1970, Becke and White 1967) and sputtered Si_3N_4 (Meiners 1978) have been used in the past for the passivation of GaAs surfaces. Despite the many attractive advantages associated with the stable dielectric properties of Si_3N_4 films, it has been reported that the GaAs–Si_3N_4 interfacial properties are dominated by a large number of surface states (Meiners 1978) and electrical instabilities caused by charge injection and storage in the insulator films (Foster and Swartz 1970, Cooper *et al* 1972). An alternative passivation scheme, involving the growth of native oxides on the surface of GaAs by wet anodisation also yields interfacial properties similar to those observed with deposited films (Meiners 1978). It has been reported (Spicer *et al* 1977) that the clean surface of GaAs has no surface states within the band-gap region, but large number of states can be created by the oxidation of the surface. Suzuki *et al* (1978) have shown the effect of thin native oxide layers underneath deposited insulators in producing the electrical instabilities of the interface that are commonly observed when anodic native oxides are used alone.

The passivation scheme reported here makes use of Si_3N_4 films formed by plasma-enhanced deposition (PED) on GaAs surfaces treated *in situ*. One attractive feature of this deposition technique is that good-quality insulators can be grown at relatively low temperatures ($\geqslant$100 °C), thus preventing the dissociation of the volatile component (i.e. As) from the GaAs surface. It also allows the sample surfaces to be treated with low-energy nitrogen ions before deposition; this was found to be an effective way of reducing the thickness of the natural oxide layer.

2. Experimental details

A commercially available apparatus was modified to accommodate the processing steps necessary for low-oxygen-content films (see figure 1). A Pyrex chamber containing

0305-2346/80/0050-0280$01.00

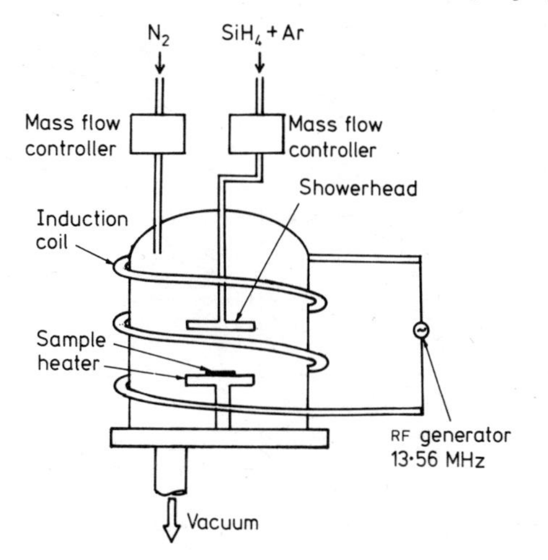

Figure 1. A schematic drawing of the reactor used for depositing Si_3N_4 films on GaAs.

a heater was used for the reaction of silane gas and N_2. Samples placed on the heater may be subjected to temperatures of 100–300 °C. In addition to the thermal energy available for the reaction, an RF oscillator operating at 13·56 MHz was used to excite a plasma within the chamber. This was induced by means of a coil wrapped around the chamber.

The dispersion of silane gas is critical for the uniformity of the film. A precision-fabricated circular showerhead placed above the heater ensured proper injection of the gas. Furthermore, the gas flow must be high enough to cause turbulent mixing with the N_2 present from the other gas port (located near the top of the chamber). The height of the showerhead above the heater plate was also found to be critical for proper uniformity.

Various combinations of operating parameters were explored before the set used in this study was selected. The quality of the deposited films depends critically on the rate of gas flow (both silane and N_2), the level of RF power, temperature, geometry of the system, and the method of evacuation of the reaction chamber. The determination of optimum conditions will be discussed elsewhere (Theis *et al* 1979). The parameters used in this study were a flow of 72 SCCM† of gas composed of 2% SiH_4 and 98% Ar carrier, 22 SCCM of N_2, a measured RF power of 10 W and sample temperature of $\simeq$150 °C. The total system pressure had little effect on the film growth and was maintained at 1 mm Hg. This produced a film growth rate of 150 Å min^{-1}; the index of refraction of the film was 2·0. Furthermore, the film contained very little oxygen, as shown in figure 2.

The quality of the interfacial properties of Si_3N_4–GaAs system was found to depend strongly on the treatment used before deposition. In our system a metal shutter was placed between the sample on the heater and the volume enclosed by the RF coils. The sample could be selectively exposed to the plasma or be protected in the pump-down phase. Before injection of SiH_4 into the chamber, the sample was exposed to the N_2 plasma for a length of time (typically 1 min). This step was found to be effective in reducing the thickness of the natural oxides which are always present on the surface of GaAs after exposure to air. The RF power and the exposure time were varied to obtain optimum removal of oxides from the sample surface. The Auger depth profile of a typical sample shown in figure 2 indicates that the amount of oxygen near the Si_3N_4–GaAs interface can be reduced to less than 1 at.%.

† Standard cubic centimetre per minute.

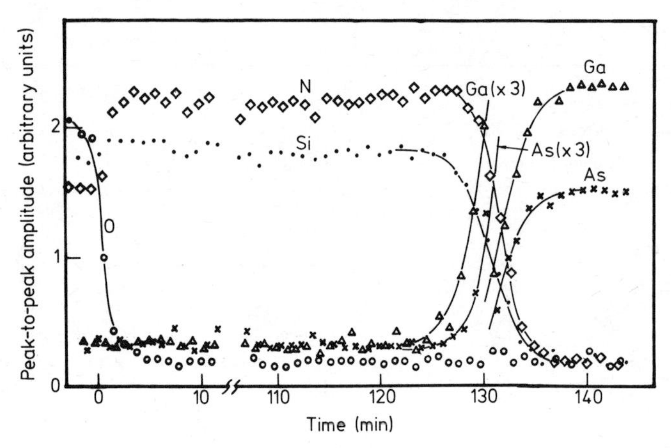

Figure 2. Auger depth profile of a typical sample. The insulator thickness is 900 Å.

All results reported in this paper were obtained from investigations performed on devices produced on bulk-grown, (100)-oriented n-type GaAs with a carrier concentration of $0{\cdot}8 \times 10^{16}$–2×10^{16} cm^{-3}. Au–Ge Ohmic contacts were formed on the reverse side of each sample before Si_3N_4 deposition. Samples were then cleaned in acetone, trichloroethylene and methanol using an ultrasonic cleaner and rinsed in 18 MΩ deionized water. Before loading into the reactor, each sample was etched in NaOH : H_2O_2 : H_2O (1 : 2 : 100 by volume) to remove at least 0·5 μm of the surface. After Si_3N_4 deposition, circular gold dots of area 4×10^{-4} cm^{-2} were evaporated through a stainless steel mask for electrical measurements. The sample was then cut into sections and mounted on a holder. Electrical contact was made to the gold dots by means of a microprobe. The electrical breakdown strength of the devices was typically 5×10^6 V cm^{-1}; thus during the C–V and G–V measurements, the applied field was kept below 2×10^6 V cm^{-1}.

3. Results and discussion

Both continuous and point-by-point C–V and G–V measurements were made to extract information regarding the interfacial properties. The apparatus used for continuous measurements is described elsewhere (Bayraktaroglu and Hartnagel 1979) and the ramp speed could be varied from 10 mV s^{-1} to 10 V s^{-1}. Point-by-point measurements were made with a General Radio 1615 capacitance bridge in the frequency range 100 Hz–10 kHz and with a Boonton 75C capacitance bridge in the range 5–500 kHz. Additionally a Hewlett-Packard 4270A capacitance bridge and Princeton Applied Research 410 fixed-frequency (1 MHz) C–V plotter were used for high-frequency measurements.

During each deposition run, a second sample of Si coated with a thin layer of Al was included. Gold dots of the same size were also produced on this sample and the capacitance of the resulting MIM structure was taken to be equal to the insulator capacitance of the MIS structure produced on the GaAs sample. MIM devices were further used to determine the frequency dispersion effects of the insulator.

A family of C–V curves obtained with a typical MIS structure is shown in figure 3. Each curve had a clockwise hysteresis ($\simeq 1$ V) which for simplicity is not shown in this

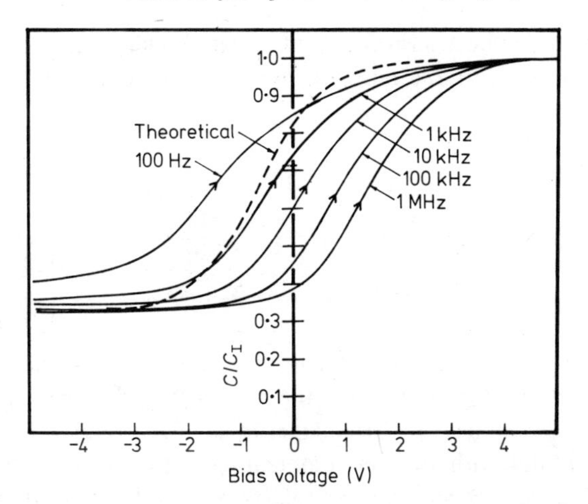

Figure 3. Capacitance–voltage (C–V) curves (with the frequency as a parameter) obtained with PED Si_3N_4–GaAs MIS devices. $N_D - N_A = 1 \times 10^{16}\ cm^{-3}$; insulator thickness $\simeq$ 1000 Å; ramp speed = 200 mV s^{-1}.

figure (only the positive half of the cycle is drawn). The frequency response of the C–V curves is consistent with the 'freezing-out' of the fast surface states. The frequency dispersion of the C–V curves in the accumulation region that is commonly observed with anodic native oxide MOS devices (Swada and Hasegawa 1976) was not seen with the present structures. However, some dispersion of capacitance in the inversion region was seen. The flat region of the C–V curves with a large negative bias was assumed to be due to the formation of an inversion layer; no further investigations were carried out to establish the existence of this layer.

In figure 3 the theoretically calculated C–V curve is also shown. A comparison of the theoretical and the high-frequency (1 MHz) curves indicates that most of the states are located in the top half of the band gap, but the exact position of the states in the band gap cannot be determined from these curves. More information about the surface states can be obtained from conductance techniques (Nicollian and Goetzberger 1967, Goetzberger *et al* 1976). The equivalent circuit of the MIS capacitor used for the conductance measurements is shown in figure 4(*a*). C_I and G_I, are, respectively, the capacitance and conductance of the insulator (as measured with the MIM structures), C_s is the surface-state capacitance, R_s is the majority-carrier capture resistance and C_D is the depletion-

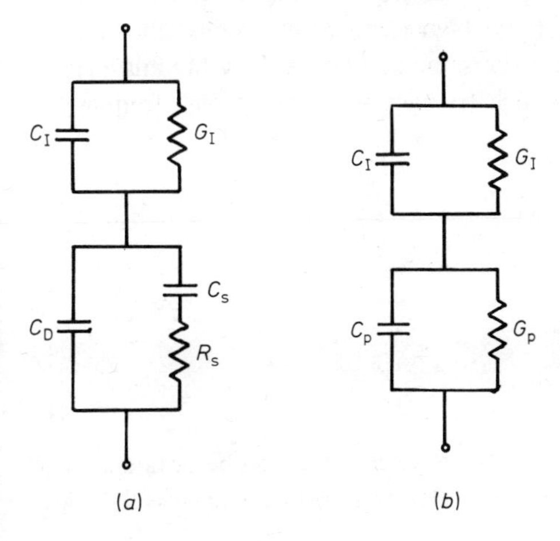

Figure 4. The equivalent circuits of the MIS devices.

layer capacitance. An equivalent circuit can be transformed into that shown in figure 4(*b*) by representing the surface-state branch of the network by its parallel equivalent (Nicollian and Goetzberger 1967):

$$G_p/\omega = C_s\omega\tau/[1+(\omega\tau)^2] \tag{1}$$

and

$$C_p = C_D + \{C_s/[1+(\omega\tau)^2]\} \tag{2}$$

$$\tau = R_s C_s. \tag{3}$$

Information regarding the density of states, N_s was extracted in three ways. In one procedure (method A), $G_p/\omega C_I$ was plotted against $\lg\omega$ and the peak value of $G_p/\omega C_I$ from the resulting curves was used to determine N_s (Goetzberger *et al* 1976). Figure 5 shows some curves obtained experimentally with the device operating in the depletion mode. Each curve shows a peak and symmetry about this peak. N_s values determined from these curves were in the range 10^{11}–10^{12} cm^{-2} eV^{-1}.

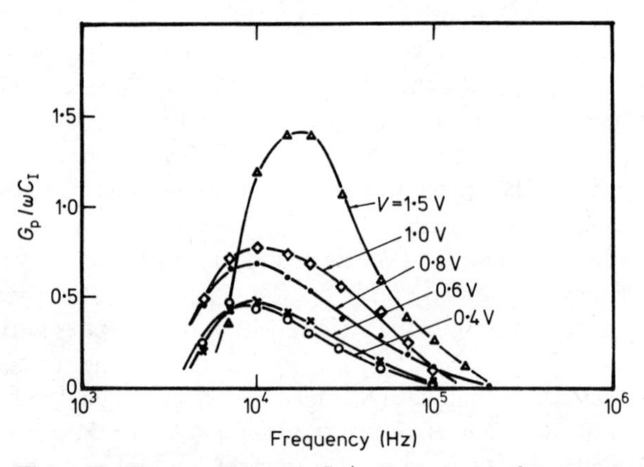

Figure 5. The experimental $G_p/\omega C_I$ against $\lg f$ curves obtained with MIS devices with the bias as a parameter.

The relationship between the bias voltage applied to the top contact and the surface potential was determined from the values of C_p at high frequencies, as described below. The second method (method B) is essentially the same as that used by Malmin (1971) and makes use of the C_p values at low and high frequencies. At very high frequencies equation (2) becomes

$$C_p(\omega\to\infty) = C_D \tag{4}$$

and at low frequencies

$$C_p(\omega\to 0) = C_D + C_s. \tag{5}$$

Therefore

$$C_s = qN_s = C_p(\omega\to 0) - C_p(\omega\to\infty). \tag{6}$$

At every applied bias, C_D and therefore the surface potential ψ_s can be obtained. The high- and low-frequency values were taken to be 1 MHz and 100 Hz respectively. Plots of G_p/ω against C_p (Cole–Cole plots) showed that, over the range of biases used, at these two frequencies $G_p/\omega \simeq 0$ and the curves were nearly semicircular in shape.

Figure 6 shows the results obtained with an alternative method (method C). This makes use of the relationship $C_p(\omega \to \infty) = C_D$ to determine the surface potential ψ_s. Experimental plots of V against ψ_s (figure 7) are then compared with those calculated from the known parameters of the devices. N_s is obtained using the relationship

$$N_s = (1/q)(\partial Q_s/\partial \psi_s) \tag{7}$$

$$Q_s = \Delta V C_I \tag{8}$$

where Q_s is the charge stored at surface states and ΔV is the difference between values of the applied and the theoretical bias that is necessary to create the same surface potential.

There is some scatter of the experimental points in figure 6, but it is obvious that the surface states have a continuum in the upper half of the band gap and show a broad peak

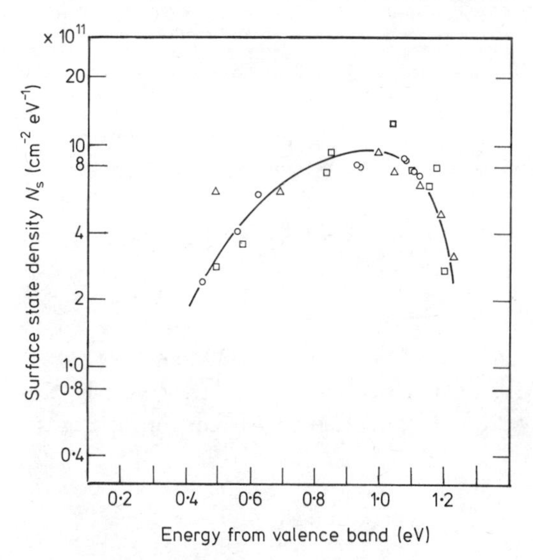

Figure 6. The densities of surface states plotted against energy within the band gap as obtained by methods A (□), B(△) and C(○).

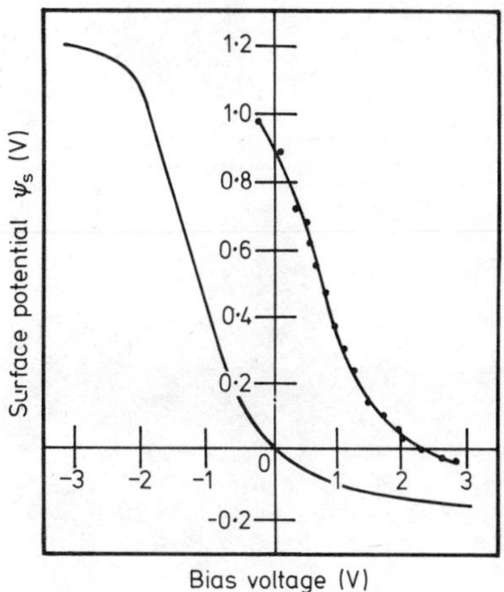

Figure 7. The theoretical (—) and experimental (• – •) plots of V against ψ_s. $N_D - N_A = 1 \times 10^{16}$ cm^{-3}; $C_I = 6 \times 10^{-8}$ F cm^{-2}.

between 0·8 and 0·9 eV from the valence-band edge. The values of N_s lie in the region of 10^{11} cm^{-2} eV^{-1} and the peak value approaches 10^{12} cm^{-2} eV^{-1}. From figure 5 the time constants of the surface states were calculated to be in the range $2{\cdot}6 \times 10^{-5}$–$8{\cdot}0 \times 10^{-6}$ s.

The tunnelling effects which seem to dominate the interfacial properties of CVD Si_3N_4–GaAs devices (Cooper *et al* 1972) were not encountered to the same degree when PED Si_3N_4 was used. This is possibly due to the thinner interface region produced after the removal of the natural oxide layers (figure 2). However, the hysteresis effects in the *C–V* curves are similar to those seen with CVD Si_3N_4 and can be explained by majority-carrier trapping at the states located in the insulator. Some of the hysteresis effects were eliminated by annealing the samples (after evaporation of the contact pads) in air or N_2 at 150–200 °C for 10 min. This annealing cycle was not able to remove the hysteresis completely and therefore an alternative method is needed to reduce the effect further.

4. Conclusions

It has been shown that the PED Si_3N_4 films can be used successfully for the passivation of GaAs surfaces. Removal of the natural oxides before deposition seems to be necessary for lowering the density of surface states. Treatment of the surface *in situ* with low-energy N_2 ions helped to produce devices with surface-state densities around 10^{11} cm^{-2} eV^{-1}. All samples showed clockwise hysteresis (n-type GaAs) and partial removal of the hysteresis effects was possible by a short annealing cycle. Before the PED Si_3N_4 films can be used for applications such as insulated-gate transistors, the hysteresis effects must be reduced further.

Acknowledgments

The authors would like to thank D L Mays, C Geesner and J E Ehret for their help in producing the samples used in these investigations and also Dr J Grant for making the Auger measurements.

References

Bayraktaroglu B and Hartnagel H L 1979 *Int. J. Electron.* **45** 561
Becke H W and White J P 1967 *Electronics* 82
Cooper J A Jr, Ward E R and Schwartz R J 1972 *Solid St. Electron.* **15** 1219
Foster J E and Swartz J M 1970 *J. Electrochem. Soc.* **117** 1410
Goetzberger A, Klausmann E and Schulz M J 1976 *CRC Crit. Rev. in Solid St. Sci.* **6** 1
Malmin P C 1971 *Phys. Stat. Solidi* **A8** 597
Meiners L G 1978 *J. Vac. Sci. Technol.* **15** 1402
Nicollian E H and Goetzberger A 1967 *Bell Syst. Tech. J.* **46** 1055
Spicer W E, Planetta P, Lindau I and Chye P W 1977 *J. Vac. Sci. Technol.* **14** 885
Suzuki N, Hariu T and Shibata Y 1978 *Appl. Phys. Lett.* **33** 761
Swada T and Hasegawa H 1976 *Electron. Lett.* **12** 472
Theis W M, Ehret J E, Geesner C R and Park Y S 1979 to be published

GaO_xN_y-based multiple insulating layers on GaAs surfaces

J Nishizawa and I Shiota

Research Institute of Electrical Communication, Tohoku University, Sendai 980, Japan

Abstract. The passivation method reported previously using single layers of GaO_xN_y on GaAs surfaces is developed further. In order to obtain GaO_xN_y films of high insulating capability with smaller x/y ratios and a density of surface states less than 1×10^{11} cm^{-2} in contact with the GaAs, two methods have been used: (i) application of another insulating film such as SiO_2 or Al_2O_3 to the GaO_xN_y film (double insulating film system) and (ii) control of the composition of the GaO_xN_y film to give a higher oxygen content near the outer surface (graded-composition film system).

1. Introduction

It has been found that gallium oxynitride (GaO_xN_y) serves as a very effective surface passivating material, suitable for use with III–V compound semiconductors, particularly GaAs (Shiota *et al* 1978). Initially the reasons behind using GaO_xN_y for passivation were involved with improvement of the quality of GaN films, as proposed by Nishizawa (Shiota *et al* 1975) for surface passivation of III–V compound semiconductors. This was based on the finding that the incorporation of a small amount of oxygen in a GaN film produced great improvements in its resistivity, chemical properties and adhesion to a GaAs substrate (Nishizawa *et al* 1976, 1977, 1978). A GaO_xN_y film was prepared for the first time by the CVD method (Nishizawa *et al* 1976, 1977, 1978) and then by RF reactive sputtering (Shibata *et al* 1978, Hariu *et al* 1978).

Highly adhesive GaO_xN_y films with a total charge density N_{FB} less than 1×10^{11} cm^{-2} and a density of surface states of the order of $(1–2) \times 10^{11}$ cm^{-2} eV^{-1} were readily obtained over a range of small x/y ratios for fairly low deposition temperatures such as 450 °C; the resistivity of the GaO_xN_y film tended to decrease as the x/y ratio became smaller. To obtain a highly insulating passivating layer while still retaining excellent surface properties, the use of GaO_xN_y-based multiple insulating films is investigated here.

2. Experimental details

The gallium oxynitride (GaO_xN_y) films were deposited by thermal decomposition of an ammonia–gallium tribromide complex containing a controlled amount of oxygen. This complex was transported by NH_3 gas flow with a flow rate of 2·5 l min^{-1}, with which NO gas was mixed to incorporate a further amount of oxygen into the film. Further control of the oxygen : nitrogen ratio was achieved by thermal oxidation of the GaO_xN_y film in the O_2 gas diluted with excess NH_3 gas. The rate of evaporation of the complex

0305-2346/80/0050-0287$01.00

was varied by adjusting the temperature between 270 and 350 °C. The original purity of the $GaBr_3$ used was 99·999% and that of NH_3 was better than 99·99%.

The Al_2O_3 film was also formed by thermal decomposition of aluminium isopropoxide $Al(i\text{-}OC_3H_7)_3$ at 400 °C in a N_2 gas flow of 1 l min^{-1}. The rate of deposition was 400 Å min^{-1} for a source temperature 160 °C. The film-forming reactions described above were carried out in a fused silica tube (internal diameter 25 mm) heated externally by a resistance heater.

The preparation of SiO_2 films on GaO_xN_y films was carried out by conventional parallel-plate RF sputtering of pure quartz and also by ion-beam coating, where sputter-generated SiO_2 molecules from an accelerated Ar ion beam (800 eV at 1×10^{-4} Torr) were deposited.

The original GaAs (100) and (111) surfaces were first etched with a solution of $H_2SO_4 : H_2O_2 : H_2O$ (ratio 4:1:1 by volume) and then dipped into concentrated HF solution; further slight etching was then carried out with $H_2SO_4 : H_2O_2 : H_2O$ (ratio 10:1:1 by volume) and the samples were finally dipped into concentrated HF again to minimise the residual oxide layer (Shiota *et al* 1977). GaAs surfaces treated in this way were finally rinsed with doubly distilled methanol.

Composition profiles were obtained with Auger electron spectroscopy combined with simultaneous Ar sputtering. The $L_3M_{4,5}M_{4,5}$ Auger peaks for Ga and As and the KLL peaks for oxygen and nitrogen were monitored automatically by a Physical Electronics multiplex system. The sputtering rate of the GaO_xN_y films was almost unchanged at 102 ± 7 Å min^{-1} with O/N $\lesssim$ 0·8 for a sputtering voltage of 2 kV and an emission current of 30 mA at 5×10^{-5} Torr. A value of 2·43 was used for the relative Auger sensitivity of the oxygen and nitrogen KLL peaks in the present study (Shiota and Nishizawa 1979), with a primary electron-beam voltage of 4 kV and a beam current of 1 μA.

3. Results and discussion

GaO_xN_y films with total charge densities less than 1×10^{11} cm^{-2} were readily obtained over a range of smaller x/y ratios and for lower deposition temperatures such as 450 °C, as shown in figure 1; however, the resistivity of the GaO_xN_y film tended to decrease as the x/y ratio became smaller.

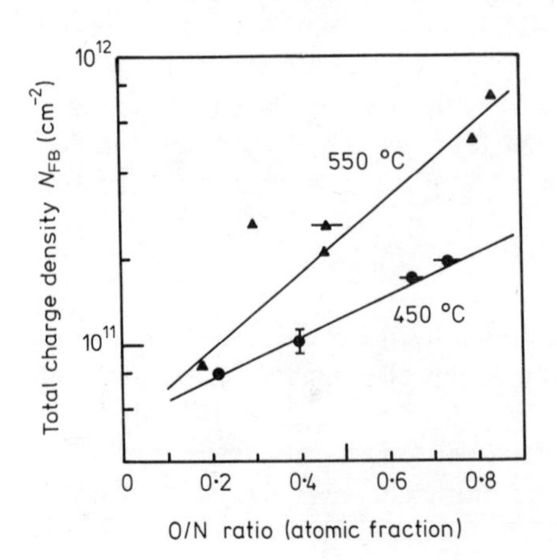

Figure 1. Total charge density N_{FB} plotted against O/N ratio for GaO_xN_y film.

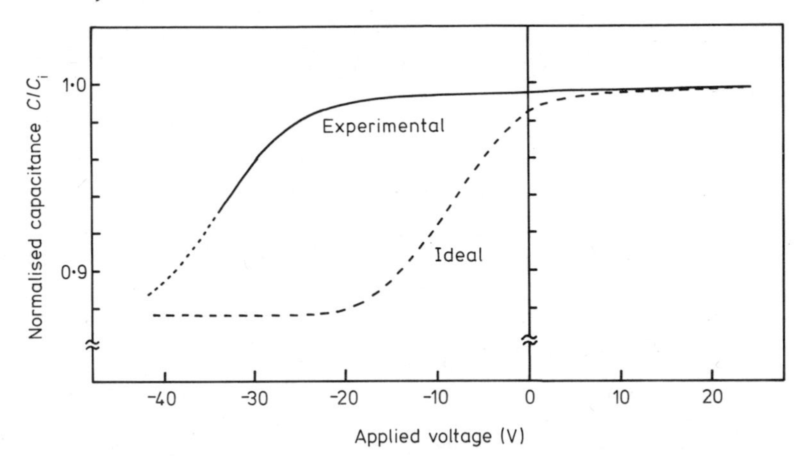

Figure 2. The high-frequency C–V curve (full curve, obtained at 1 MHz) for a MOS capacitor with an RF sputter-coated SiO_2–GaO_xN_y double layer. The broken curves are theoretical results.

Based on these results, we have attempted to produce a highly passivated GaAs surface with the use of GaO_xN_y-based multiple insulating layers, which have a high resistivity as well as a low density of surface states. To accomplish this, two methods have been used: (i) application of another insulating film, such as SiO_2 or Al_2O_3, to the GaO_xN_y film (double insulating film system) and (ii) control of the composition of the GaO_xN_y film to give a higher oxygen content near the outer surface (a graded-composition film system). For method (i), the two types of physically deposited SiO_2 films and the Al_2O_3 film prepared by thermal decomposition of $Al(i\text{-}OC_3H_7)_3$ in N_2 atmosphere, as described in §2, were used as the second insulating film on the GaO_xN_y film.

The RF sputter-coated SiO_2–GaO_xN_y shows a large negative shift in the capacitance–voltage curve (measured at 1 MHz) and has a total N_{FB} of the order of 10^{12} cm^{-2} as shown in figure 2, although the breakdown strength of the double layer is improved. The total charge density was determined from the flat-band voltage ΔV_{FB} or the depletion approximation ($(C_0/C)^2$ against V) from the C–V curve determined at 1 MHz. For these experiments, SiO_2 was deposited at a rate of 40 Å min^{-1} to form a layer 2500 Å thick. The control MOS capacitor with no interfacial GaO_xN_y appears to give a much larger value of N_{FB} than does the double-layer system as little variation is observed in the capacitance for the same range of applied voltages. On the other hand, the ion-beam-coated SiO_2–GaO_xN_y double-layer system seems to give a small shift in the C–V curve, and in some cases even a positive shift of the curve is observed, compared with the theoretical curve, corresponding to approximately $N_{FB} = 5 \times 10^{11}$ cm^{-2} or less, as shown in figure 3. The C–V curve shows a small ion-drift-type hysteresis as well as that illustrated in figure 2, whereas an MIS capacitor with a GaO_xN_y film as an insulator shows a carrier-injection-type hysteresis. These results indicate that the additional fixed charge and movable ions are introduced when the SiO_2 film is deposited. A major cause of the larger value of N_{FB} in the RF sputter-coated SiO_2 system may be the fact that conventional diode RF sputtering tends to cause a specimen located directly in an excited plasma to heat up during film deposition because of the electron current flowing across it. With ion-beam coating, the increase in the temperature of the specimen can be safely

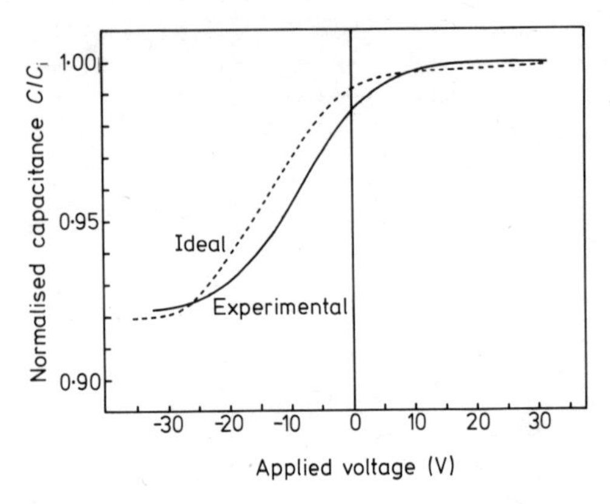

Figure 3. The high-frequency C–V curve (full curve, obtained at 1 MHz) for a MOS capacitor with an ion-beam coated SiO_2–GaO_xN_y double layer. $C_0 = 4{\cdot}7 \times 10^{-9}$ F cm^{-2}; $N_D \sim 4 \times 10^{16}$ cm^{-3}. The broken curves are theoretical results.

ignored and, in addition, the operating vacuum of 1×10^{-4} Torr is better than the pressure of 2×10^{-3} Torr used in the present RF sputtering experiments. It is therefore obvious that an SiO_2 film formed at a low temperature may serve as a possible second insulating layer for the GaO_xN_y–GaAs system, except when the system is required for use as a thermal diffusion mask or in encapsulating film for annealing because of the rather high solubility of Ga atoms in an SiO_2 film.

The Al_2O_3 double-insulator system also results in a higher breakdown strength and a low total charge density, but the C–V curve exhibits a rather complicated hysteresis, as shown in figure 4. The inversion region shows a minority-carrier-injection-type hysteresis and in the accumulation region an ion-drift-type hysteresis is observed. For these studies the field-plate was made of evaporated gold film. The control MOS capacitor with just an Al_2O_3 film shows a large positive shift (ΔV_{FB}) in the C–V curve and also a larger carrier-injection-type hysteresis, as reported by many authors, so that negatively charged species are formed inherently in the CVD Al_2O_3 film itself. The hysteresis caused by the drift of positive ions in the accumulation region of a C–V curve increases when the field-plate is an evaporated Al film. This may be attributable to the chain reaction associ-

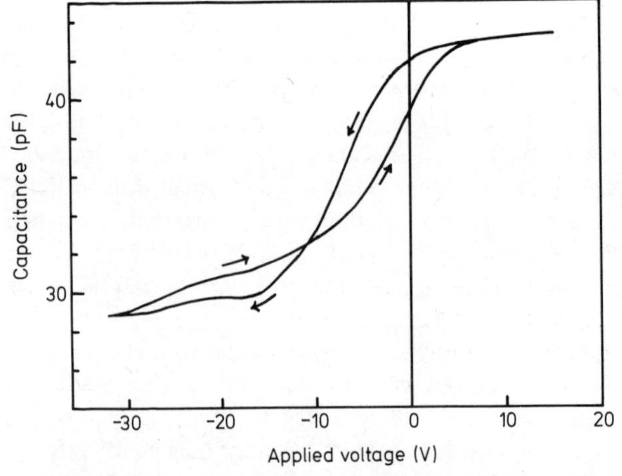

Figure 4. The high-frequency C–V curve obtained at 1 MHz for a MOS capacitor with an Al_2O_3–GaO_xN_y double layer.

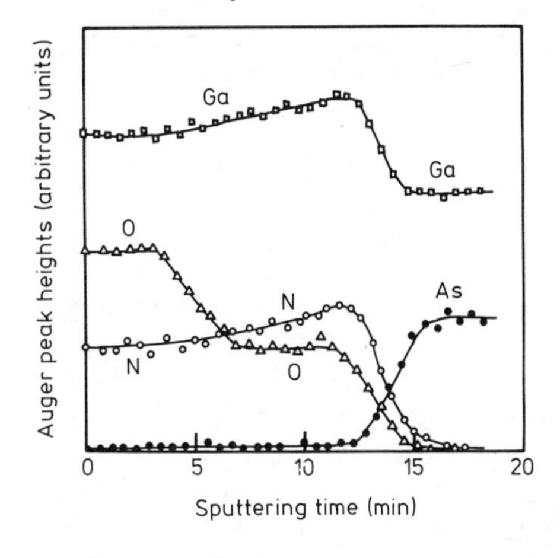

Figure 5. Depth composition profile of the GaO_xN_y–GaAs system obtained by the Auger–sputter method, in which the x/y ratio for the GaO_xN_y film varies stepwise from 0·31 ± 0·02 to 0·79. The thickness of the GaO_xN_y film is about 1400 Å.

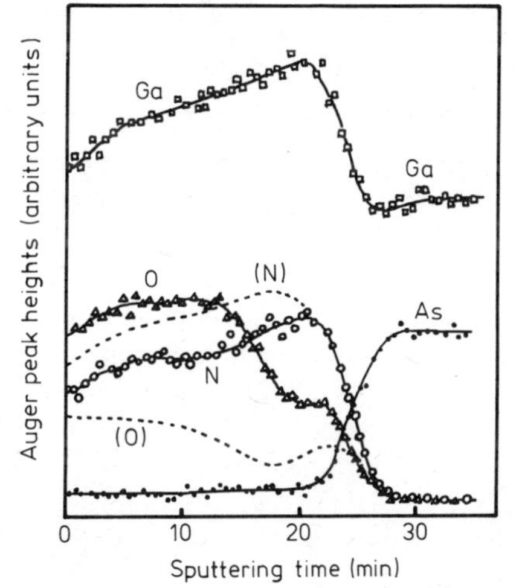

Figure 6. Depth composition profile of the GaO_xN_y–GaAs system thermally oxidised in NH_3 (1 l min^{-1}) + O_2 (0·1 l min^{-1}).

ated with active Al and initiated by reduction of part of the Al_2O_3 to Al at the Al–Al_2O_3 interface. Further reduction of part of the GaO_xN_y by the resulting Al produces positively charged Ga atoms which are less reactive towards oxygen than are Al atoms. It is thus concluded that an inactive metal such as gold should be used for the Al_2O_3–GaO_xN_y double-layer system in order to prevent the generation of driftable constituent metal ions. It is also noted that the deposition of the Al_2O_3 film should be carried out in an atmosphere containing as little oxygen as possible, in order to avoid generation of an oxide layer at the interface between the GaO_xN_y and the GaAs substrate.

The structures (ii) described are easily fabricated by thermal oxidation, by addition of an oxidising gas, preferably NO, to the carrier gas NH_3 or by varying the distance between the source and the GaAs substrate: the O/N ratio increases as the source–substrate distance L becomes shorter. For example, the O/N ratio is 0·6–0·8 for L = 14 cm and 0·4–0·5 for L = 22 cm. Thermal oxidation should, however, be carried out in an O_2 atmosphere diluted with excess NH_3 gas; an environment comprising pure O_2 or O_2 + inert gas including N_2 should be avoided because of the undesirable generation of an oxide layer between the GaO_xN_y and the GaAs. Figure 5 shows the compositional distribution of the GaO_xN_y film as the x/y ratio is varied in steps. The structure was prepared at 450 °C by mixing NO gas ($0{\cdot}9\ l\ min^{-1}$) into a flow of NH_3 gas ($2{\cdot}5\ l\ min^{-1}$) during the later stages of film formation. The O/N ratio (x_1/y_1) of the inner GaO_xN_y is about 0·31 and that of the outer material (x_2/y_2) is about 0·77 ~ 0·79. The surface-state density N_{FB} resulting from this double layer $GaO_{x_2}N_{y_2}$–$GaO_{x_1}N_{y_1}$ (where $x_2/y_2 > x_1/y_1$) which is formed continuously is a very low value $8 \times 10^{10}\ cm^{-2}$; the breakdown strength is also improved

Figure 6 shows another example of variation of the x/y ratio in steps. This film was made by thermal oxidation of the $GaO_{x_1}N_{y_1}$–GaAs system at 550 °C for 30 min in NH_3 ($1\ l\ min^{-1}$) + O_2 ($0{\cdot}1\ l\ min^{-1}$). Addition of N_2 to the O_2 gas flow had little effect in preventing the production of the interfacial gallium oxide, which brings about a large increase in N_{FB} to a value of around $1 \times 10^{12}\ cm^{-2}$ and also increases the number of defects at the interface with energy levels 0·8 and 1·2 eV below the conduction band. These defects are, however, greatly influenced by the temperature used.

4. Conclusions

The passivation of GaAs surfaces by GaO_xN_y-based multiple insulating films has been carried out by improving the resistivity of the film and maintaining the standard of its interface properties. An ion-beam coated SiO_2 film deposited at a lower temperature has been found to serve as a good second insulating film on the GaO_xN_y film, although there may still be problems for applications at high temperatures. It is more desirable therefore to choose stable films such as Al_2O_3 or Si_3N_4 that are resistant to the interdiffusion of the constituents of the system at higher temperatures and to other mobile harmful ions. The GaO_xN_y film with the x/y ratio varying across it is thus very promising, with or without the use of another insulating film.

References

Hariu, T, Suzuki N, Matsushita K and Shibata Y 1978 *IEDM Tech. Digest* 598–9

Nishizawa J, Miyamoto N, Suto K and Shiota I 1976 *Proc. Mtg on Special Research Project Surface Electronics, Tokyo* part a pp79–82

—— 1977 *Proc. Mtg on Special Research Project Surface Electronics, Tokyo* part a pp9–12

—— 1978 *Proc. Mtg on Special Research Project Surface Electronics, Tokyo* part a pp9–12

Shibata Y, Yamashina T, Hariu T, Adachi H and Komiya S 1978 *Proc. Mtg on Special Research Project Surface Electronics, Tokyo* part a pp5–8

Shiota I, Miyamoto N and Nishizawa J 1978 *Int. Conf. Solid Films and Surfaces, Tokyo Collected Abstracts No.* B190 (1979 *Surf. Sci.* **86** 272)

Shiota I, Motoya K, Ohmi T, Miyamoto N and Nishizawa J 1977 *J. Electrochem. Soc.* **124** 155–7

Shiota I and Nishizawa J 1979 *Proc. Int. Conf. on Surface Analysis, Karlovy Vary, Czechoslovakia, 1979* No. 4

Shiota I, Yamakoshi S, Miyamoto N and Nishizawa J 1975 *Paper presented at 2nd Research Mtg on Surface Electronics, Sendai, Japan* unpublished

Author Index